An open letter "TO SELECTED ACADEMICS # 5" Part 1

ISBN-13: 978-1499531688

ISBN-10: 1499531680

http://www.titius-bode-law-explain.co.za/index.html

Author Peet (P.S.J.) Schutte

Can you raed this?

Olny srmat poelpe can.

I Cdnuolt blveiee that I cluod aulaclty uesdnatnrd what I was rdanieg. The phaonmneal pweor of the hmuan mnid, aoccdrnig to a rscheearch at Cmabrigde Uinervtisy, it deosn't mttaer in what oredr the ltteers in a word are, the olny iprmoatnt tihng is that the frist and lsat ltteer be in the rghit pclae. The rset can be a taotl mses and you can still raed it wouthit a porbelm. This is beuseae the huamn mnid deos not raed ervey lteter by istlef, but the word as a wlohe. Amzanig huh? yach and I awlyas tghuohot slpeling was ipmorantt!

If you can raed this psas it on !!

An open letter

TO SELECTED ACADEMICS
ISBN 0-9584410-9-X

Introducing as well as acting as
THE ACADEMIC PROLOGUE TO
MATTER'S TIME IN SPACE
THE THESIS
ISBN   0-9584410-8-1

TRANSLATED BY
PEET SCHUTTE
FROM THE ORIGINAL AFRIKAANS:
"MATERIE SE TYD IN RUIMTE"
I. S. B. N.   0-620-27041-1
WRITTEN BY PEET SCHUTTE

© **KOSMOLOGIESE EN ASTRONOMIESE TEGNIKA**

TO WHOM IT MAY CONCERN,
RE: AN OPEN LETTER TO ACADEMICS **ISBN 0-9584410-9-X announces**
The Four Cosmic Pillars
In this letter to Academics I, the author, introduce a new method whereby **the origins of the solar system can be proven**, if one make use of the natural phenomena which I named **The Four Cosmic Pillars**

They are as follows:
The Titius Bode Law
The Lagrangian System
The Coanda principal
The Roche Limit.
First one has to prove how these four Pillars bring about gravity, and that I do as you are about to find out. Sit back and enjoy the ride because you are going to stray from Mainstream Physics as far as you have never strayed before.

THIS LETTER IS THE ANNOUNCING OF THE BOOK IN SEVEN VOLUMES CALLED
MATTER'S TIME IN SPACE: THE THESIS ISBN 0-9584410-8-1 VOLUMES 1-7

This letter was the letter that was sent to near enough eighty Universities through out the world in regard of announcing a new cosmic theory. It now is turned into a separate and individual commercial book.

TO WHOM IT MAY CONCERN,

An open letter **TO SELECTED ACADEMICS** ISBN 0-9584410-9-X Is THE
ACADEMIC NOTIFYING OF
MATTER'S TIME IN SPACE: THE THESIS ISBN 0-9584410-8-1 Written by PEET SCHUTTE

Dear Professor,

I present you again with my work and this time I compiled the work into three books. In the three books I introduce for the first time ever into science an explanation for the phenomena not yet explained. They are The Roche limit, 2) The Bode law, 3) The Lagrangian points and The Coanda effect.

These above mentioned phenomena work in harmony to form the principle called gravity. They work as a whole as an assembled unit to compile gravity. Each one has its individual role but unifies their combined principles to form gravity. Each fills a function that finds a position in the unit where it participates in forming that what we regard as keeping the Universe glued. Mass therefore, has nothing whatsoever to do with the generating of gravity. In the books I purposely explain my conclusions in minute detail, which I did not do the previous time when I offered my work for academic investigative evaluation. This time the information is now so much simplified it is to a level where I believe it will probably be understood by any child still at school on the condition that that scholar has developed an understanding about physics to the level of high school physics standard. In the previous presentations of my work that I submitted, my work was rejected at the time by your institution amongst and including other institutions equal in stature to that of yours. At that time I could have been guilty of accepting that there were facts I presented that the reader was unfamiliar with, but I did not realise that because I saw those facts as general knowledge, which at the time I personally would regard as simple everyday information. In the end those facts did prove not to be general knowledge after all and that complicated some of the issues. One such example is where to find singularity. It is so obvious allocated that at the time I could not dream any person did not see it at first glance. The other possible mistake is not to explain how I conclude how the value of singularity comes about because that too, is so simple to see. One more example I may mention is to explain why singularity has developed space-time by specific measurements according to time and by using time. I might have failed to present the limits of time, which I clearly make an issue of indicating this time round. I took this as general knowledge being located in an obvious position where all can immediately see the presence of eternity and infinity but it turns out not to be that obvious to others. (This is only naming a few.) However, I have corrected such mistakes and this time I have explained every aspect in such great detail that more explaining of any issue with the use of more detail would not help explaining more facts but would become tedious and monotonous. Therefore my explaining at present is so simple a scholar should understand it.

I am under no illusion that you suddenly will find any inspiration to read this letter this time round since you never made any attempt to read any of my correspondence in the past. I am under no illusion that those of you that have incisive influence about recommending what the criteria should be about printing material of published material in major University or stand in as an adviser for any other Publishing Houses which may or may not form a department in your institution, those then will suddenly read my work with inspired interest and ruminate on matters I refer to. I do not envisage that you would this time round find a change of heart where you suddenly would enthusiastically devour every word in this letter, let alone get around to actively read any of my work in any of the books. I am totally aware that when you receive any letter such as this, you would immediately delegated the task of reading the letter to a lesser office since you have no time to deal with such small issues as this letter holds. I am well aware of all your dismay which I am about to release where your response is dumping this onto a much lesser desk and that is the reaction that this letter charges where it is causing the rejection that stirs in you and I am equally aware that there is a poor chance that any of your members of staff would even care to read this letter to its end. You may well wonder why I then would address this letter to your office. Your reasons for not reading my work in the past was that you were too busy and that did not allow you the time to read it, but it is more to the point of simply being because I reject Newton and you counteract that by rejecting my work from which your reason for doing the rejecting is that in justifying your action as that you simply could not generate any interest to do so. Sir, Madam, by my sending you this letter is in motivation that I can find solace because you are about to reject what I say again but your doing so notwithstanding my writing you this letter will be my consolation in my conclusion about an academic conspiracy. Reading this statement at this point might seem ridiculous and execrating but if you think in such terms do yourself the favour and keep on reading. If you are surprised in the statement then be even more surprised when I challenge your decency to prove there is no academic conspiracy going on to the effect of being deliberately instigated or otherwise being protective of the status quo because you feel vulnerable about academic issues. I see no need to introduce the books in this letter because of the simplicity of the information as it is presented for reading purposes in the books. In the books I am again showing what is incorrect about Newtonian gravity. That is not my personal perception but a fact I established without a doubt, but irrespective of whatever degree of undeniable evidence I bring, notwithstanding the degree of correctness that I deliver, my statements are never good enough, not even to justify an unbiased evaluation thereof. I know from past experience that this statement where I suggested anything concerning Newton and Newtonian views as being incorrect nailed my coffin shut before you even opened the book. In the past and at present I am aware that such accusations about Newton already bring automatic disqualification to my work. It spurs immediate and total academic rejection of my work by all academic's concerned and I provoke the resentment they experience towards me in person in the most intensity any Academic can experience any resentment. They immediately feel offended by my challenging the system and I am also aware that much of the anger is as a result of their feeling personally offended in their position as guardians of physics. While they are the caretakers and Masters of the physics they guard, I come along and show flaws in what they see as being more perfect than God. Even those amongst you that declare them to be devoted Christians think more of Newton than of Christ and don't fool yourself on the matter because I have a means to prove this statement. In defence of my condemning Newtonian gravity I ask you which is more important, the ego of the Masters or the truth of the work? In the past it seemed the ego of the Masters about the stature of the absolute Master being Newton carried supremacy above the truth while overshadowing logic and reason. Therefore I say this in total confidence that I am more than aware that I am about to evoke the very same response as I have done in the past in my rejection and disputing of all the Newtonian gravity views. In my saying this which is commanding the anger that I evoke I challenge you when you go onto reject my work again, to have some degree of honesty about why you again reject / ignore or dismiss my work. Be honest to your conscience as to why you are not prepared to analyse and to scrutinize Newton with me and in the manner that I do. However, when you do go on to reject my work once more, then I charge your honesty and your sense of fairness as the bearer of an academic pillar, to go on to prove how mass does bring about gravity. After six years of continuous trying and after eight books were presented on the subject and with no clear response yet, what else must I conclude than that there is some conspiring going on to suppress my findings.

Sir, Madam, I have reached a point where I am beyond diplomacy and I have taken off the gloves of hypocritical politeness. I reached a point where I call a spade by name. I now shout fraud where I detect conspiracy. I will do so even when this is present in the highest circles of the Physics paternity. I feel tested to a point where I am no longer prepared to use cotton wool as a means to avoid confrontation with the highest in all academic intellectual circles in order to evade embarrassment by hiding honesty, decency and respect behind a cloak of hypocrisy just to save the face of a supposed respected paternity.

In the very same breath I also will admit: Sir, Madam, I feel like Satan incarnated in person while I accuse you because I do not know you in person or even as an individual person. I can only consider you as my superior intellectually and on all other levels and by all other norms and with that then what you represent what I have to regard as being above and beyond reprimand in any way. Yet when no one takes notice of what I say when I show what I say I have to come to a conclusion that there is some conspiracy going on. Why would the members of the academy ignore me rather than act surprised when I show a clear mistake that Newton made. Why trash my work regardless without reading it in serious consideration when I present a new clear definition about views concerning the basics in physics. Your ignoring comes in spite of the fact that no one knows what forms gravity and that includes even Newton in person. I bring a new suggestion to the table that holds all cosmic phenomena dearly and the phenomena are a fact whether academics attribute any reality to their being there or not. From that I define gravity in a new sense but from persons of your status I find rejection because the new definition does not involve mass. If anyone presents an answer to a problem the least one would expect, as an honest reaction is that the party which is informed would be surprised and appreciative of any new suggestion that may solve such a problem and I would think not knowing what mass is would constitute to a problem. Why would any one dismiss the problem by dismissing the suggestion and ignoring the solution that is presented? The more the lot ignores me the more my suspicions mount because I prove Newton and his mass has no validity. If you are not guilty of what I come to accuse you of when I accuse you of conspiring to suppress evidence about Newtonian incorrectness, then why ignore what I say. Why do you then not just read what I have to say? What I have to say puts the cosmos out there in a completely new perspective and if I may add, a logical and understandable perspective. I bring a perspective where one does not have to go Bohemian to show singularity and to show space-time. It is there for all to see and even a child can understand what there is to see. So far my attempt to use a respectable avenue going through the official channels and following the guidelines on offer got me precisely nowhere and it took me years to get there with all the aid I got from the Academic paternity through the years. I am now at the point where I chuck all the niceties out the window and see what I came to suspect because of the treatment I received. I made excuses on behalf of the Academics and tried to find reasons for their behaviour but the only and last conclusion I could arrive at was that I was very naïve about the honesty and unwavering integrity in which I regarded the Academic world in its complete devoted sincerity to which they strive to achieve about the factual correctness in science.

I feel like the crook in the fairy tale while I accuse you of dirty dealings but what other choice do I have. I prove that Newton does not pan out and you brush me off by telling yourself and me that I do not understand Newton. That is complete and utter rubbish. I understand Newton better than any other person I have come across. Newton does not make sense at all. Mass only has validity when gravity is restrained because gravity is the motion of space that is filled with matter in $a^3 = T^2k$ and when space does not move it cannot act on gravity or motion where it becomes immobile and frustrated. Being immobile and frustrated results in that the space that is filled with matter is forming mass just because it cannot move independently any longer. However gravity is the motion of space in time and mass is the restricting of such independent motion. Read my work and you will see! If you are not prepared to read my work but always delegate the evaluation thereof to a person that is clearly not up to the task, then what must I conclude from that? If you do not even try to address the shortfall of Newton then what must I conclude? Then it becomes obviously apparent that those academics being in the status and position that you are either detest what I say or do not believe what I say. If you do not believe what I say, take my challenge serious and prove to your mind why you accept mass has gravity as a result while you scrutinize my ideas. When you detest what I say then find the reason for your detesting me including what I have to say.

Again I have to admit how embarrassed I am in my behaviour by accusing a person in your position but what else is there to conclude. The Critical Density theory makes a mockery of common human intelligence that reminds me of a bunch of drunks going on in an argument about matters they do not understand even when they are sober and in their state of drunkenness they are going mad. I just cannot believe the world's most intellectuals can reason in such an obscene manner that suits the likes of high school boys arguing about their fantasies rather that persons in such position as they have. It fits boys arguing to impress girls much more than it can be associated with the worlds supposed best minds there are. At the end of this letter when and if you have the courage to read it to the end and simply also on the condition that you will read it to the end this time round, then just be honest to your mind and think what I said and how I say what I prove in my books. I honestly do not wish to be insulting by shouting conspiracy but can you find another way that I am supposed to reason when I gauge the way you react?

The following four phenomena are real and are evidence found in the cosmos where they stand undisputed but alas also unaccepted since the phenomena does not match Newton. Newtonians rather accept Newton than would they accept the phenomena since the phenomena do not apply in the manner Newton explained the cosmos works. The Phenomena are as real as outer space is but because it does not match the concept of Newton and his mass by gravity, therefore the Newtonians discard them with many wide ranging excuses. The phenomena are amongst so many other fitting proof of Newtonian misconceptions and I can explain every one as much as I can fit the lot into the forming of gravity. I can ensure you they have nothing to do with mass and so has mass nothing to do with gravity where gravity has everything to do with mass since mass announces the end of gravity. Mass forms gravity's grave. The one phenomenon is called the Titius Bode law and because it is named a law most astro physicists scorn at the fact it is referred to as a law. I prove that this is not only a law but this is the Universe. This is gravity compiling a Universe. This is space-time and without this applying, stars collapse as we find Black Holes do. To this day I must still find one Newtonian that even had a glance at this proof I bring where they did not tell me to my face the phenomena are coincidental and my proof therefore is also coincidental. Should these apply then Newton cannot apply because all the planets relate to one another as well as the Sun in this precise fashion, which rebuts mass differences. The relation is there and that Newtonians do not dispute but in order to avoid the explaining they rather put it down as a coincidental insignificant phenomenon. The fact that all the planets show exactly this formation and that all the planets are distributed (all nine of them excluding not one, however Neptune does not precisely fit the order but Pluto does and I do explain why Neptune is slightly out of line) is approached in a manner that would rather be degraded as a fanciful thought best left outside the mentioning of good conversation. They ignore a mountainous discovery because of what disgrace Newton's claims would suffer. The true disgrace is the rape Newton committed about the work of Kepler. Newton removed Kepler's formula and suggested that the rotary motion would nullify the radius since a circle does not develop any drive $\frac{dJ}{dt} = 0$. The concept is mathematical fraud

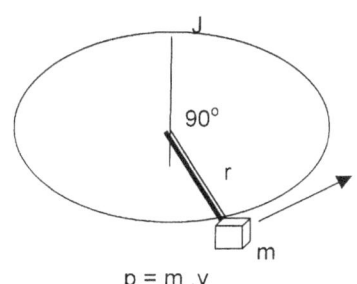

$p = m.v$

as big and as wide as any scam can go. In Kepler's formula the one holding time (T^2) compliment the other factor in time (**k**) to allow space to duplicate as space moves through time. They are not in division as Newton claimed. Secondly there is no mass involved in outer space and in that we find the Titius Bode law that provides us with the evidence of unanimity where all planets are at a specific ratio, which removes any concept of size differentiation from the

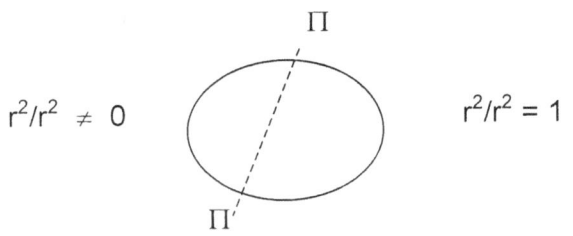

possibilities present which should be forthcoming where mass is present. One cannot divide the radius to the smallest point and find the radius would diminish to zero as Newton suggested. That is just a mathematical impossibility. The very essence of the Phenomena renders its result onto form. When a line is shortened by dividing the line will keep on reducing until the line becomes infinitely small and the line would have all point that forms the line positioned on one specific spot.

That is singularity where 1^0 have no sides. If Newton did suggest that $\dfrac{dJ}{dt} = 1^0$ then that statement would ring true because I can prove singularity being present at that very location and that is correct! Hubble proved the universe is expanding. Then by backtracking we have to set about reducing the sphere constituting the expanding universe. If r in the circle is growing we have to reduce r to backtrack. When the circle reduces, the value located to r will become implicated because r determines specific size. Not so in the case of Π, because Π in the true sense only indicates that the circle is a square without corners and therefore Π dictates form and not size. By reducing size only r comes into contest and will point to such reduction. By reducing the circle radius r by half continuously will lead to an infinite small circle but Π will remain because the circle as a form remains even being infinitely small. In answering this question about presuming the Universe to form as a sphere, one will arrive at the point where we can begin in finding answers to the cosmos. It is the spontaneous choice the humans make by selecting a sphere when thinking about the cosmos in an overall picture that we should address. From logic comes the answer about the sphere being the form that inspires the cosmos. To find solutions we must then investigate the cosmos in that direction. You're visiting this web page may bring along answers about the Life story of the Universe. When one draws a line through the centre of the sphere it would connect the most outward points of each opposing side. Taking such a line horizontally will draw the longest possible side from edge to edge through the very centre. Another line coming from the from to the back will run also choosing the longest line by connecting the most furthers points and the line will cross the other line at the centre at an angle of 90^0. Then a third line with the same characteristics will also cross at a centre point at 90^0 to any one of the others running from back to front. The lines will have a specific line in length as a common delineator as well the angles crossing at 90^0 and most important the lot will cross at an infinitely small precisely centred point. That point is singularity because when all the lines are evenly reduced they will all share one point in the centre that disregards size.

Finding the shortest line is also the process in which we will be able to find the beginning of the Universe. Light gives information and light travels by a line from point to point. Light uses the Universe to travel and if we wish to find where it started we must find where that which parts material, started. We must see where

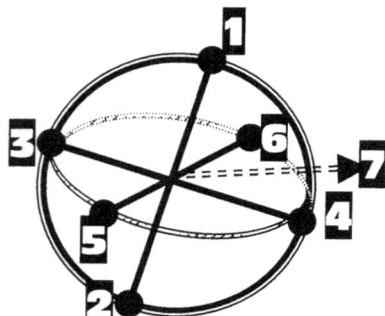

the line starts to find where the Universe started because the Universe is a system of combining lines that is running all over between points rotating.

In the very centre of the sphere the form dictates that the shape will relinquish all grounds in space that it can hold and the form will finally be without dimension. Being without dimension means at a point in the extreme centre of all spheres there are a point that holds singularity because this point with no space has a mathematical position although it is

invisible since there is no sides to such a point to give that point any dimensions.

When holding the strength of the shape of the sphere in mind as well as taking into account that all cosmos objects of importance is in the form of planets or stars and they are all in the form of a sphere, we therefore may contemplate that it is where gravity originate. We now only have to find the reason why gravity will hold a base in a space less ness as Einstein predicted. It is clear to be seen that gravity is in the centre of the sphere controlling from the centre everything that is outside the space less centre. We can reason with confidence that gravity is the strongest where space is the least. We can further reason that it is gravity that is holding the sphere in true form and since the sphere allow gravity the best working opportunity, gravity can form the sphere in as strong a shape and form as the sphere seems to have. From every point on the surface of the sphere is a point where that point connects with the other side on the surface of the sphere by a line that runs through the spot which is space less at such a centre point in the sphere. Such a line also connects by an angle of 180^0 as but also it holds other lines 90^0 where those lines show exactly the same characteristics than the original line. These lines are between six other points forming three lines that is running from top to bottom, right to left, and back to front, where all join and cross in the centre of the sphere. There are therefore six points at 180^0 forming three lines at 90^0 crossing at one centre crossing that has all the lines at 90^0 to one another. That shows the law of Pythagoras. All those lines are connecting through a centre from any given point on the surface of the sphere where one spot on the surface will have three lines covering six points on the surface all together and cross at a seventh centre point. Such points connects in total six surface points on each side of the sphere while they all support one another through the space less centre. In that absolute space less ness in the centre holding singularity we find gravity supporting and controlling all space within the sphere as well as space connected to the sphere. That is where gravity control and guide the space, which falls in the parameters as well as under the influence of the form of the sphere. In the gravity centre space goes singular meaning space becomes space less or flat. That is where Einstein's Universe goes flat because that is where gravity is at its strongest.

Also it is true that the entire form that is the sphere has, is controlled from a centre within the sphere. That centre holds the sphere in form and shape. Therefore the strong form the sphere presents is dictated from that space less centre where there is no space and no form left. The natural inclining is in the form of the sphere. It is part of the roundness that the overall shape of the sphere represents and this structural strength is carrying down to the very centre. Because the circle is forever reducing that reducing which is inherently part of the form of the sphere becomes a tool in distorting of space in the sphere and is eventually removing all forms of space from within the centre of the sphere. The very centre ends up as having no space because of the reducing that continuous down to become the space less inner centre. The all roundness is the ingredient that forms the backbone of the absolute strength that the sphere has and that is the component that the sphere is so famous for. The form the sphere has allows the sphere to have a control that is coming from the centre deep inside the sphere where the space vanishes and being without space seems to keep the entire structure rigged. From the centre the sphere shape shows strength that the shape as tough as it is. How does it work in its most basic analyses? Every point on the sphere surface is in regard to every other point on the sphere surface by a space less centre.

By using mathematics the rules dictating the use of mathematics will apply which when correctly interpreted. It is about analysing Kepler's message mathematically correctly. If it is done, then such interpretation of the formula gives new meaning to claims coming as a result of analysing Kepler's formula. This is also without Newton meddling and without Newton telling Kepler what he (Kepler) should have found and whereas instead he (Newton) should have seen what there was to see when Kepler said what there was to find. Through Kepler's investigation we now know what the cosmos tried to tell the human species if only there were some sort of skilled mathematicians since that time to decipher the true meaning of the message. Newton should have been looking at instead of telling Kepler (and in actual fact the cosmos) what, and then he (Newton) could have been able to see what gravity is. He (Kepler) said the cosmos told him that in the cosmos we would find that $a^3 = T^2 k$. He (Newton) was to be recognised as the everlasting expert on matters in motion, but his personal arrogance disallowed him to see that the cosmos told Kepler that gravity forms when space is duplicating space by the motion of space. A close study reveals that space a^3 is held in check by the motion T^2 of the space at a specific relevancy k of the motion of the space. The space a^3 will be on the move T^2 and the moving T^2 will be equal to the space a^3 in motion T^2 at a relevant distance k. There is a specific ratio of space moving in relation to a specific centre at a very pre determined distance from that centre.

The line reduces r by any point to the centre and r remains connected to the centre for all purposes that may play any role. The line will reduce until such a line is all but vanished forming singularity at 10 because that is singularity by value but the line can never become zero just because such a line always will connect material.

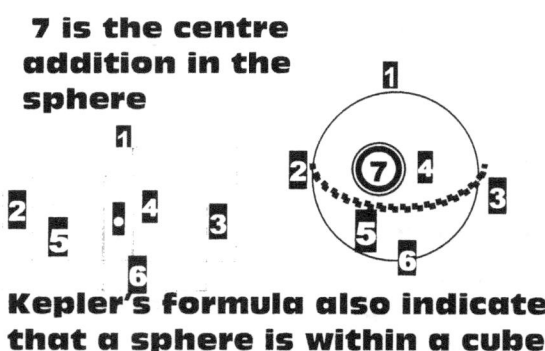

7 is the centre addition in the sphere

Kepler's formula also indicate that a sphere is within a cube that is holding a sphere

Taking the outlook from the point the sphere is holding from that centre out into space there are ten points connecting to the centre. In that are the dimensions of singularity connecting to space where five connects to space in the second dimension of singularity, and five connect in the third dimension of singularity. On the other hand does the cube show a very different characteristic, which involves only six sides (at least) being connected.

$a^3 = (T^2 k) = a^{3 +2 + 1} = 6$ with the sphere presuming the position of singularity as part the of $k^0 = 1 =$ **singularity**. Einstein proved that at the point where space reduces and such reducing reaches a point where space as a factor in the third dimension disappears into the single dimension (space going flat) gravity is overwhelming. Einstein interpreted this, as the complete Universe going flat while the Universe going flat, that can only be within singularity since singularity represents the Universe as flat as it can get.

Humans (including Einstein) interpretation of the Universe is faulty but the faulty aspect does not include the fact that the Universe is going flat but only which is the flat going Universe referred too. According to Einstein he proved that the Universe is alternating between going flat and holding space but his lack of studying Kepler lead to his spontaneous misinterpretation collected from our culture and his incorrect interpreting of what the Universe actually is. We all have a faulty perception of the Universe because not only he (Einstein) as an individual Scientist but all humans throughout has also never asked the Universe what the Universe is. Kepler did and the Universe answered using the mathematical equation $k^0 = a^3 / T^2 k$, which when interpreted means singularity placing space-time is the Universe. No one ever thought about this statement in sincerity because from a Newtonian aspect it seems silly. But rethink the silliness presented by the Newtonian Universal centre and compare that thinking about what the Universe told Kepler then decide what is silly. Newton's never acquiring the effort to do a study of Kepler's work withheld him (Einstein) from reading his very own mathematic translation accurately because apart from Newton Einstein must be considered the second most important Newtonian ever. What Einstein saw was that space disappear and he then jumped to the conclusion that the space he saw in his mathematical equations was outer space referring to the space falling outside the parameters of the material occupied space secluded by dimensional borders. In the sphere placing the borders that the sphere holds there are deliberate and very distinctly placed edges or points forming a specific distance from the centre. The centre is also proven beyond any debating.

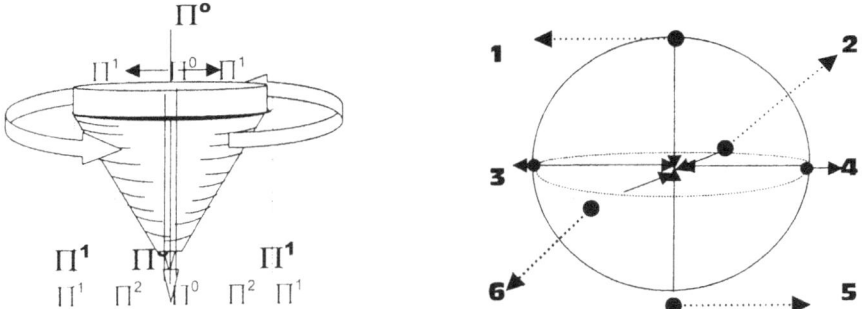

With Π^0 being a little more than the figment of an intelligent person's imagination, there are actually two values connected to singularity at Π^0 measuring Π^1 that is facing each other from both sides of the divide in relation to Π^0, and by moving from Π^0 to Π^0 by going around Π^0 the total combining value is $\Pi^1 \times \Pi^1 = \Pi^{1+1=2} = \Pi^2$, and with two sides being the very same that is opposing each other the movement will conclude as Π^2, in relation to every point that holds Π^1 and is facing Π^0. This is no more so visible when a circle is spinning which unveils the presence of a centre line that becomes a factor when an object spins in a circle.

Locating zero

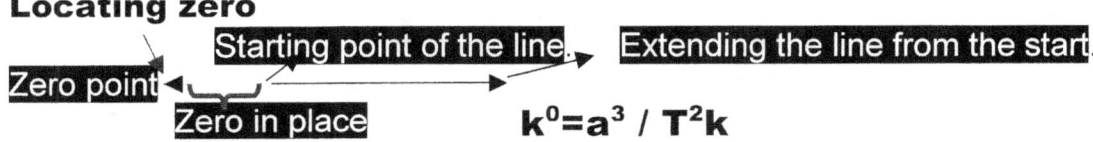

A line starting at zero has to start with zero and starting with zero removes the line altogether. Starting the line with infinity the line poses eternal possibilities.

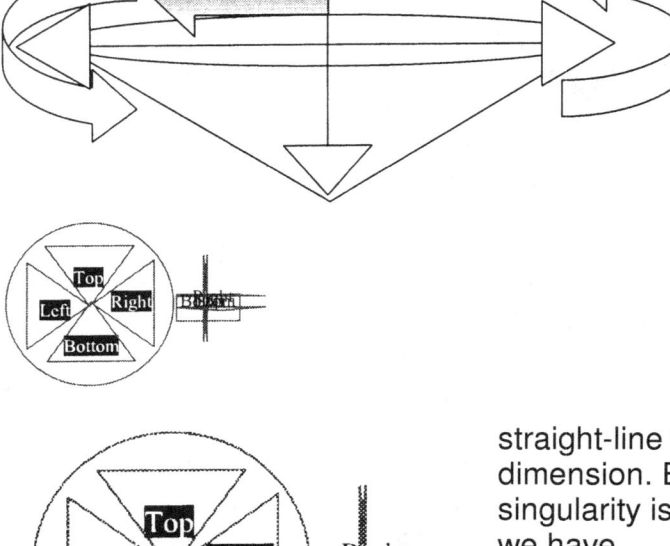

Tracing the centre of the Universe is still possible by any one wishing to find such a centre. The centre falls outside the accepted Universe since it cannot be mathematically accounted fore but that centre is in control of every thing in its influence.

The centre changes motion to gravity by diverting the straight line to an immediate circle. By tracing the line back to where the circle is no more a straight-line will uncover singularity plus one dimension. But in the entire centre forming singularity is still locatable within the Universe we have.

Reducing the radius r from all angles possible, throughout the circle will bring about that all possible direction will eventually land on the very same spot with no more dividing possible. Yet zero cannot be a factor since the sides still hold value. Whatever value may progress from such a point notwithstanding there are that many possibilities holding all the value that can rise from such a spot is concentrated on one point. This is at a point arrive where more reducing will land the one side on the opposite side of the line but it will not bring about zero in the equation

But keeping Π as one ($\Pi^0 = 1$) we keep the Universe in the first dimension.

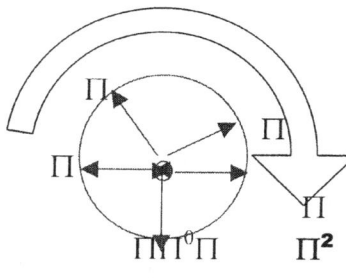

This point, which I now am referring to, is the point where Π is a fully appreciated value while the diameter D still remains a dimensional factor of one. This is the dawn of the second dimension where space was there but space was sparsely shared in some cases. It was Π^2 when Π^0 shifted to become Π for the very fist time.

What this argument further proves is that the circle reducing must then come from all points because the radius might be a line but that line represents a circle through 360^0 coming from and accounting for all possible directions. Taking that into account it is important to recognise that notwithstanding the size of a line, which any radius of any size is there is another line (or dot) eternally bigger as well as eternally smaller than the line in question. While we are in the third dimension being part of the third dimension such being in the third dimension then allows that all parts of the third dimension forever can be divided once more until the line in the third dimension is no longer part of the third dimension. When such a line leaves the third dimension it is still dividable because it might not be part of our dimension any more but it can still reduce further as part of the second dimension. By that time it has left our scope by miles but that does not mean that it end there because from our perspective that is where it ends. But our perspective does not represent reality. Yet, even then it can still reduce infinitely more until it has left the second dimension and then at last forms part of the first dimension. Only then when the line reaches the first dimension no further dividing of that line is longer possible. We can never grasp the size of a line that the first line in size that came about when the first motion broke the eternal stranglehold on space. According to our big and small conceptions of what we perceive as large, ultra large, small and microscopic small is just mere words describing thoughts totally unrealistic in the context of what the cosmos sprang from as the cosmos moved out of the spot and formed a dot. Even by the standards of forming the dot, which was eternally bigger that the spot T, as the dot and all the many dots that came from the spot. The size differentiation only between those two exceeds all limits and divides we wish to create forming borders that we can appreciate.

Being part of the 3D we have the inclination to think of something and then we also includse in that something the space we experience. Thinking of the spinning top we will think of the edge (A) forming the end of the line, the position of seven. This cconcept we have is part of the 3D Universe. To understand cosmoogy we have to rteturn to the biginning of cosmology. The relations then was when Π^0 formed Π.

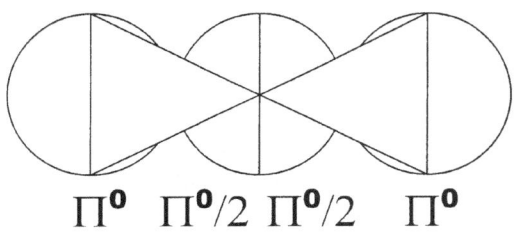

$\Pi^0 \quad \Pi^0/2 \quad \Pi^0/2 \quad \Pi^0$

Let us start telling the story as it was.

At the very first sign of any of the sides departing from the centre shared by all, all other points must also show signs of a willingness to depart. There will be one point where r still is one coming in as a factor but pi moves out from only being a factor of $\Pi^0 = 1$ and at that point pi will become a full factor of Π.

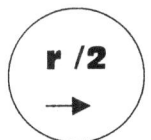

By dividing the radius r by the half of the value that then reduces r to a point where the left edge of the line reducing will be at the very same place the right hand edge of the line that is reducing will be. At one point the spots that formed the two ends of the line will be at the same spot where the original centre between the two points were. The two points would have moved evenly towards and in the direction of the centre by reducing all the space on both sides of the centre.

Then by moving towards the centre they will at some point have to reach such a centre point notwithstanding cultural concepts favouring nothing to be filling that spot because reaching that centre point will land all the sides on the same side and because of the presence of all possible sides such presence of all possible sides removes nothing out of any further possibility.

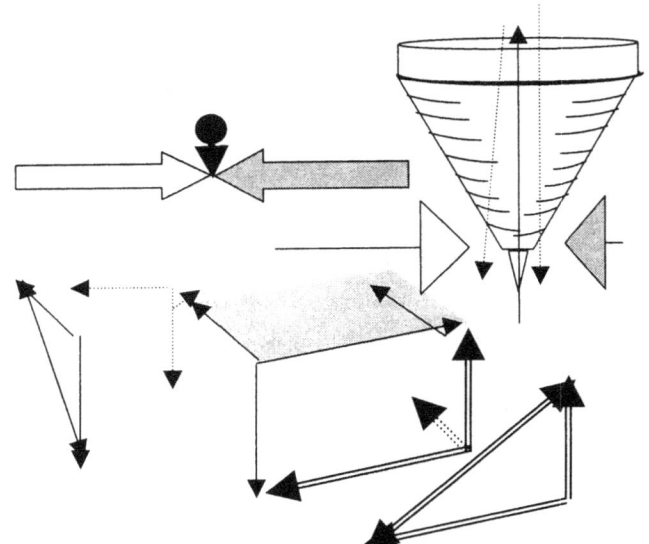

Any further dividing will land the left hand spot past the right hand spot in the opposing half where it then will grow once again but in the opposing direction that the specific spot previously represented. All possible dividing then ends on one spot where such a one spot that represents the perfect centre point and that divide the left side from the right side and the top from the bottom and the front from the back will land on one spot. At that spot all the sides just mentioned shares a location with all other possible sides. The centre that then is holding all the previous points in one spot then physically is in the single dimension applying as one spot to share a location for all sides. At such a point there is no further dividing possible. That point cannot be zero because that point represents an eternity of possible growth in an infinite number of directions available to grow. The line starts in infinity and not in zero or nothing as is taught to scholars by teachers' worldwide. Trace that centre while the top is spinning and one will find a centre that favours no side since the centre divide equally all sides while spinning. That centre proves to be no specific side because such a centre proves to be all possible sides. On several occasions in the past I have been accused of manipulating the argument to produce none-existing or overrate facts. That is not the case. I am not manipulating facts to create an argument as some intellectuals in the past accused me of. What I am talking about is a mathematical fact that any one can prove by calculating.

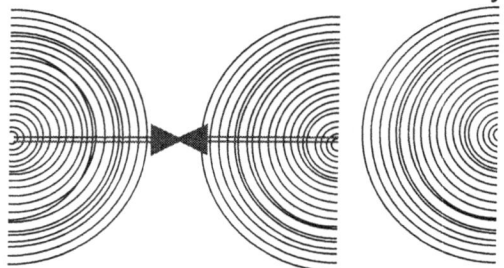

By following a very simple procedure it is within any person's reach to detect the centre of the spinning top, which I am referring to, although there is no such a centre to detect, the centre is there for all to detect. A child is capable of using the two times table and the dividing by two every time that is the most simple form in which mathematics may be used. It is a mathematical fact that a line will reach a point where all sides are at one spot and as such the line cannot divide any more. I have been accused as being dubious about my arguments while it is Mainstream science that dubiously found a way to get to zero as a mathematical starting point of any and all lines. Then they put the double standard blame on my arguments where it is.

At such a centre starting point all sides share one specific spot but that spot holds all further future possible growth in any direction of all sides and since everything is in there, there is no room left for zero to be there. That point is filled with all possibilities which prevent zero becoming factor since the sides share one spot and in that sharing they are present and their presence prevent zero from becoming a conclusion. While the different sides are in one place the factor and value is one to all without allowing zero any part to play.

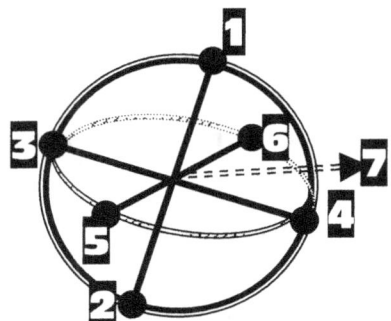

In the centre forms a point that connects every possible point to every other possible point in the sphere and all sides run through that centre to connect to one another. The point is no more than a dot and has the infinite value of 1^0.
Motion is required to activate such a point to be more than just a spot and to evolve from the spot to the dot.
That point confirms the line that comes about when the rotating motion of the top releases the time components. Inside the top is that which has no beginning because it is the beginning that holds the entirety there are on one spot and from such a spat motion releases eternity. Eternity isd the time on theoutside of the tio because that which is on the top runs as long as it is inconlusive and that which is outside the top may concentrate as atmosphere but never end anywhere.

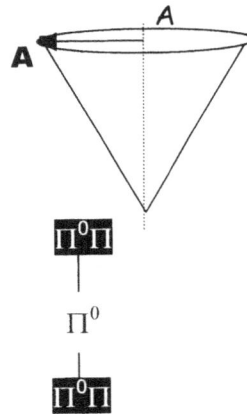

The moving of Π^0 to Π involved relegation and not motion as we consider motion. It was Π^0 getting a side and that is all. There was no true side but only a form that came into place. Singularity (A) received singularity (**A**) and no more of anything but the shift to comply with having a relevancy forming in relation to singularity. The dots had no sides, had no length or diameter. There was not measurable space or measurable time involved. The time could have been a micro, micro second as much a trillion millennium because time had no relevance. It was eternity interrupted by infinity, as it still is the case, however the line that eternity followed was no line because there was no space to hold the line. The line was momentarily interrupted by infinity, however with no one there, there was no one to notice. The lines were not lines but relations to sides being formed.

The relevancy that had the power to set Πapart from Π^0 is the only relevancy that still has the power, to set particles apart or join particles. It is heat in variation from cold. In order to excite singularity, singularity must establish a basis of heat that sets such a heat basis apart from cold. From there the form the atom will take on, however, the atom was still enumerable eternities to the development side.

Where they are equal in value we must test the reason why this then is valid.

Locating and finding the presence of singularity

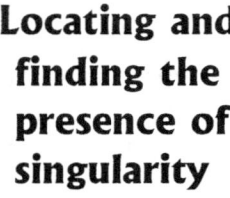
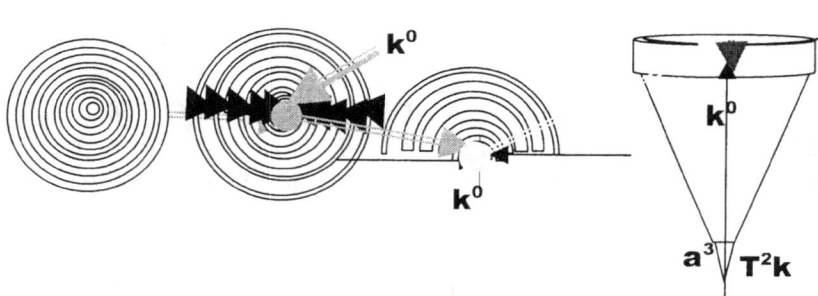

What is in the Universe is spinning. In the **precise middle** of all **objects in rotation** is a precise centre dividing the object in sectors that will **start the spinning initiation** from that centre point.

$k^0 = a^3 / T^2 k$ **states that whatever is, is also spinning in order to be present.**

Thus, the spinning object **will have a middle point,** a very specific **centre point that does not spin** and only holds Π as a specific value because no radius can apply. But also the one value such a line **cannot have is zero** because the line **is there and holds contact** to the rest of the material bringing about that **zero does not start any** line and therefore the **value of the line must be infinite,** just as described in accordance and by **the definition of singularity.** As I am introducing a very new idea, I wish to explain in better detail what I try to convey. While the top is spinning, one will find a line that formed in the centre where no line can form. It comes from spin but can never participate in spin.

That line must be singularity because if one moves any point on that line one position on, such a movement will land the point that then form on the

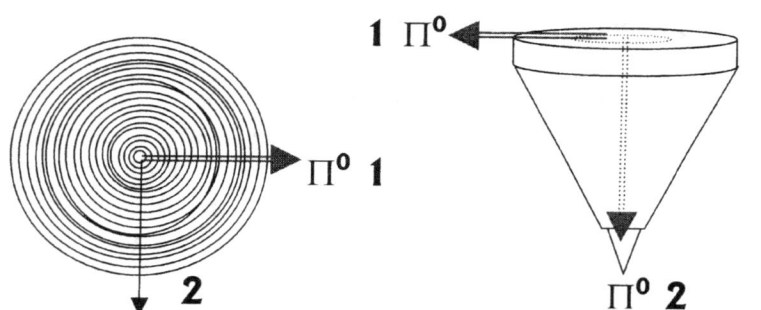
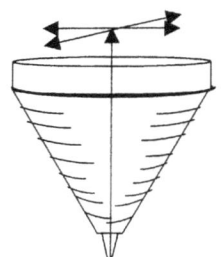

line, on the other side of the line. The line is where the radius ends and starts because the line divides what is spinning in innumerable sectors and when reducing the radius progressively towards the centre of the spinning top at the centre where no line can be there is a line dividing the entire spinning top.   At that centre point all further reducing must end because the next movement however slight will fall on the other side that is completely contradicting the one side. One movement further will change whatever is, so completely every aspect of that characteristic will contradict what it was before. There is one point that is neither left nor is it right but any point next to that point must be either left or right. The only value that point may not have is zero because albeit so small that it is not part of our Universe, still the point is there for all to witness and that point is a reality as much as the entire Universe is a reality. Whatever one attaches to the top either in the line of being material or a concept, such a concept or material has to start at the spot in the centre of the top because every aspect of the top changes in contradicting from that point onwards in all directions. **That** point albeit hypothetical, is also as much a reality none the less and is placed where that point **must be standing still** because every line **running from that point in opposing directions** is also **in opposing directional spin the other or opposing side.**

As the rotating direction moves inwards, the rings holding Π will become smaller and smaller. The reducing of the radius r will eventually end at r^0 but the top does not end there because the top still then is Πr^0. The form we attach to the spin still applies as Π and the top finds directional contradicting change at a point that never moves because it can never move being in the centre where the spin direction ends at Π^0. **It is the only aspect in the entire Universe that can be and still be motionless because it is not within the Universe. It is the centre of the Black hole because it is the centre of the Universe. It is 1^0 and only the centre of the Universe in singularity can have 1^0** However that point where the directional spin ends is the point where the actual spin does not takes place because if its immovability. It is at a point in singularity $k^0 = a^3 / T^2 k$. It is where space ends because motionless ness ends space there. The spinning is on the precise location where the point is not spinning because the Universe ends in its not spinning there.

That line running through the centre of the spinning top divides every possible side from the opposing side in innumerable points that are divided by angles and degrees. Moreover, it changes the future position of the point from the present and from the past as it redirects every point every time **k** moves to reposition in a location where T^2 ends. In the end it proves that both **k** and T^2 confirms a^3 just like Kepler stated before Newton interrupted with dishonesty.

Another huge factor that favours the use of Π as a measure in singularity progressing is that any expanding by any mathematically sympathising method will have to use Π since Π is the only route that a spot of no significance can develop into a dot that represents a Universe of development in waiting. Only by promoting through the measure of Π can all possible sides progress on equal terms in all directions simultaneously without bias. The development must progress by measure of equality to the smallest indication and that purpose only Π can serve. The progress must be generated so that it can flow equally to all sides in all directions spontaneously where not one side will favour the growing process as such. Time in it's flow does form a bias but explaining that at this stage will also involve too much other concepts which I would rather leave to the books. There is only one way to permit such a flow and have a mathematically correct outcome and that would be using Π for such expanding. The use of Π would ensure that a dot rises from the developing that comes from the first spot. The dot would have to form a sphere and a sphere is Π in relation to seven. This bring as back to the Titius Bode concept where ten is one half of a sphere relating to seven points forming the sphere where the seven with singularity puts form to the double ten that totals (including singularity) 21.99999 and that is the overbearing dominating issue.

When the cosmos came into motion, motion was not yet defined. When the cosmos brought about motion, the first motion was relevancies. Cold parted from hot. Eternity parted from infinity. Motion parted from motion absence. Infinity broke the laboriousness of eternity for the duration of infinity. The spot became and grew into the dot.

From what the spot was to what the dot now is might be just a mathematical implication of going from 1^0 to 1^1 but in reality that first motion was the creating of and establishing of an entire Universe which was with all possibilities that now is it.

Never again can that much growth become a reality, although to us the growth is beyond what we ever can notice. But it is because the growth is so massive and we are so small that we are unable to notice such almighty growth. When the spot Π^0 became functional and established all relevancies possible, heat parted from cold as eternity parted from infinity. The expansion was not clear motion but more a parting of relevancies where a centre formed a relevancy because the centre could not provide motion. Without being capable of motion, the centre established four points, which also served singularity. From the inverse square law we know that the centre doubled by producing the four points holding singularity. We have to presume there is a time line because the Universe has this as evidence. The fact that light travel from there to here and from here to there proves of such a time line because there is no distance in outer space except in the Newtonian's misconception they have about the cosmos. Any line shows direction and the direction implicates positions according to the line having dimensions in the Universe we have in space and time. The line brings in Pythagoras and Pythagoras implicates mathematics.

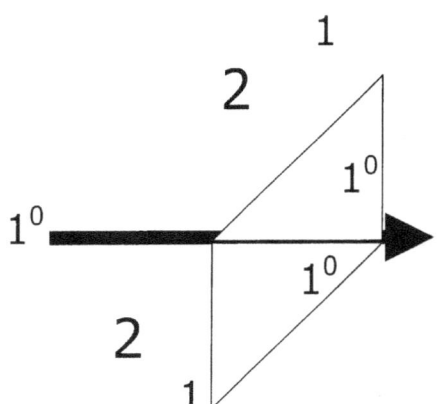

1^0 going 1^1 where $1^0 \Rightarrow 1^1$ If there were progress that developed from singularity in the form of the first spot and we are the evidence of such progress, then the mathematical conclusion must be that a line formed where the line developed two sides and we have the evidence of that still present in our Universe. That brought about that three markers formed in relation to one and by admitting to the law of Pythagoras we find that what formed was 3^2 in relation to singularity 1^0 that became 10 on the one hypotenuse and 10 on the other hypotenuse which forms the square of space. Therefore mathematically space has ten positions and material has seven.

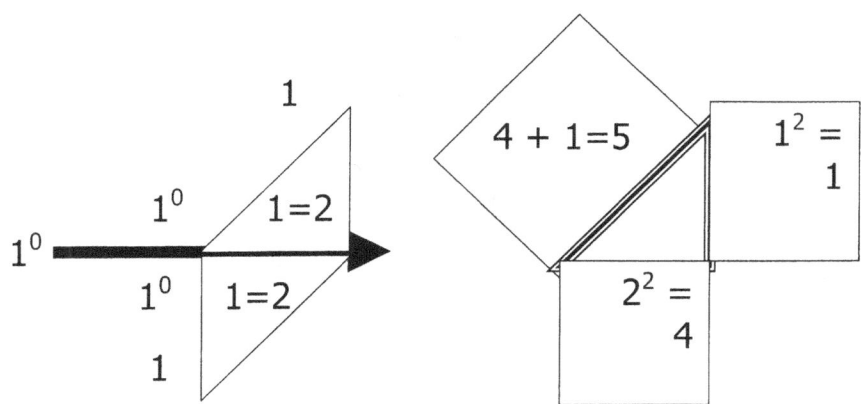

From the three the four (2 on both sides of singularity allocating the cosmic divide) the two in square developed as a mathematical consequence and that brought about the five.

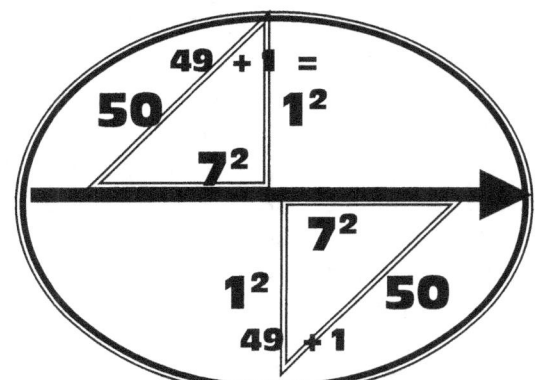

The five was duplicated as a response on the other side of the divide and having five as a result of $(2^2) + (1^2) = 5 \times 2 + 10$ we find that the square of space holds the value of ten in place.

At the very beginning there was a spot. How do we know that? The spot is still with us and holds a value of 1^0. This spot is in the centre of all spinning objects. Then time came into motion and 1^0 moved to 1^1. Since from where we stand we see 1^0 and 1^1 as being the same and therefore with the moving of time in the very beginning such moving must contribute to an increase of space. Therefore 1^0 and 1^1 has to have a difference where the one is 1 and the other is one point in singularity smaller making the 1^1 coming from infinity and rising into eternity 1^0, making infinity forever one point smaller than what we find as the value one will associate with the one we find as a measure in infinity. Therefore we can judge that singularity combines   to   have   a   total   of   1   + .9999999999999999999999999999999999999999 999, which then becomes 1.99999999999999999999999999999 or whatever because the one going smaller is running into infinity and since infinity is one less than eternity we are in eternity 1^1 looking at infinity 1^0 which is one point reduced in infinity. However this moving from 1^0 to 1^1 involved 1^2 as well as 1^2 on the other side of the divide. As a result of the form the sphere holds, there is a centre connecting the sides and the centre holds singularity. However, by presenting a centre where all lines cross on a point that cannot distinguish sides since that point has no individual sides, the centre holding singularity is inactive. Motion makes it active and the motion of space in time activates singularity to charge gravity that we find as a factor in the Coanda effect. That motion that establishes the purpose of space a^3 as a result of motion **k** through time T^2 was what Kepler presented as a formula. Gravity is $k^0 = a^3/T^2$ **k** and to install k^0 the motion of space-time a^3/T^2 is required to complete the task.

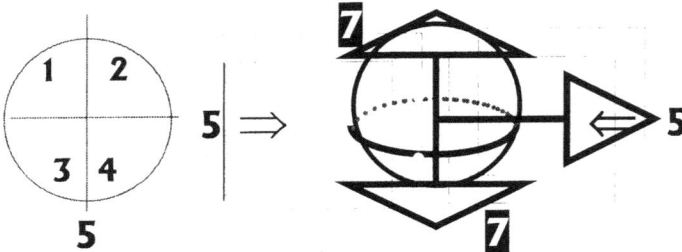

In the Roche limit the **straight line** forms part (1) and the **half circle** is part (2) and the **triangle** forms part (3) to singularity (4) Holding 5 points outside singularity

By rotational motion, the top creates a line confirming singularity running down the line and by generating the line the line charges gravity. The gravity is what drives the top as the top and as long as the top spins. There is an influence generated by the spin of the top that keeps the top upright while the top is spinning.

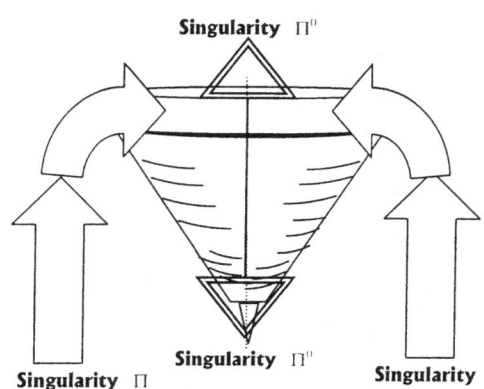

The line is generated but the line is far from magic. The line is where the centre of the Universe is which the Universe is then that what the top filled by particles from the line to the edge of the sphere. The particles in motion generate motion by electing a centre from the centre of every particle in the spinning top. Such an elected centre becomes the centre of the Universe as far as the top relates to a Universe because all the atoms in motion elect the centre of the Universe.

In this, it is clear why the Titius Bode ([10 + 10 + 1 + .991] / 7) and the Lagrangian 5 \\ 1 systems part their ways when applying the different processes they hold. With all the differentiating, the observer must also consider the dual message that light uses in travelling through the vastness of universal space. The thought of nothing is just what it is, a thought of nothing and although it is in the human mind common nature to present nothing as a value in the recalling of something, nothing is a presentation of the figment in the human mind. There can be no number such as nothing and that was (possibly) Newton's biggest error. Nothing represents non-existing and that is just what nothing is, it is non-existing.

When taking Kepler's formula as a valid criteria we find that singularity indicates a sphere with seven positions $k^{0(1)}=a^{3(3)}/T^{2(2)}k^{(1)}$ and $^{(1)}+^{(3)}+^{(2)}+^{(1)} = 7,$ which confirms singularity within the sphere but where space-time without the aid of singularity forms a value $a^3= T^2\ k$, it also proves the absence of a charged singularity not generated into a position of control outside the sphere $a^3= T^2\ k$ and where the sphere $k^0 = a^3/\ T^2\ k$ meets space $a^3= T^2\ k$ the factor k becomes the common denominator that is shared which leaves a ratio of seven to five ($a^3= T^{2\ (3\ +\ 2=\ 5)}$ where through motion one part of space (5) lands in the past and five lands to the future as well as five to the past leave a distinct space square of ten. To form space-time in terms of singularity the singularity holds 1.99999 in relation to ten on the one side of the Universe and ten on the other side of the divide where the total then is 20 + 1.9999 = 21.999 in relation to 7, which gives the sphere validity.

Since the Titius Bode is only proving a relation from the centre to a point and from the point to a centre and where such a point can only be at one position on one side of the divide in all events only half of the space factor takes up the given value while the object claims singularity by (1^0 = .9999 standing in for the Sun 1^0 and 1^1 standing in for the planet 1^1 making the seven the rotating action of space-time forming motion as the structure of the orbiting planet. The Titius Bode is not proof of the cosmos being in motion but the cosmos being in motion serves as being proof for the Titius bode generating gravity by the measure of Π^2

When the seven occupied by material rotates, it removes the ten in contact from influencing space-time and by rotation introduces a very new ten points in relation by the act of rotating. Such a process also comes about on the other side of the divide because it is the divide that cannot rotate and forces space-time to comply with the frequency of breaking down space-time as space-time releases from one side and if necessary shift or only apply new rotation alliances as the seven points become a part of the new ten.

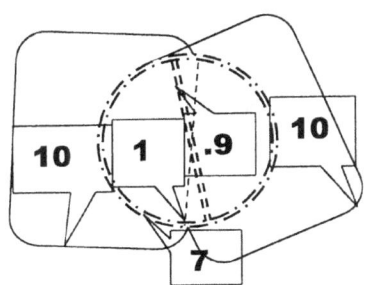

The motion duplicates space (10 + 10) by rolling over singularity (1) from the one side by reducing space (.09991) on the other side. All this motion of space 10 duplicating is directly related to matter 7. The total value established from this motion is the interaction of seven with ten producing Π^2 as the half of singularity Π /2 on the one side interact with half the singularity Π /2 on the other side and in this process Π the duplication of Π by means of matter and space produce gravity in the measure of Π^2

The conditions proving singularity is that the circular form will produce a centre point from where gravity will dictate the reproducing of space. The gravity part is the fact that motion must contradict the centre point around which the motion will produce space. The space part is proven by the motion that produces a running line of space created and followed by the liquid producing the space in the motion of the liquid. It shows that relevancy comes from space shared by space and motion separates space shared by space. It shows gravity coming from motion separating shared space.

When further consuming is not possible, when finding any singularity that challenge is the consuming, the Roche limit comes into effect.

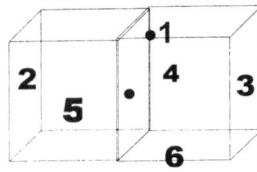

There is one more point in the sphere in the centre forming an addition in the sphere. That point holds gravity secure.

The motion that establishes gravity secures space by duplication thereof.

The only absolute constant we have in the Universe is that when something is part of the Universe it can go nowhere but remain in the Universe. That very first instant where the

Universe came about is still applying and it is still happening because nothing can remove it from the Universe and where nothing might be a concept in Newton's head nothing is the only thing that is not anywhere in the Universe.

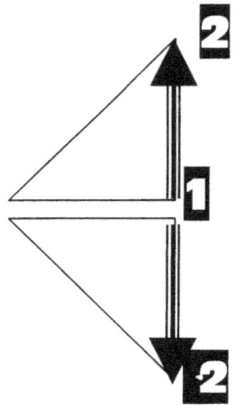

As time committed in moving along a line the rest of the motion was desiccated to produce space since a line is in one direction but that does not include the other six of the seven direction of space. One must keep in mind that singularity is the line producing the flow of time from the past through the present and into the future and therefore connects to the six sides by taking such connection in a change of position to the following time instant. On a dimensional level the flow of time progressed from a position of one to two but also it expanded into space from one to two. By flowing with singularity then it is singularity that takes the six dimensions of the sphere along and in changing newly allocated positions singularity secure the space to flow through time.

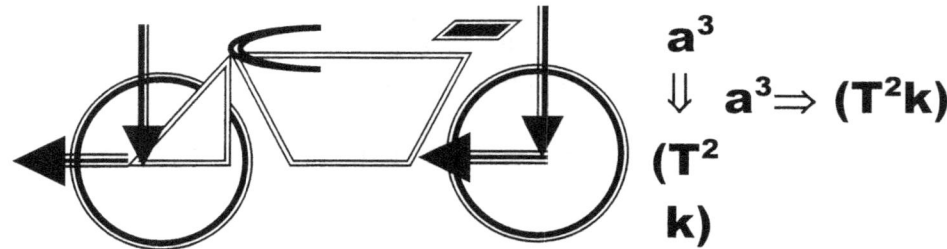

It is the movement of a^3 during the period of T^2 in relation to the relevant change in position k that brings about the movement. But while the bicycle has it's motion of $a^3 = T^2k$ the Earth too has its motion of $a^3 = T^2k$ it is in contrast to the bicycle's motion of $a^3 = T^2k$, which is at an angle of 90^0. In this lot mass has no part as far as cosmic relevancy goes. Yes mass is a factor but mass is a factor just because humans invented mass as a measuring unit in order to establish some common denominator to calculate but that also applies to distance as it applies to temperature as it applies to time. It is what humans may use to calculate human perception but human perception finds no valid grounds in the cosmos. It is because the bicycle time in motion exceeds the time in motion the Earth insist on that the time in motion of the bicycle find a means to keep the bicycle moving and upright. It is time in motion coming from the past, going through the present and onto the future that keeps the "momentum" higher than the moving of the Earth by duplicating the entire structure of the Earth from time in motion coming from the past, going through the present and onto the future which control what we find the Earth to be.

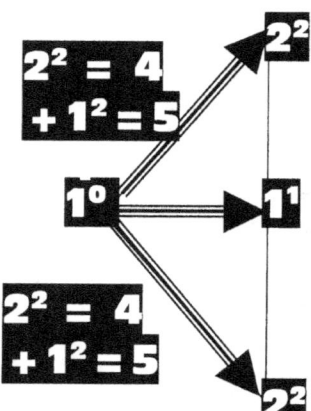

One dimension up and 10 = 10^2 = 100

Newton claimed that $GMm/r2 = m (\omega^2 r)$, which I guess in itself is good and true. Then by replacing $(\omega^2 r)$ with $2\Pi/T$ we get $T^2 = a^3$.

That is rubbish because for one, one cannot exchange (ω^2 r) with 2Π/T because (ω^2 r) clearly is $\Pi^2\Pi$ (moving from Π to Π in relation to Π) or relevancy T^2 **k** or r where one must remember that it involves movement and movement enforces a triangle because the motion depends on a centre. The fact that Newton put ω^2, he clearly admitted a centre and the movement T^2 of a^3 is clearly by distinction of **k**. One cannot substitute that which is in a square (meaning a flat dimension) with a double Π. What was the man thinking and was the man thinking at all? He then corrupted Kepler's formula to the tune of putting $a^3 = T^2$ or $T^2 = a^3$. That is not what Kepler said! Kepler said $a^3 = T^2$ at the rate of **k** or if you will at the distance of **k**.

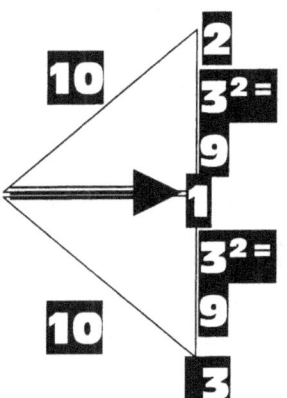

If he persisted that **k** has a value of 0 then the space was zero and the time was zero because in outer space distance **k** has no relevance because **k** is a result of time and there is no distance in the Universe. Distance is a measure we on Earth attach to an unknown, which we wish to calculate. Distance is manmade **(k= a^3/T^2)** and **($k^{-1}=T^2/a^3$)** because the cosmos devised time developing instead and mass is manmade $a^3 \neq T^2k$ because the cosmos devised gravity $a^3 = T^2k$ instead while temperature is manmade $T^2 = a^{3/}k$ because the cosmos devised time instead.

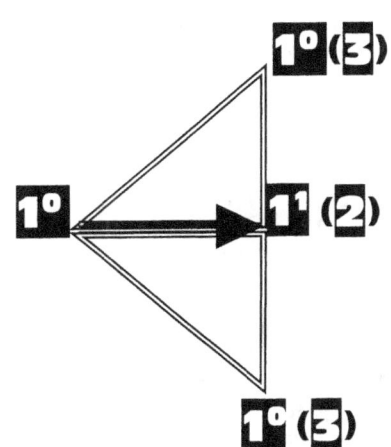

The distance between the Moon and Earth the fluctuate while what Newton suggested was that $a^3 = T^2$ always proved that the time factor we see as distance which is using the symbol **k,** is in fact a time continuance that always remain the same. That is the measure of time because the rotation of T^2 puts the space a^3 in a relation to the duration (distance) of **k** and where the rotation of space is $T^2 \times T^2$ that motion would duplicate space in its full contingent a^3 by the measured relevancy (distance that time progressed or developed) **k,** and therefore the Big Bang came about by $T^2 \times T^2 = a^3 \times k$. In that regard we find that $(T^2 \times T^2) / k = a^3$, which is what the Lagrangian point system is. Time moves one point away from the value of $4\Pi^2$.

However the following explaining involves the smallest area where singularity immediately commands space to motion. It is where Einstein found the Universe to go flat and where the smallest form of material functions. It is where space is time by mathematics.

We are in the zone that is still active because it is where material increases. One thing that is another part of the Newtonian dysfunctional concept about the expanding Universe is that in the Universe there can be no expanding at all.

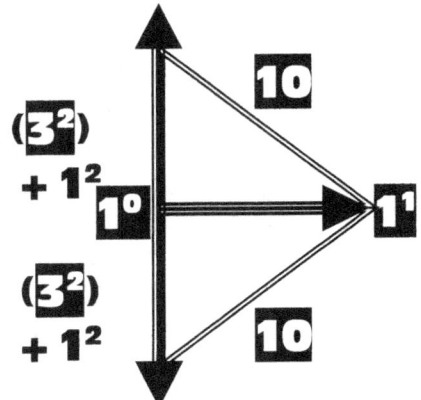

Every person agrees that the Universe is as big as it can get and there is no limit to the Universe. The universe has no end and has no borders because the

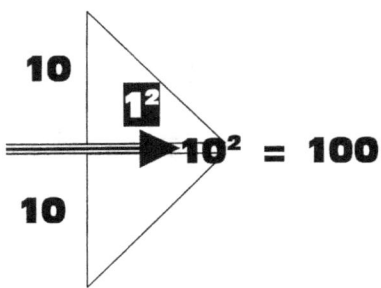

Universe is everlasting. So how can a thing that is eternally big and has no ending and presents the eternal entirety of what there is, still grow bigger and become more! How can that which has no limit still grow into something with more limitlessness? The universe is expanding by shrinking into the oblivious. That what was previously too small to be a factor becomes a factor by the shrinking of everything becoming relative to the reducing of the space. If the Universe is truly limitless on the outside then the only limit must be to the inside and it then must be in that direction that the growth of space is flowing. I explain much more of that in the book. The growth in motion concerns the scope where no Newtonian can see and that is why they have no evidence of smaller aspects growing. It is because they simply can't see that. To apply motion we have to start at the smallest there is because the smallest has to shift as a unit of independent structures all working in a coherency in order to move. Motion is the most complex issue there is in the entire Universe but motion involves not the biggest as a pretty lump but the smallest acting together as a group to eventually participate as a lump of material in time.

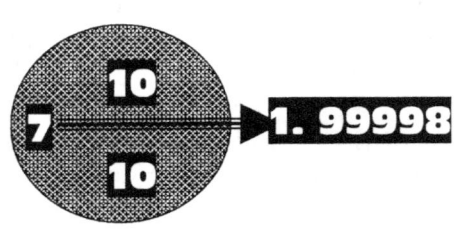

In the sphere there are 10 on both sides of the divide as time is pushing matter by seven along through the instant of singularity on both sides of the divide which is 1 + .999998 = 1.999998

To understand the process that involves motion we have to argue about where the smallest are and that will be just outside infinity because infinity is the part of the cosmos that has no start leaving eternity to the part that has no end. Space-time is the space filled with time that parts infinity in singularity (1^1) from eternity (1^0) and where the smallest is there is that is the growth of the space-time. That is where gravity is. It is where space disappears into motionlessness and not where the Universe draws flat. Gravity is where the smallest particle meets space less ness and space disappear into singularity because singularity cannot move. If one

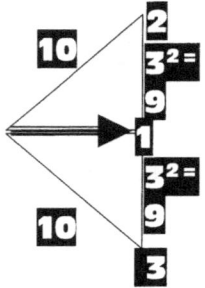

wishes to locate gravity it is there that one should be starting and not at some stupid mass idea that resonates from the dark ages. It is where space starts because it is where space disappears. When the first line moved (I choose to refer to time in beginning because although it still apply it puts a dimension to the occurrence that stands as correct because I am in time delay many moons away from where that which I refer too, is taking place at the present time) it moved by acquiring three points on one line and the one line was on one side of the Universe but since motion is the duplication of what is at present from the past into the future, it has to be on both sides of the divide in a cyclic manner.

Therefore it was three points (1^0 going 1^1 on the one side and coming from 1^1 on the other side) as it involves three 1^0 to 1^1 and 1^1 and since motion is a triangle by dimension the triangle implicates Pythagoras in the square which is $3^2 = 9 + 1^2 = 10$. As this is applying on both sides of the divide the total shift by dimension is in cyclic perspective 10 X 10 = 100 but in shifting along development it is a shift of 10 + 10 = 20. While this is in the shifting that singularity develops it is 1,999999 plus 20 and that is a sphere. This is happening beyond the dimension of the proton where the motion is still beyond space, as we know space is. On both sides of the divide we find that 10 X 10 = 100.

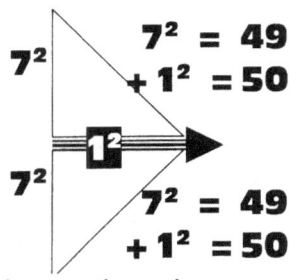

$$7^2 = 49$$
$$+ 1^2 = 50$$
$$7^2 = 49$$
$$+ 1^2 = 50$$

Also directly linking mathematically we have seven in the square plus singularity in the square in a Pythagoras motion association forming the triangle where then the hypotenuse is valued at fifty and on both sides of the singularity divide it totals 100. There is the connection in motion. The sphere holds seven and in the square with the adding of 1 in the square we have fifty on the one side and fifty on the other side.

In motion the square of seven fills the square of ten and in that where space has no legal status all motion is by mathematics because all motion exceeds the

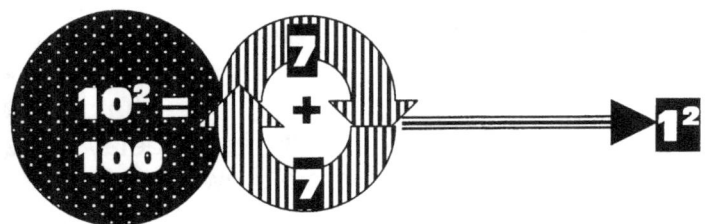

speed of light by a measure of many light years on end.

Material in the sphere holds the measure of seven and in seven circling the motion has to instate the triangle where the triangle enforces the use of the law of

Pythagoras. That is as basic mathematics as one may get and more simplistic than this there cannot be.

On the side of space the motion comes from the five sides in the past making contact with the sphere by the five sides in the present and continuing into the future by commanding another set of five sides.

In that manner we find the curvature of space-time honouring the value of Π. This is where the Lagrangian system becomes part of the space – time that generates gravity. There is four points in a circle that is completed and the fifth point is where time progresses to a new location. It is one fourth of the sphere in space where time completed forms the other four to come to a total of 21.99999 to the

seven that forms the material sector as the sphere. The motion involves the dividing of space into units comprising of six blocks where the material annexes one side leaving five sides to act on behalf of all material because where the material connected to space, that zone is not defined in connection as the material is by having specific borders and edges. The area that the five sides cover

goes on indefinitely limitlessly and with no borders to form limits or endings that might remove a certain part as an excluded area. The two blocks on one side becomes a double of five, which are ten and the same value of on the other side also of applying in the same manner.

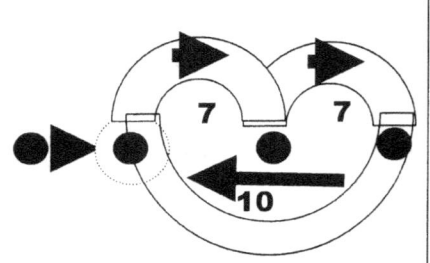

> In absorbing one in four duplicated points in singularity, the centre singularity shows growth by dismissing in doubling the adsorbed heat. The centre established a density growth. One must keep in mind that space was not yet an issue there fore the half circle was equal to the straight line. Whether the growth was seen as a half circle or whether it is seen as a straight line it is the same thing at that stage.

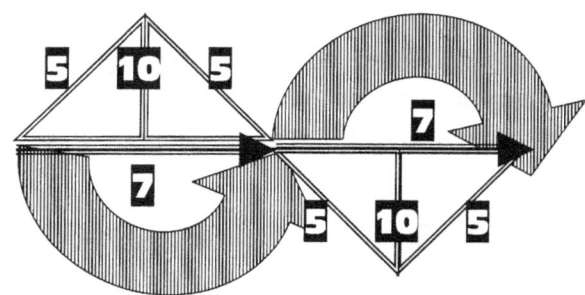

That results in the sphere and by the pin of seven points running through ten on both sides of the divide the sphere takes shape as material duplicates. Again I wish to remind that we are dealing with the very most inner space of material where space-time forms an apparition of mathematical proportions. That is where the duplication results in material and material is the result of time delay of heat that is confined to time in space. This is where the Universe draws flat. This is where time is as current as it can be. It is where gravity keeps the entirety connected to reality and the time delay is reversed.

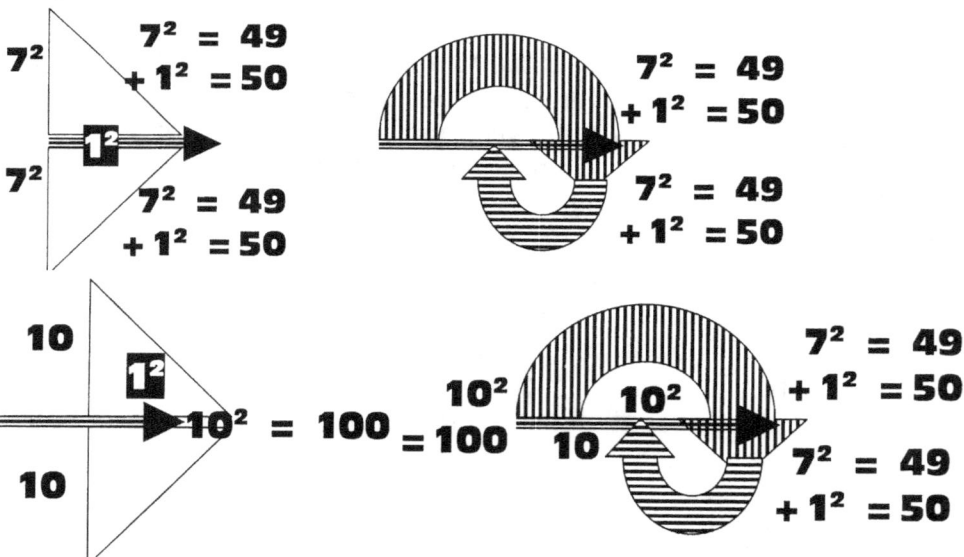

Material with seven points forms a joint value of doubling on both side of the divide relating to the filling of space. It is where the sphere fills space totally not because a square can hold a circle and have no space left but it is where 50 + 50 can fill a hundred and have no further figures unfilled. The filling of space is a mathematical consequence of the cosmos duplicating what is within the cosmos at the time at that point in time. Yet there is no way I can describe in this little space available the full extent of what motion involves because every spot is singularity and singularity find value by performing as a point in time delay.

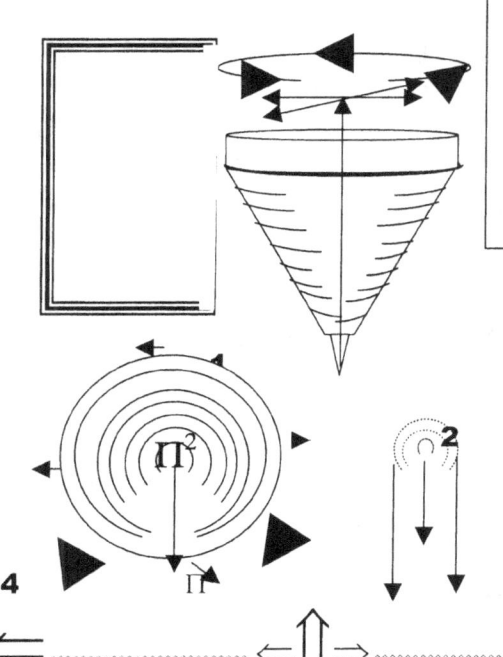

The sphere has seven points. The cube without truly being a cube but is just in consideration of having a cube in form holds five points to singularity. In the centre runs singularity to the value of Π^0, which means that which surround Π^0, holds a position of Π^2,

The spinning sphere activates the seven points, which places gravity in relation to a centre. Outside the centre there are five sides by dimension. The sphere has seven points of which four is spinning. The four spinning stands related to the gravity of spin, which are Π^2.

Since the square of space is directly involved with the dimensional depleting of space from ten to Π^2 and singularity divide the affect of gravity being Π^2 into two sectors forming one double of the four quadrants of time, the affect of Π^2 may not be ignored since that totally dominated gravity from even before the establishing of gravity in the form of the Roche limit.

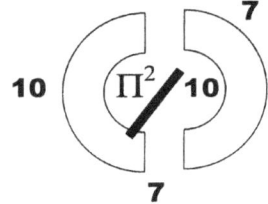

$$A1 + A2 = .49 + .49 = .98$$

$$A1 + A2 = 5.033 + 5.033 = 10.066$$

The result is the dimensional dismissing of space from a relative 10 to Π^2.

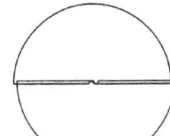

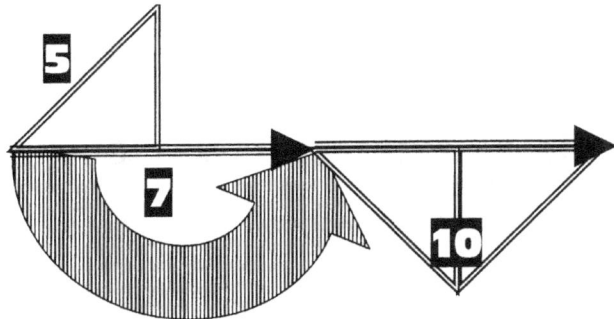

In it self the points are serving as singularity and the motion is a result of generating singularity by gravity to an elected common centre. Every point activates to fill a role as the centre is excited in being generated for the task. Every point is there as every point is time delay or time lagging behind. As space moves the singularity established in the centre is static.

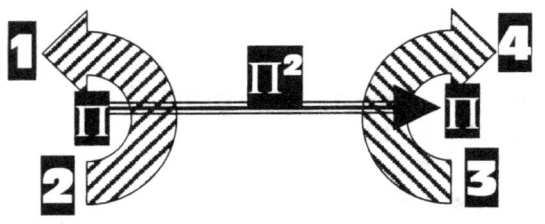

To get space to move, the time delay activates and charges a common point and the common point respond by activating time within singularity in the centre. The common point serves as the stabilizer unit around which the spin of gravity is generated that provides a spot from where space can serve material as time and material can be within space serving as time. The point holds gravity Π^2, where the gravity divides the time into four points as the four points are generating the time delay. Every point holds a spin and from the spin the duplication forms a relevancy to what is duplicated but also in as far as the total unit that is duplicated. The unit charges a single point that serve as the divide of all the duplication that comes as a result of motion. That charging of a centre by the motion of all the atoms within the top and the centre being able through

7 / 10 that is interacting with 10 / 7 across the divide or the Roche limit at $\Pi^2 / 4$

motion to generate gravity is what keeps the top in spin. That is extremely briefly explaining the essence.

It is the amount of spinning material in motion within the unit that forms the unit in motion, which is duplicating as a unit that determines the rate of duplication of the entire structure. The Sun would duplicate much more frivolously than the Earth and the Earth more that the bicycle which puts the three in different time domains. The amount of atoms spinning together as a group renders the motion that is carried on to the singularity that the group is generating.

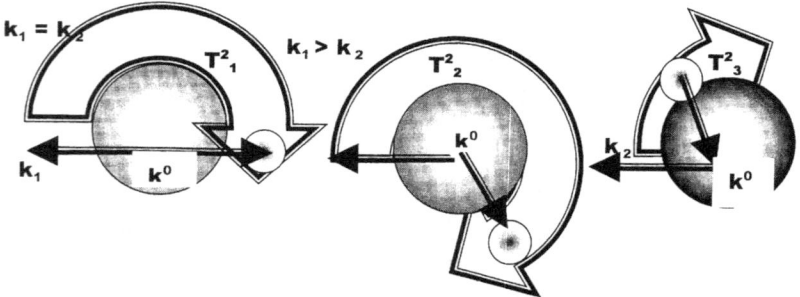

When the top starts to spin eternity is divided from infinity by such a spin. The spin establishes a cosmic identity by parting that which has no start from that which has no end and the parting of singularity in the two forms is what keeps the top in an upright stance. Through the motion the top may serve a centre in singularity that the motion charges into service. This motion identifies a Universe by attaining the standard of a Universe, as it establishes space-time by motion of space in and through time or have the entire Universe being the top in representation thereof while spinning or when not spinning have the top form a

Black Hole where that Universe which was the spinning top, then by not spinning, allow the entire top Universe that was, to collapse all together.

In outer space the bicycle will willingly orbit as a satellite as long as the orbit speed supersedes that of the Earth. This has nothing to do with mass. The second the orbiting speed lags behind that of the Earth orbit velocity, the satellite bicycle will fall to the Earth and that too has nothing to do with mass. Galileo proved that when falling the bicycle would show no indication of mass influencing such a fall since all object descend at an equal rate. Again the proof is on the velocity discrepancy and has nothing to do with mass. The small object will fall as easily as a large object when speed discrepancy allows such a fall to commence. Not at any stage does mass come about to render the object with more mass superior in any way or to be at a disadvantage in any way. As one can see an object such as the Earth in close proximity has more duplication as that of what a natural bicycle in orbit would have, therefore both objects in the duplication process must set a trend to match the duplication that the Earth establish or become a part of the duplication process of the Earth in the event where such duplication of a less prominent object cannot sustain the relevance which the Earth in total establishes.

$$k=k^{3-2}=k^1$$

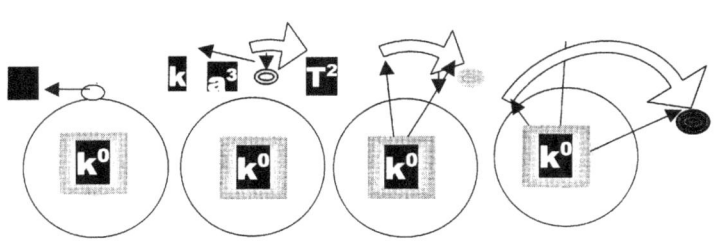

By being in motion there is no evidence that such differentiation is brought about by mass being a factor but everything depends on the relevancy

of duplication or sustained orbit velocity that establishes the trend or relevancy of

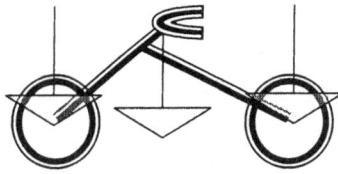

motion applying. Where duplication does not presume in relevance a discrepancy in relevance in terms of motion and not mass will bring about that there has to be changes in the relevancy that the motion provide. Mass is no factor to be descried anywhere in the whole assembly of motion creating space by duplication of space -time.

By accepting the mass that the Earth offers the dropped object and the relinquishing of independent motion, the object is offered a far better duplication

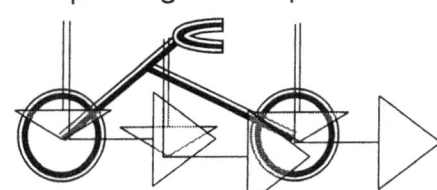

by being part of a larger unit and the duplication as a unit is able to sustain a larger relevancy in time in the period of spin. The bicycle becomes a part of such duplication where it absorbs the duplication by being part of such duplication. The motion persists as equal to all until it hits a

blockage where mass comes about and there the equality changes to identifiable differentiation as a result of mass. However the mass secures the objects new identity as part of the Earth.

Then while being on Earth and with a desire to move above and beyond that what mass killed, it then has to establish a higher rate of duplication than what the Earth provides in duplication to find such an ability as to move with and in mass. Only life can provide that feat but life is only on Earth.

In order to move while in mass it requires a source or a supply of energy that will extend the duplication of its atoms by increasing the accumulative heat they

assemble. Under normal cosmic conditions that is not possible and the only place known where this might take place is on Earth where the qualities that life may render can accomplish that. Mass has killed individual motion but with the aid of life and where the object has such motion inherited from the Earth life can make use of the qualities of duplication that the Earth provides and extend additional motion to the already inherited motion the bicycle receives from the Earth as being with mass part of the Earth motion. Considering the number of atoms the bicycle as a unit has to reproduce to

substantiate its motion in space through time and comparing that to the space and time involved where the Earth perform the very same task one can

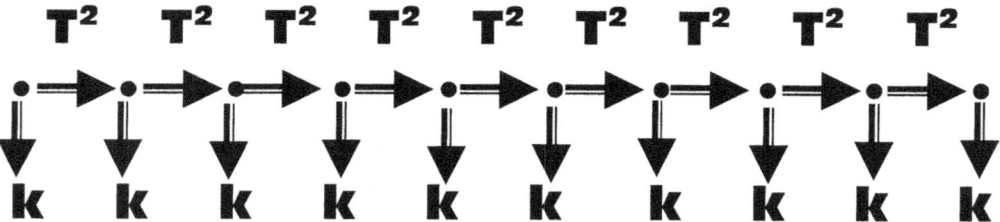

see the relevancy that applies to the duplication of the Earth and that of the bicycle. Just duplicating shifts the Earth through time by an innumerable larger relevancy, which is completely beyond the duplicating possibility of the bicycle, where the bicycle has to rely on individual duplication and therefore it is very unlikely that the bicycle can sustain an even greater pace than the Earth does. The space a^3 that applies to the Earth demand so much more generating of gravity in the circular factor of T^2 as well the linear relevancy k and it is clear that the Earth's relation to time will overwhelm and overall the bicycle's time component many times over.

$$T^2 \quad T^2 \quad T^2 \quad T^2 \quad T^2 \quad T^2 \quad T^2 \quad T^2 \quad T^2$$

$$k \quad k \quad k \quad k \quad k \quad k \quad k \quad k \quad k \quad k$$

When considering the layout of the sphere and the fact that both are just opposing factors in the same sphere the one would regard the other by 90^0. What is T^2 to the lesser will be k to the more progressive because in both cases a^3 is directly the linear factor to the other a^3. As soon as the Earth relevancy k supersedes the orbit motion of the bicycle T^2 the factor k will override the motion in the relevance factor of the bicycle and the bicycle will start to "fall".

The falling is the fact that the superior relevance of the Earth by factor of k will overall the motion of the bicycle by T^2 whereby the Earth factor k will render the bicycle factor k of no sustainable value. At that point the Earth relevancy factor k

annexes the bicycle motion but the bicycle still has no mass. The bicycle only starts to capitulate its time factors in favour of the total domineering Earth time factor and the Earth take position of the bicycle by supplying a time factor the bicycle in orbit cannot compete with.

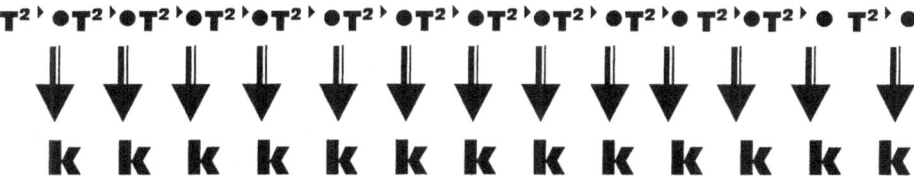

The Earth then strives to liquify the bicycle by the measure of $\Pi^2/2$ and the bicycle loses its time factor of T^2 in favour of accepting the Earth T^2. The bicycle still has no mass but the motion the Earth renders to the bicycle leaves the bicycle totally dependent on the Earth to provide motion in the time the Earth domineers.

When the bicycle capitulate it's motion and completely accepts the motion that the Earth provides, the bicycle will continue at a descending rate of $7(3\Pi^2)$ notwithstanding what mass Newtonians may connects to the structure. If that is not the case Galileo is completely wrong and Galileo is everything but wrong.

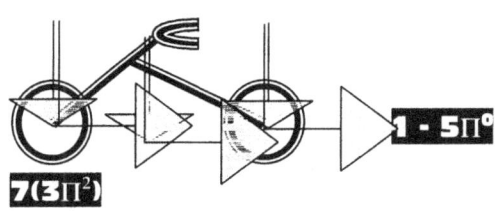

After landing in mass can the bicycle again move but only by the intervention of life will the bicycle again find motion and all motion thereafter will extend the Earth motion $7(3\Pi^2)$ by an extending factor of Π^0 to $5\Pi^0$. The motion will go from $7(3\Pi^2) \times \Pi^0$ to $7(3\Pi^2) \times 5\Pi^0$. This fits any and all criteria notwithstanding mass of what magnitude is involved. When the density caused by motion exceeds a limit of $\Pi^2/2$ in relation to the material the ratio is so thinly spread the air cannot even support sound waves travelling through the thin air.

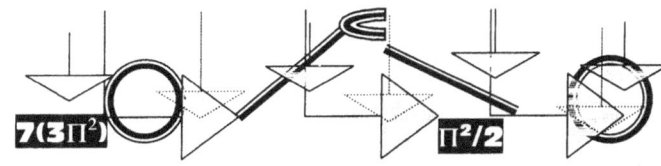

Again all of this is relying on motion, which puts a relevancy between the moving material and the space, or the time that it moves thorough. When mass is a factor such relevancy is predetermined by the Earth and the ratio the Earth maintains with air or space or time which it really is.

However where life does create artificial motion because life is artificial motion we find the density between material and space redeploys new standards applying to mass. At a point where the bicycle has artificial motion to the value of $7(3\Pi^2)$ times $2\Pi^0$ (this figure does vary considerably in correlation with the structure profile and layout and therefore it is only a suggested figure given on my part) we find that mass will tarnish and the bicycle will at one point become airborne. This has nothing to do with wind

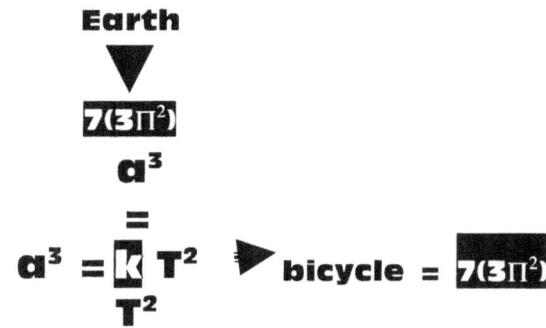

rushing through underneath the car because to counter the air rushing through underneath the car there are even more air rushing over the top of the car so in that sense we find mechanical engineers are as Newtonian as the rest are. The increasing in velocity will establish a changing density factor between the material and the air it connects to in a specific time period. More air makes contact with the same material, which makes that less material makes contact with the same space and in ratio with more air less space starts to fly. It is exactly why water skiing has a person glide on water and not sinks into water.

The moment the bicycle accepted mass it adopted the Earth motion and therefore all other motion is an extending of the Earth motion. That is why Galileo found all

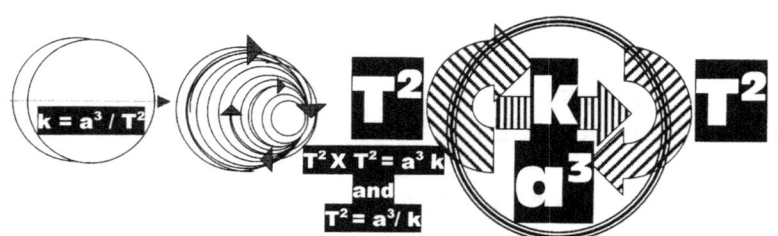

objects fall equal because when falling all objects fall in relation to the Earth singularity which, is $7(3\Pi^2)\Pi^0$ and mass does not yet apply.

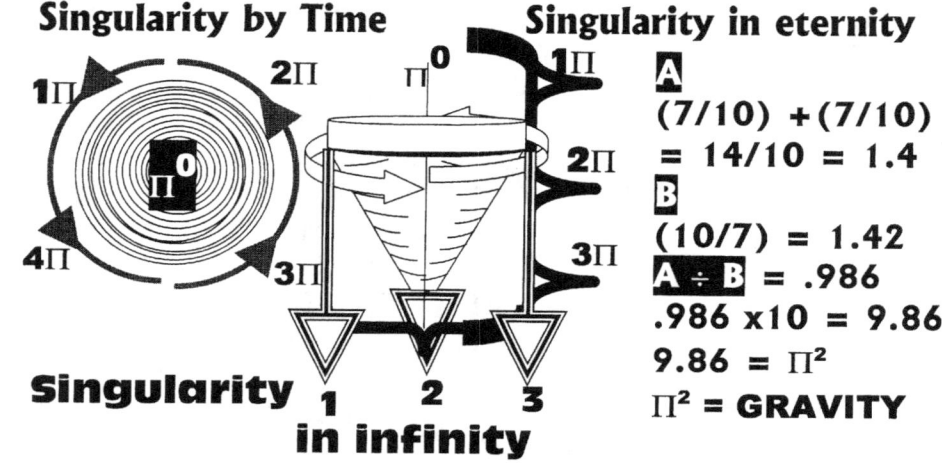

Singularity by Time

Singularity in eternity

A
$(7/10) + (7/10)$
$= 14/10 = 1.4$
B
$(10/7) = 1.42$
A ÷ B $= .986$
$.986 \times 10 = 9.86$
$9.86 = \Pi^2$
$\Pi^2 = $ **GRAVITY**

Singularity in infinity

When entering the Earth atmosphere the descending object must either adapt to the Earth motion and adopt the Earth or become liquid with a bang as Mir did. The Earth holds more atoms that has to establish more atoms from the past to the future by way of duplicating every point that is confirming the presence of material and since the task of the Earth involves more duplication than the bicycle has, the time in effect of the bicycle will have the Earth absorb such time if the bicycle of the bicycle cannot find a means to duplicate more ferociously than the Earth does. It is all a ratio with heat that science never detect and is even part of life. When we breath we breath oxygen but that also is oh so Newtonian because what do we do with the oxygen. We don't eat or drink the oxygen but we do take heat from the oxygen and it is the density of the oxygen carrying heat to our blood for use of our fibres that we need oxygen. We need the Heat the oxygen associate with to use to build our vessels and the building we call aging. Only motion can turn singularity from the sot it has in the sphere to the dot to form a line that takes charge of the independent object, which is in motion and is in relation to another centre. This confirms Kepler's $a^3 = T^2k$.

The time activates a line from which space begins. The line forms a space that turns the top, but just as important is the fact that the line extending singularity activates a dimension in time in which the top can spin.

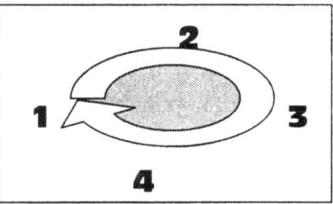

It is the time that relates to the spin of the space filled with material that allows the position the top has when spinning. When the co ordinance between the space spinning and the timeline keeping the balance falters that the top looses independence and fall.

The issue is that the rotation activated the time component where four points in singularity moves about a centre point in singularity and the four points moving around such a centre point extends the centre point by establishing a line of three points in the centre of the top. This line can never move. This forms the basis of the Coanda principle as the rotating four points forming a circle around the top then becomes the motion that acts as liquid, although the wood is solid.

That also provides a line in the centre that can never move and that line forms the solid part in the partnership. The point around which the rotation turns becomes the solid. This is evident by the fact that the line holding three points is activated by the motion of the circle holding the four points while not participating in the spin. As the spin reduces in dynamics the turning still present a fight to the last to preserve an individual gravity that sustains a cosmic reality. The one position in time that is turning the four points around 1^0 which is the original singularity, represents eternity in time, the part that forever turns because it is time.

Infinity is the part on the inside, which represents time where time has no start since infinity is always part of eternity until space parts infinity from eternity. The instant eternity shifts from infinity through the evoking of space by motion separating eternity and infinity space-time comes about. But we are dealing with the cosmos and what was in the cosmos at the first instant is still in the cosmos at this moment because that which was part of the cosmos at the start has no where to go but to remain in the cosmos and be part of the instant of the cosmos in eternity.

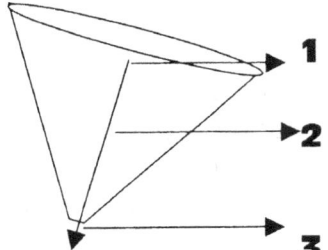

The cosmos did not get bigger from the first moment. The cosmos is not expanding but is reducing. That which we cannot see where 1^0 parted from 1^1 is not to the inside of where we are but is so big it now serves as home to all there is within the cosmos and for what which was to become, to be able to be where it became, the cosmos had to reduce to accommodate that which now is where it is located in its allocated position in the cosmos. That which parted when the cosmos came about is now so big it is to big for us to see because we are too small to bear witness to something that large. The line was at one stage one half of the entirety there is and the other half is the entirety we are able to see and the rest we are too small to see. That what now is. Was once so massive it could only be parted from that which is so small it clearly is not.

The contracting of space diminishing as a result of singularity established

a^3
$\Downarrow$
T^2
k

The motion provides new space and the new space provides motion and in the midst of all of this a point forms through the motion where space will flow towards and disappear within that point of space less ness because of motion less ness.

$$\Rightarrow a^3 = T^2k \Rightarrow \Downarrow \qquad k^0 \qquad \Uparrow \Rightarrow a^3 = T^2k \Rightarrow \Downarrow \qquad k^0 \qquad \Uparrow \Rightarrow a^3 = T^2k$$
$$k^0 \qquad \Downarrow \Rightarrow T^2k = a^3 \Rightarrow \Uparrow \qquad k^0 \qquad \Downarrow \Rightarrow T^2k = a^3 \Rightarrow \Uparrow \qquad k^0$$

Space duplicating is creating motion and from the motion space is created.

The establishing of motion is creating space and providing a centre point of singularity. This proves that the atom links with singularity and that the value of singularity extending forms a relevance with the space-time it influences up to a point that it controls the space as much as it controls the motion of the space. That proves singularity extend way beyond the space of the material it holds and establish duplicating points of singularity within singularity by merely applying motion to space within space.

The presence of the Coanda gravity principle is portrayed throughout the entirety of the Universe and every galactica holds the Coanda principle as proof that this is the way gravity applies. Yet no Newtonian ever bothered to see the indicating of the Coanda effect that produces gravity by providing motion in space and why gravity is used as the glue that is bonding the all of "the everything". But it has no mass and therefore in a thinking manner the Newtonian brain switches into a denouncing mode.

That means the second motion coming from the object travelling through space which arrange its atoms in the same time duration as that it would have used while being stationary on Earth pushes the electron circle out of the normal sequence that electron had while being in a steady state of motion on Earth and under the control of Earth time. The principle, which I described in the motion of the spacecraft travelling through time, and space in space-time also apply in the same manner in the Coanda effect. It is the same motion producing gravity by applying a space relevancy that changes in the same duration as when not in motion but through motion the space changes to, a new motion that comes into play. In the motion producing a relocated centre a new centre will play its part in the compensating motion that will force upon the relevancies a new space-time dispensation.

When the motion comes about we find three positions in time moving on the one side of the divide where such divide is formed by singularity and we have three positions on the other side of the same divide and because it is motion such motion transforms the line to a half circle as much as it translates to a triangle.

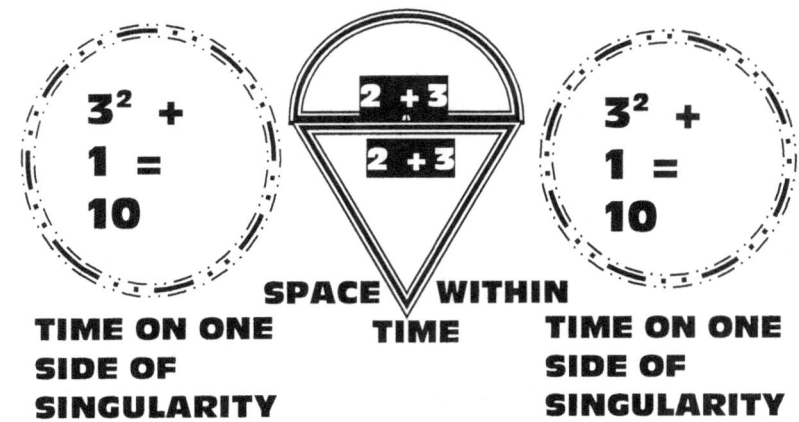

TIME ON ONE SIDE OF SINGULARITY   **SPACE WITHIN TIME**   **TIME ON ONE SIDE OF SINGULARITY**

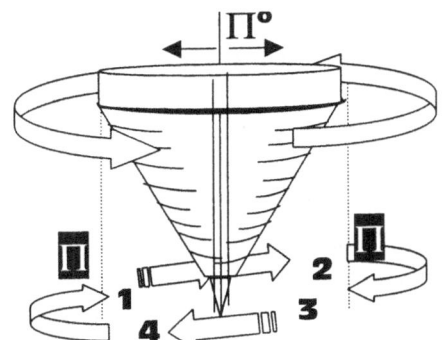

The motion goes around a centre that forms the half circle and is commit ting one side of the motion to oppose every value of the other side in a relation.

On the one side we have T^2 forming the space in relation to a centre and also on the other side we find the rotation committing the space defined by the half circle in relation to the centre.

We therefore have $T^2 \times T^2 = a^3 \times k$ and that positions the completer relevancy brought on by singularity to complete what the motion provide that space to have in relation to a centre established by such motion. It is the spin on both sides of the circle that establish and define the space in regard to the dividing centre and from the centre to the outside is space in time This is

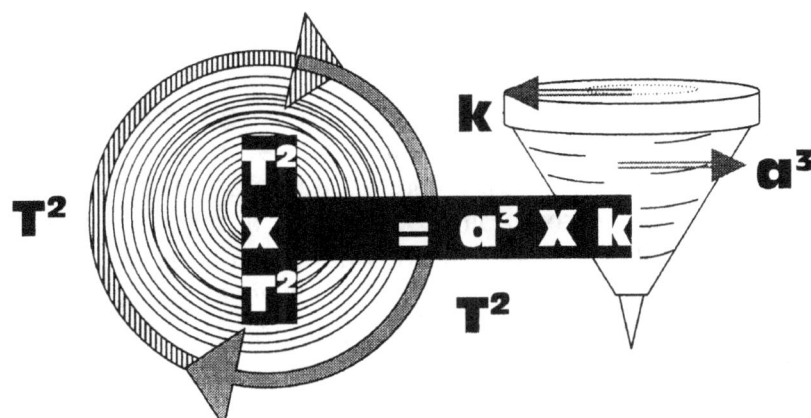

what Kepler's formula confirms when one take out all the hideous misinformation Newton connected to the concept. The motion in rotation T^2 confirms the space a^3 in relation to a specific centre X k or mathematically equated as $T^2 = a^3 \times k$ that one easily derive from $a^3 = T^2 k.$

Mechanical engineers propagate that it is air flowing underneath cars that causes the lifting of the cars at high speeds. That is as Newtonian as blaming it on insufficient mass or some other polony story they come up with. We find the top doing the same at high peed as the car or the high-speed boat does in water, where the top tries its best to lift off from the ground. If they are correct about the wind underneath the car the next question that comes to mind is what wind is lifting the top when the top starts to move rapidly?

One should remember the wind in the case of the top must lift a needle if it is wind that is lifting the top. The process is identical and the top indicates to what incorrect surmising the engineers are capable of. The top is in motion because of it duplicating its space position in time in relation to the time it is in contact with during the instant as well as the duration of combined instants.

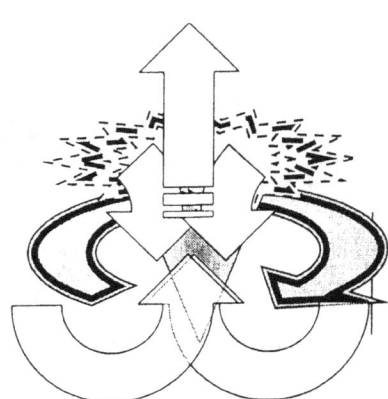

The top has the material it holds in space becoming more because of more duplicating through rapid motion and as the electron wants to extend the orbit in relation to the time factor when the electron becomes more active, the top also wants to extend the relevancy of k because the atom material becomes more in relation to the time aspect by which the duplication is controlled. With all the spinning and the commotion that the additional heat brings about by which the governing singularity extends, and by extending it is promoted to a higher time zone in the atmosphere. The singularity suddenly is in charge of more space-time by which the space is duplicated in time. This is the line that science always shows, which runs from where it grew as a motionless point in the centre's centre.

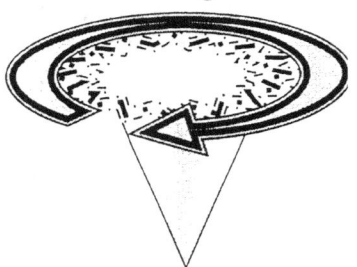

With all the excitement and nowhere to take it, the extending goes down to singularity keeping the whole top erect. That is why the top is spinning in the first place. We know the spinning is attributed to the will of life but in the cosmos where life does not apply the spinning would be a result of gravity accumulating and concentrating heat in a small area. The more assertive the spin is, the more reaction there is from the lines running towards and extending outwards. It should in real terms expand the top because the motion creates more space per time unit filled and the space has nowhere to go but to excite the newly established centre singularity. By duplicating more space a^3 the rotation of time T^2 provides for a larger extending of time k^1 to extend the boundaries of a^3, which is the result of a more ferocious spin in T^2

The spin under normal conditions can only come about as a result of heat being more concentrated by space being rapidly reduced. That in short is one part of gravity. With that aside the spin normally comes about by the exciting that brings on rapid movement, as is the case when water falls on a hot plate. This proves that material grows by removing heat from the time sector where the time component $k^{-1} = T^2 / a^3$ reduces its value and material is gaining from such motion $k^1 = a^3 / T^2$ by the measure of k as material grow.

This is consistent with the Big Bang as well as the Hubble shift because time removes its position by allowing more space to form in between particles of matter. The linear component of time k^1 removes heat from time to singularity being in $k^1 = a^3 / T^2$ by introducing $k^{-1} = T^2 / a^3$ as the flow of space-time from outside the atom to inside the atom. That means liquid heat is removing from time in the space between particles $k^{-1} = T^2/a^3$ and is adding to material $k^1 = a^3/T^2$. That puts the Universe in as much contraction as it is in expansion and secures the perfect balance we would expect from a harmonised Universe.

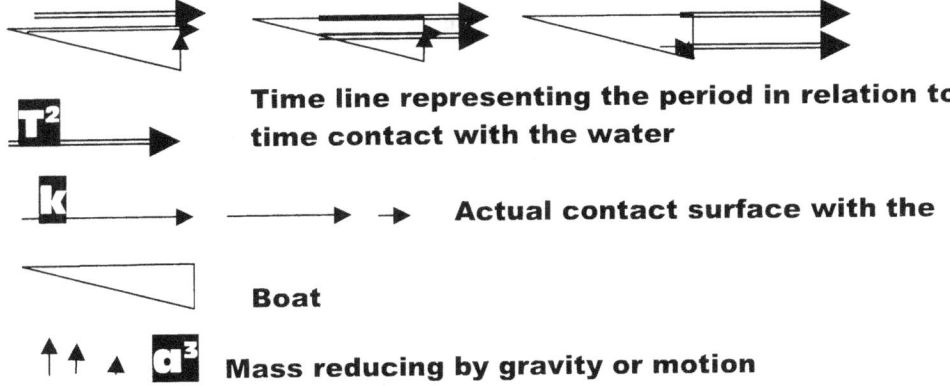

T² → **Time line representing the period in relation to time contact with the water**

k → → → **Actual contact surface with the**

Boat

↑ ↑ ▲ **a³ Mass reducing by gravity or motion**

More spin increases both line directions $k^1 = k^{-1}$ that produces gravity by applying rotation T^2 where the rotation is the result of the extending of k where k^1 removes heat from time $k^{-1} = T^2 / a^3$ in order to add it to space by the measure of $k^1 = a^3 / T^2$. By heating space exceeds its boundaries which the liquidity of time allows through the Coanda effect and in the motion that results from the expanding of material by acquiring additional heat to sustain form the motion allows heat to extend by the margin that the liquid provide such extending to come about. Therefore the extending is both $k^1 = k^{-1}$.

The racing car getting air born and the aeroplane lifting off and the water skier gliding over the water and the top spinning about its axis is all the manifestation of Kepler's formula where Kepler (wittingly or not) predicted that by moving space a^3 through time $T^2 k$ the movement will produce space a^3 by ratio of time T^2 allowing the frequency k of such duplicating $a^3 = T^2 k$.

Motion is the result of space duplicating $a^3 = T^2 k$, exactly as Kepler said before Newton interfered with work that was totally above his (Newton) level of comprehension. A boat lifts from the water by the increase of velocity because the increase of velocity reduces the mass the boat has. The more motion there is the less mass the boat will have because the more duplication the motion allows the less water displacement the boat will produce and the water displacement is the mass the boat delivers in the water.

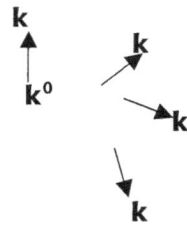

By applying motion the flow of the liquid around the round solid which represents as much as rein acts singularity the motion brings about a new space-time as the factors have to compensate for the motion and the motion in itself brings about a new controlling k^0 in the position

elected by all atoms forming part of the relevant motion. The motion introduces a securing of the position that is elected by all the atoms and is forming the new part that will serve as a controlling singularity by substance producing the motion or the substance of the solid securing the position of the newly elected controlling singularity. By applying motion the Earth gravity is interrupted by the space producing motion that through motion takes charge of the space and turns the vertical flow of Earth space into a

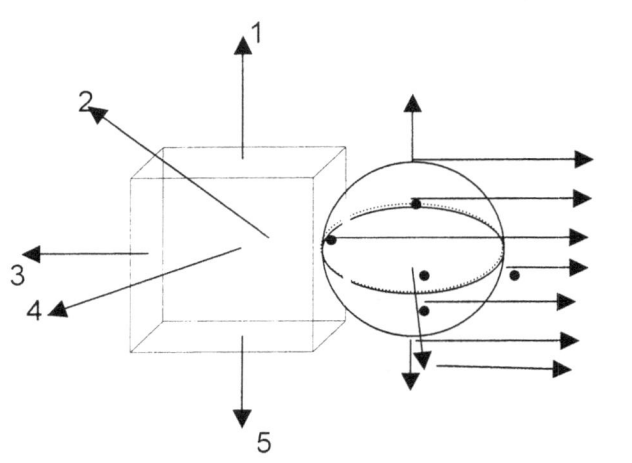

circular motion although only for a short while. This is how electricity is charged but the motion then involves the reducing that the flowing electrons undergo to compensate as much as producing speed of light. Other elements and their characteristics are also involved and in the charging of electricity the Coanda principle personifies the gravity that comes about in the condensing of space-time around singularity.

By reducing the space towards the centre in the using of the Coanda effect re-insures the correctness of the Big Bang theory. A newly established centre will create a newly controlled Universe and the motion of Iron $_{56}$ will force space towards the newly established centre. By forming the flow of space in relation to copper the space will break down and form the inside we find the space to flow on the inside of the Earth centre. We have to remember that the relevancies, which applied during the Big Bang is present in a minute capacity on the inside of the Earth.

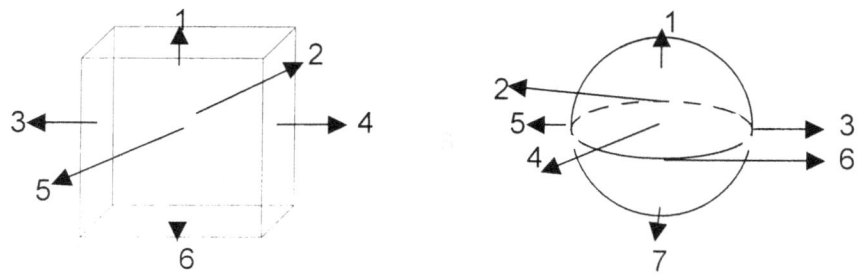

In the cosmos we find two forms available to use. The one is a six-sided cube that connects at the edges and can be found in whatever name one may attach to the angles that connect the sides but after all it still remains six sides with three sides facing three opposing sides. Then we have a sphere, also with six sides with three in opposing to three others but in this case all the sides connect via a precise centre and the centre gives the value of a seventh position that secures the precise location of all six edges and that keeps the sphere true. In the case of the sphere there are seven points where the six points representing the sides connect through the centre and the centre connecting is charging the form of the sphere to Π and with the centre forming singularity the centre is $\Pi^0 = 1^0$ that holds the form in charge and in truth.

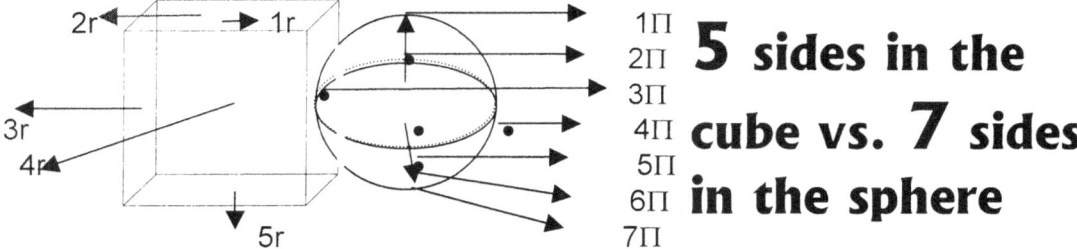

When the cube, which is having six loosely connecting sides comes into contact with the sphere with six sides but also with a centre that charges gravity by its positioning of singularity, the cube is destined to lose the one side that forms the contact with the sphere. The centre in the sphere removes the cube side that has the contact and the sphere allows that which is on the edge of the cube to become committed to the gravity of the sphere by descending to the centre. The centre will charge a reducing of space through implementing motion and the evidence of that we find in the Coanda effect which I shall explain later on. Only where such descending is no longer possible because material that forms the edge of the sphere will not allow further moving by the descending object to the centre. Then at that point where motion becomes a tendency to move, does the object receive a position of mass.

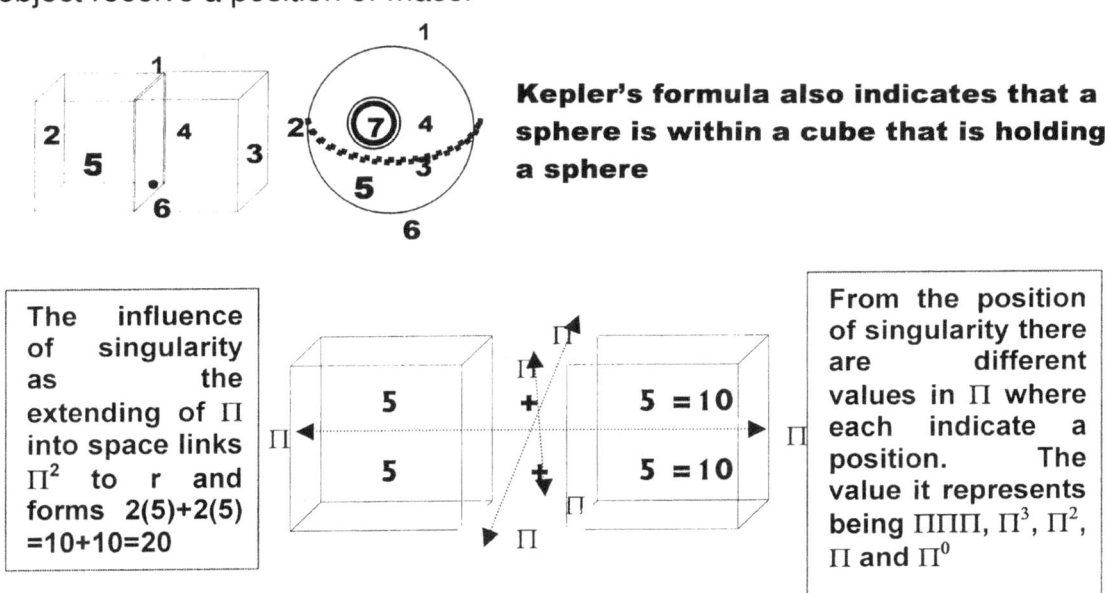

But it only becomes mass after the motion or gravity that it had became restricted and the motion becomes frustrated and confined. The motion ends where mass confines the object to the motion that the larger object confirms. As the sphere

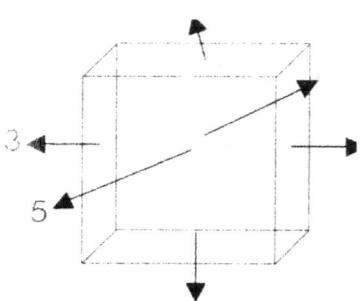

spins it have seven positions in the sphere that is eternally part of the form the sphere has, acting as one while there are six in the cube but with the demanding sturdiness of the sphere, the sphere removes the one side of the cube where contact is made. The six sides in the cube then become five sides and five in the cube then has contact with the outside leaving one point in contact with seven points in the sphere.

The centre will dominate because the centre charges seven points to gravity while the cube cannot secure the position of any particle that is in the cube. The cube with five lines that loosely connect and that don't form any firm position to secure material within the cube will be conformed by the spinning centre of the sphere, which generates gravity. The fact that the cube lacks any ability to confirm material within the form it secures shows no structural bonding in the cube.

Therefore, where the sphere and the cube touches the sphere depletes the space the cube holds and remove any material to a position that relates to the centre of the sphere.

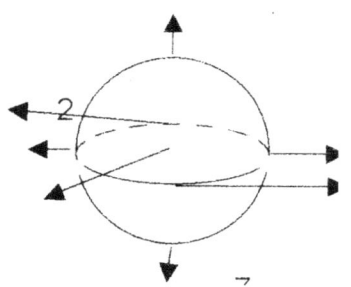

The loosely connecting sides that concludes the cube as form will lose one side to the more overpowering structural form the sphere has and the sphere is going to capture one side from the cube. Then with five sides standing related to six sides and a seventh in the sphere centre, which also is commanded by singularity in the centre, the pure domination is going to affect the motion of the material that is going from the cube to the sphere because of the six points that is spinning around the seventh point.

In that the cube is represented by five sides forming a form plus five on the side also forming the cube where both are constructed by the loosely connected cube in form that stands in relation to seven points connecting in the sphere and the sphere has seven to the double cube having ten and that is why the Titius Bode holds a form of ten that stands in relation to five on the one side and another five which is part of the following direction. Space –time is built in such a manner. As the movement propels it will move from a cube of five to a cube of five and the total sides being affected will be ten on the side of space.

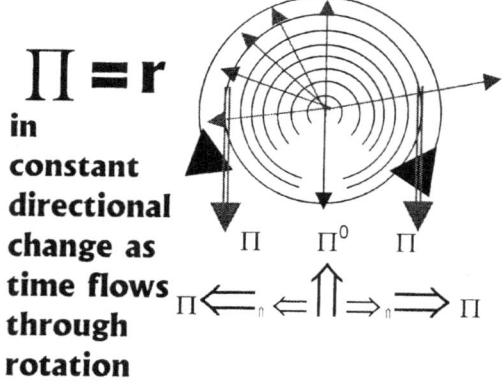

$$\prod = r$$

in constant directional change as time flows through rotation

Pinpoint positioning of singularity $\prod^0$ with $\prod$ positioning space to either side forming the border set by singularity

The new direction pointing to a new location in relation to the previous point will oppose the previous point it had in relation to direction considering the centre point.

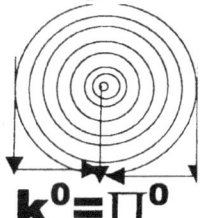

The figuration found in the planetary and has taken the name of the Bode or the Titius Bode system such a system is the mould that remains in space as part of space after the concept establishes the relation and as such is then detected by observation and growth leaving the mould in space as the departing of planets from a space concept in relation with the centre. It is a manifestation of the evidence what Kepler brought from his investigation of the cosmos. Gravity is where space vanishes and as such space is in the centre of the sphere. The sphere controls not only the space included in the sphere but also the space excluded or on the outside of the sphere.

$$k^0 = \Pi^0$$

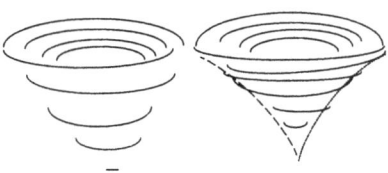

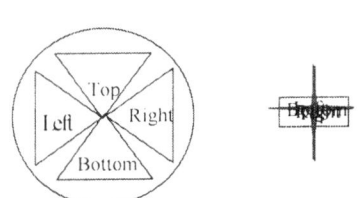

To find the invisible I had to locate singularity. I realised that my effort to locate the point holding singularity enabled me to backtrack the exploding universe to its origins

By reducing r indefinitely to the tune of half each time, r would become infinitely small, beyond human calculating means, however as mentioned in the case of the smallest dot holding one spot, r would become insignificant beyond human comprehension even, but never reaching zero and still Π would remain intact and dictating form. I believe one can begin too see where my suspicions are heading because the flaw comes about in the manner mathematics are practised for thousands of years. But before coming to the mathematics I would first like to bring your attention to the practical side. I am promoting a theory in which I am able to prove there is as much contraction going on in the cosmic universe as there is expansion and the contraction is as much part of the expansion. The universe rides on a balance and we have to locate such a balance. To prove my theory I firstly had to locate the centre of the universe.

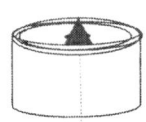

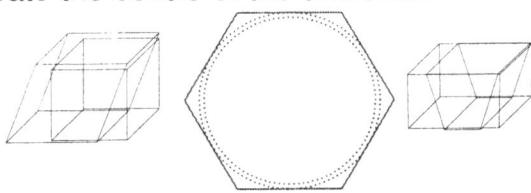

Even admitting to such a notion sounds like madness, but please give me a chance to explain in more detail. If I wish to achieve success that would depend on my ability to convince all that outer space comprises of material and as such we can locate such material even if we are unable to see such material. By applying some basic effort I have located the position from where all movement came and the direction it took moving forward in time.

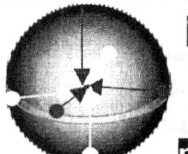

The sphere holds six sides in relation to form as unit and as does all other shapes and forms. All forms have to have at least six sides indicating different exposures to the Universe. But with gravity having a free choice, gravity always chooses the sphere. As I shall prove later on gravity is the strongest where the form produces the least evenly distributed space.

We traced the line back to the spot and so we know where it all started there was a spot. The miracle part comes in the fact that eternity shows no change. Any change, even the slightest change represents the end of eternity. With everything motionless and locked in singularity everything remained eternal because even today singularity is locked in eternity. In the centre of all objects spinning there is a space forming a divide. It divides every aspect of space spinning into sectors of space having motion as much as changing direction of initial motion. However, since it was part of the cosmos, and although it is outside the cosmos, it still is part of the cosmos. A spot holding what ever is and can be into a dimension so small it did not even have sides. Then for a reason, which at this time I do not wish to go into some miracle happened and the spot showed motion. The motion was deliberate but the motion was eternally small. The spot had one specific value, which it clung too…it had all sides on the same side and from that rotating diverting came about. The forming of $\Pi^0 = k^0 = 1$ had all possibilities available but it was only one possibility in the end. There was no space and then space expanded producing space $\Pi^0 \Rightarrow \Pi$ and from that motion comes gravity by the value of the motion creating contracting Π^2. The motion brought along Π but even Π was subject to relevancy. From this come the most basic principles in as much as forming the ground rules of the law of Pythagoras.

When drawing a line such a line then starts of with a dot serving the spot that holds all sides equal. That means the line serving as the future radius will be equal to the half circle which is then Π. The only aspect of the point that stands in for the end of the single line forming the radius of the circle is that we then mathematically reach the single dimension. We decreased the line to where a circle being Π formed on the single dimension. This dimension also hold the circle dividing line because from there the radius must once again generate a value and by such a gesture that the extending would form the circle that forms the sphere that eventually lead to the formation of particles.

Where the radius reduces by a dividing of two such dividing can never remove the radius and since the radius determines the walls of the sphere the sphere may shrink but can never become zero. All lines have to start at infinity and cannot mathematically start at zero because in cosmology as in mathematics zero has no place.

In $4\pi^2 a^3 / T^2 = G(m+m_p)$
$a^3 = T^2 k$
$a^3 / k = T^2$ but
$k / a^3 = 1 / T^2$
$k = a^3 / T^2 =$ singularity
$a^3 / T^2 = G (m+m_p)/4\pi^2$
and $a^3 / T^2 = k$
then $k = G(m+m_p)/4 \pi^2$

All Newton's changing was possibly done with good intensions but even that I doubt. The end result however was in some cases far from good, as it does not do such great credit to Newtonian insight into cosmic affairs. Only Kepler and only Kepler unaided without the intimidation and interfering of Newton can explain the Coanda effect. I grant the fact that the Coanda effect was discovered before Newton saw himself fit to change Kepler, but only Kepler can explain the Coanda gravity effect when Kepler is without the attentions of Newton.

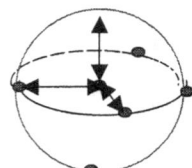

In the sphere there are never only one direction implicated in movement. Movement are always in relation to the centre position because as a line goes up it also goes in or out. When a line goes north or south, it also comes towards the centre or going away from the centre.

What Kepler saw was more of a dimensional nature than the practical mathematic symbols and values. On the one hand was a value to the third dimension, which equalled two-dimensional values one the second dimension, and one to the first dimension.

There is always relevancy present in movement. As this moving indicates direction it also apply Π^2 for indicating value forming the time factor.

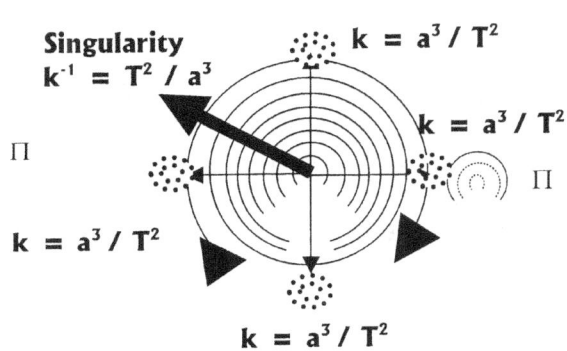

Singularity
$k^{-1} = T^2 / a^3$

Π

$k = a^3 / T^2$

$k = a^3 / T^2$

$k = a^3 / T^2$

$k = a^3 / T^2$

Π

The points duplicating is four moving around a centre by the square of gravity. The motion is the sources of heating because the heat is bringing about the movement. The heat growth therefore provides the action because the action is what energises the points to provide the motion. The motion is purely is space-time duplicating and the duplicating is feeding heat to the centre from the four points overheating thus the points that shows expanding.

The TITIUS BODE Principle Outside the sphere

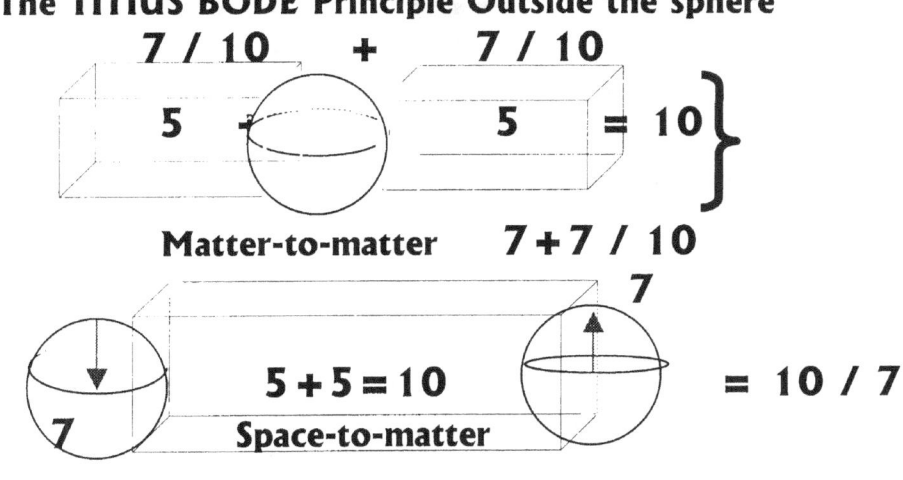

7 / 10   +   7 / 10

5   +   5   = 10 }

Matter-to-matter   7 + 7 / 10

7

5 + 5 = 10   = 10 / 7

Space-to-matter

7

In spite of all the overwhelming evidence to the contrary of Newton's mass performing gravity you persist that Newton is very much commendable and as sureties that the cosmos provide as clues to show the terms applying to how the cosmos truly works is swept under the carpet in favour of Newton's where it is almost overwhelming evident that the hoax Newton created is not working.

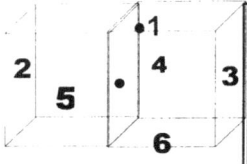

There is one more point in the sphere in the centre forming an addition in the sphere. That point holds gravity secure.

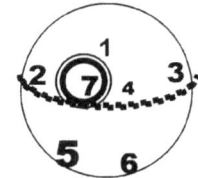

If the cosmos had any other shape the might be available contraction in the end as the final conclusion would then not be possible. Only in the sphere and more so in the circles all forming a sphere is singularity present. Singularity comes as part of the construction we find in the sphere. By singularity forming the base of the sphere the Coanda effect will forever be present and with the Coanda effect gravity applies.

As is evident from the experiment of the Coanda effect there will always be an attraction by singularity controlling the space-time from the centre of the sphere. Such control has the name of gravity, but it is no force. It is a combining of liquid (10) and solid (7) in the presence of three in time that will produce such control by motion. But why by motion?

The spinning object has seven points serving precise equality by rotating and holds the sphere in contact with the space outside by the measure of coming from a cube that has five connecting lines while the seven points that confirm the sphere in time holds seven points in the instant of the presence and those seven points are moving onto another five points serving time in space.

Let us find the smallest possible line first. We already have reached the conclusion that by reducing the line , the reduced line will eventually leave all sides on the same spot. Such a spot must be round in form. With the line being the smallest line, such a line will start off as a dot that moved away from a spot. With all possible sides being in precisely the same spot we have all possible sides onto one spot.

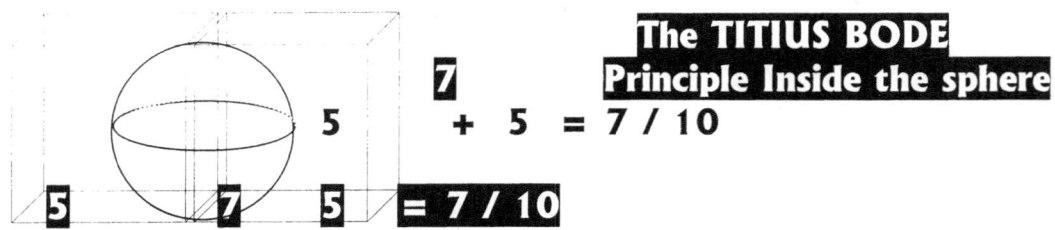

The TITIUS BODE Principle Inside the sphere

$$7 + 5 = 7 / 10$$

Mathematically the spot is in the single dimension where the space is one and exponentially zeros. There the space moved over to form the dot. We now are reaching into areas only the human mind can venture by understanding and nothing more. The understanding of this concept demands our reaching the point where the mind of the animal cannot reach. If it starts with a line that line only represents two sides being one and as such that is rather a flat Universe. The spot is not yet round because being round are requiring a shape or form and this lies beyond or before a time when any form of shape came into the cosmos scenario. It was in a period where shape and form was a part of the distant future hidden in and beyond eternity. In that time the line must have been so small it had reached a point not yet dividable in any way. If any further dividing took place such dividing would have brought growth because there then would form space between the sides going in the opposite direction. The dividing brought all there is having all sides literally on the precise same spot, and I have located singularity in just such a spot.

I came to the conclusion that the spot I found had to be singularity purely on the grounds that that spot holds only one side to serve as a start to the starting point of all directions possible. In that side is only one spot where there is only one side applicable and one dimension present. With all the factors given one can only come to one conclusion and that is that there can be only singularity. In such a case more dividing by two will land further positions on the other side of the divide. That point is serving as a position for all possible points and cannot allow further dividing as it is in the smallest line or spot there may ever be. This spot is the result of a most basic process of reduction as the Hubble constant is a most basic process of expanding during a matter of time. By reducing the line constantly the only value that will eventually remain without dispute from any party arguing about the facts is exponential zero. By only having exponential zero instead of a numerical zero and a radius as one in the square (the radius effectively becomes one holding any and all sides on one point) such a point might become any value of any significant measure implicating anything but zero as the radius. By expanding the line, it will be an evenly spaced structure growing into the most perfect round dot ever possible anywhere at the point when it starts to grow.

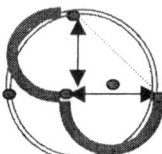 Because every moving line represents one quarter of the sphere in relation to the rest of the sphere and the line also indicate the relevant position between the point indicated and the point in the centre it is a relevancy of singularity in progress. By connecting the line, as Pythagoras will suggest the singularity within the sphere become a specific value indicated representing one half circle.

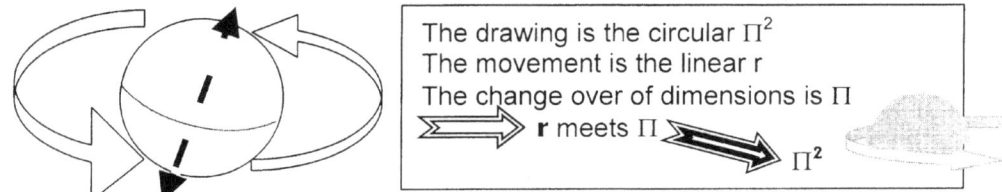

The drawing is the circular Π^2
The movement is the linear r
The change over of dimensions is Π
⟹ r meets Π ⟹ Π^2

The result of the five to one relation comes the Titius Bode principle.

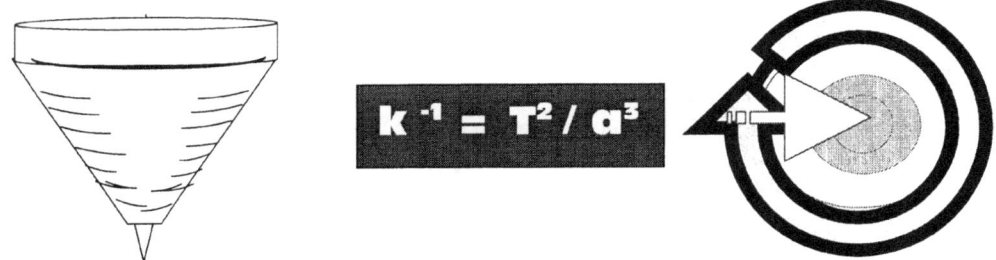

$$k^{-1} = T^2 / a^3$$

The motion established by singularity results in the implicating of the Coanda effect as much as the motion establish the Coanda effect. The spin realises the space limit while the space limits attaches the motion onto the space in the time within the time.

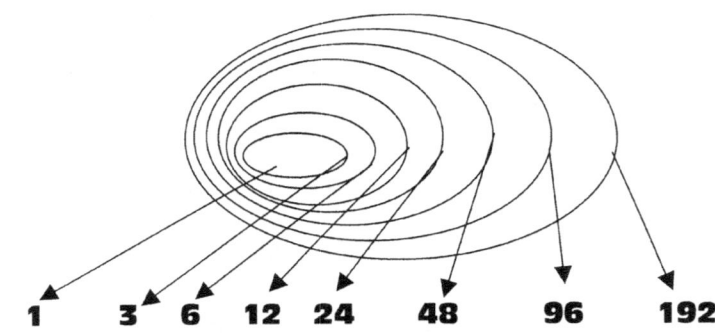

1   3   6   12   24   48   96   192

Planet	Mercury	Venus	Earth	Mars	Ceres	Jupiter	Saturn	Uranus
Bode's Law dist.	4	7	10	16	28	52	100	196
Actual dist.	3.9	7.2	10	15.2	28	52	95	192

The Titius Bode law is form where space holds a relation with material and space finds form in the cube being unconnected while material finds form in the sphere and is connected through singularity that is activated by motion. As singularity spins space converts with material and the Titius Bode law of 7 /10 and 10/7 conjuncts with the Roche limit value of $\Pi^2/4$ which finalises the value of gravity Π^2. The sphere holds seven points and the centre by dividing proves to form singularity, however the singularity can only activate into representing an active line in singularity by motion that charges singularity to take control and form the centre line. With the accepting of singularity as a point where the Universe started and that everything grew from such a point we must first allocate a value to such a point. The only value that such a spot can have while being is 1^0 because from being a spot and growing into a dot would take the development of coming from 1^0 and developing to 1^1.

Being singularity and progressing into singularity could only be at a level of one because singularity means 1. The singularity is still with us and we can see it every time a round object spins. It's a line that is but also is not in the centre of every rotating object where 1^0 are flanked on every possible side by 1^1 that has one opposing the other to every possible side. The main thing is that I can point to my claim that suits all the criteria astro physics prescribe that singularity should have but also I can say that a person is to go and look at the centre of a spinning object and find such a line there. It is a mathematical point that forms a line with the aid of spinning. By the motion it gives the object spinning a freedom that the object is willing to fight for before relinquishing such a freedom. It allows the top to stand up straight and spin as an independent Universe amongst all other independent Universes there might be. It promotes singularity, which is inside every sphere to become a line, which generates independent space a^3 by the motion $T^2 k$ thereof. The space holds autonomy as the top shows while it spins and the spinning charges independence and the independence is only achieved through inflicting independent gravity onto an existing time format. The Earth is 1^0 to the Moon being 1^1. The Sun is 1^0 to the moon being 1^1. The proton is 1^0 to the electron being 1^1. There is an eternal link in singularity and all four phenomena prove the distinct reality thereof.

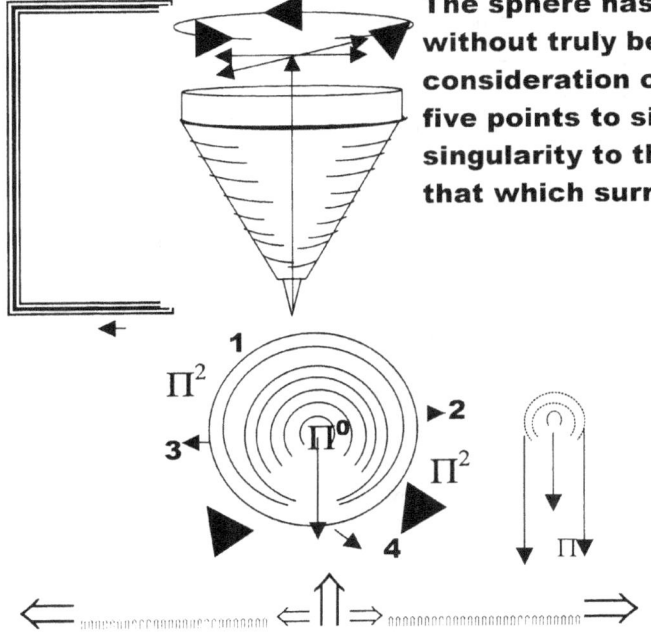

The sphere has seven points. The cube without truly being a cube but is just in consideration of having a cube in form holds five points to singularity. In the centre runs singularity to the value of Π^0, which means that which surround Π^0, holds a position of Π^2

The spinning sphere activates the seven points, which places gravity in relation to a centre. Outside the centre there are five sides by dimension. The sphere has seven points of which four is spinning. The four spinning stands related to the gravity of spin, which are Π^2.

1^0 going 1^1 1^0 ▶ 1^1 had to bring about 1^0 going 1^1 1^0 ▶ 1^1, because the eternal repeat of duplicating while contracting was not relieved from the Universe. Before the contracting was equal to the duplicating because by measure the heat was identical to the cold. It was eternity that was interrupted by one cycle of infinity and was in repeat of eternity. Once something is part of the Universe there is nowhere else to take it so it has to remain as a part of the Universe.

The sectors provide individual singularity a means in sustaining governing singularity by which provision comes through maintaining governing singularity the required spin in maintaining cooling. If this process did not apply, there would be no connecting individual singularity to major singularity

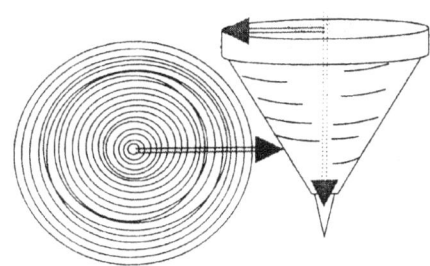

In the **precise middle** of all **objects in rotation** is a precise centre dividing the object in sectors that will **start the spinning initiation** from that centre point. As the rotating direction moves inwards, the rings will become smaller and smaller. Thus, the spinning object **will have a middle point**, a very specific **centre point that does not spin** and only holds Π as a specific value because no radius can apply. But also the one value such a line **cannot have is zero** because the line **is there and holds contact** to the rest of the material bringing about that **zero does not start any** line and therefore the **value of the line must be infinite,** just as described in **accordance** and by **the definition of singularity**. There must come a point where the ring is infinitely small, where it can reduce no more, where it reached its ultra limit, but at that point it cannot be zero, because the point is there

The centre cannot spin. The centre divides every aspect of the spin from the next and the previous where every aspect changes its relation it had with time and with the rest of the Universe in total. Everything that connects to the centre to the one side is defined by a line that is not present in our three dimensional Universe since we can't accommodate a line with the value of 1^0 going to 1^1. Being what the line is and where the line is the line has no sides but leaves all possible sides on one centre location. That is singularity, which is the smallest line or point that can ever form.

Having edges where Π^0 duplicate to present the edges singularity lost the value of Π^0 to the value of Π^1 with the same value singularity had being Π^1 to the one side and Π^1 to the other side, Π^0 must be the point splitting singularity into two parts of eternity, the eternal value of the first dimension outside eternity. It was the square of Π^1 being Π^{1+1}. That was the first dimension outside singularity Π^0 where singularity has a value of Π^1 in the form of $\Pi^{1+1=2}$. The first claim to space had a value of Π^2. This applied to both sides of the claim to space outside singularity, and the double proton became the dominant factor on matter.

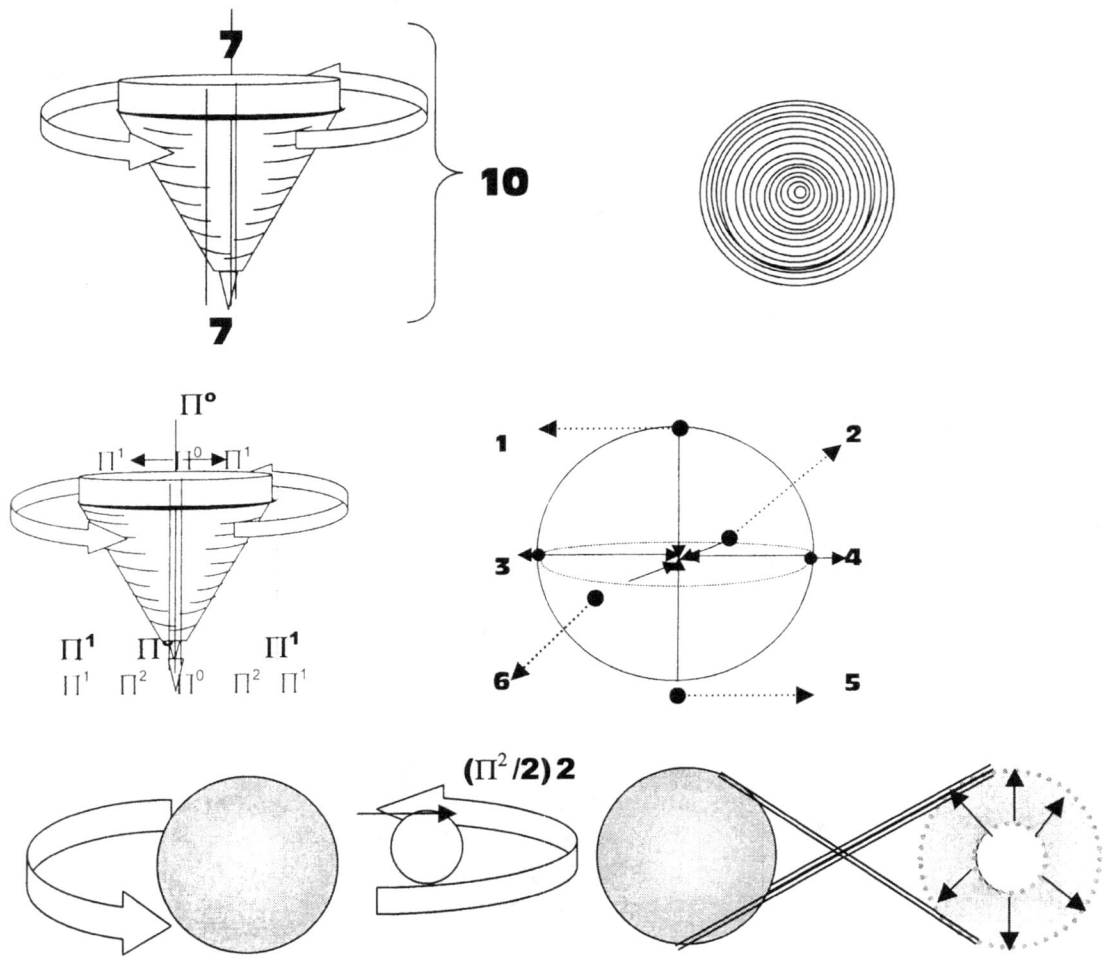

Where any object is closer than $\Pi^2/4$ it becomes part of the motion of the other structure, which is reserved for liquids. The Coanda effect becomes where the material will be turned into a liquid and the liquid becomes part of the Coanda effect action.

The Roche limit is evidently widely applying throughout the entire cosmos but Newtonian rules do not explain the phenomenon:

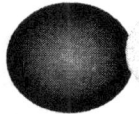

Any person taking Newton seriously should at least take on the challenge and find the comets colliding with the sun, find how much the planets moved closer to the sun since the days of Newton and indicate where there is unprecedented collision between stars. Yet the closest the Universe comes to that is to show " how stars blow bubbles" in space and that is to use the precise words.

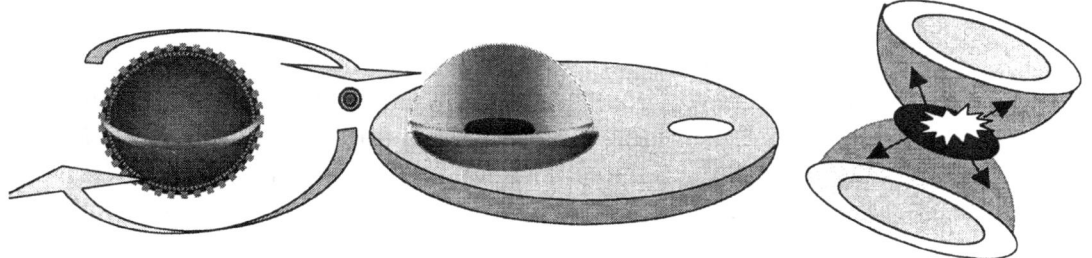

What this confirms is that no cosmic object will collide with another cosmic object and no mass draws mass closer. It is a relation where space-time relates to space- time.

5/2

Five sides divided by two spheres.

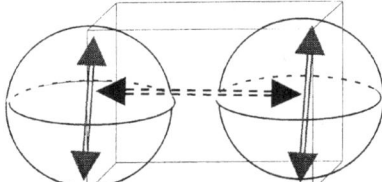

Where Π extends to lock onto the next sphere's extending indicator, Π has to connect to Π forming the square of space and translating that to the half of Π being $(\Pi/2)^2$.

The Roche limit 5/2 becoming = $(\Pi/2 \times \Pi/2)$ = 2.4674 as singularity interferes

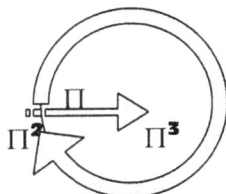

The Π^3 is matter in singularity
The Π^2 are motion or heat.
The Π is space in singularity.

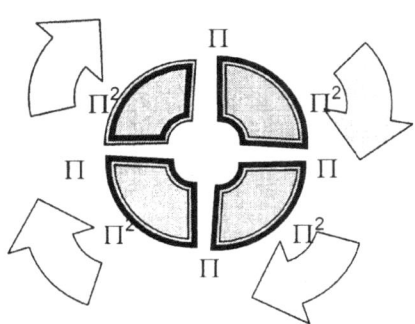

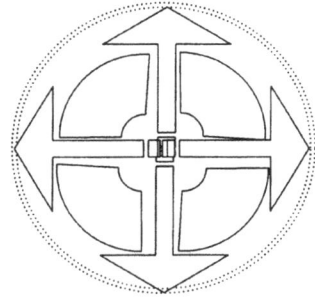

Gravity is the conforming of a centre that holds singularity and the maintaining of such a centre by rotating as well as contracting towards the centre. It is confirming the centre by motion and that forms time but the time is what keeps the space in place. It is the motion in time and the motion by time that confirms space from the past to the future through the present. I even explain what time is and that it is time that keeps space valid as Kepler said it would $a^3 = T^2 k$

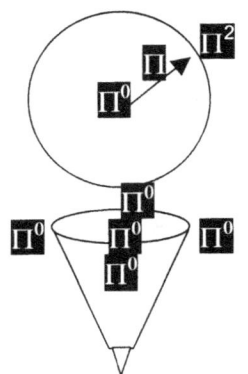

When the radius of a sphere or circle is reduced to its most infinite value it would become a factor of 1^1. That would leave the form Π as value of the sphere. In the centre where six lines cross such crossing will cut through a point where the six sides connect and that point will leaves value of singularity. Singularity can mathematically only have a value of 1 to the exponent of whatever comes as choice or of whatever symbol symbolises the radix puts the value of the exponent at a value of zero.

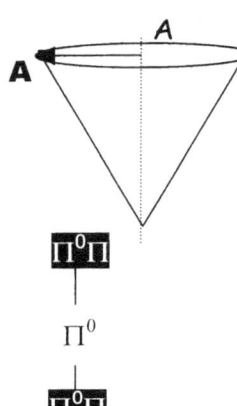

That point is there regardless of any other argument put to the contrary. If the is volumetric space the point in singularity is at that centre and it is a point that is so small it falls outside the parameter of measured space in a three dimensional form. It is not part of out Universe but it divides out Universe into immeasurable sectors. That line can only be defined by the value of 1^0. Those line cuts whatever are on their outsides into sectors of infinite measurements and only by a mathematical regarding of the fact can intelligence come to be recognised. It is only the intellectual mind the will find recognising in such a point that control the Universe without being part of the Universe. Try to explain that to a gorilla and the ape will not understand the explaining. To the immediate outside of 1^0 are two (at least) two bordering values of singularity that has to stand in value as 1^1 because they still maintain singularity although they are hardly if anything bigger that 1^0. That the lines are there is as true as the fact that the cosmos of in place but detecting the lines can only be through an intellectual apprehension of the fact that the cosmos starts at that point. Those points are there eternally in infinity and the only manner to release eternity from infinity is to provide motion to eternity outside infinity.

Because space is ten and half of the space should be five one would expect that

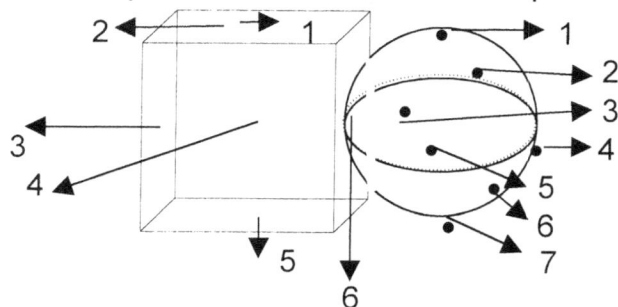

the Roche limit would rather establish half of ten to form half of space in the square by half. Since that is not the case but gravity is involved one may presume that this value reflects of a condition that applied before all space, as we know space was a factor and everything rested in motion.

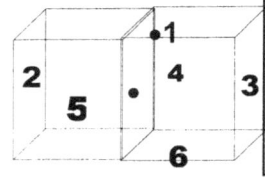

There is one more point in the sphere in the centre forming an addition in the sphere. That point holds gravity secure.

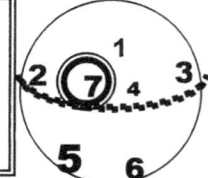

The only time only motion was about and space was excluded was when the Big bang was not in place yet. By tracing the phenomena and studying the values and positions of the phenomena that will allow investigation to go into the era that is pre Big Bang.

The generating of gravity is not confirmed by the Titius Bode law but the Titius bode law perform the generating of gravity in relation of bridging singularity restrictions by implementing the Roche limit at $\Pi^2/4$. It is the turning of material (7) through the dimension of space (5+5=10) and from that relation, which is motion that becomes the generating of gravity. It is true that it is immensely more complicated but how the hell would you find out how the process truly works when all the different element's functions join in the process when you never read my work!

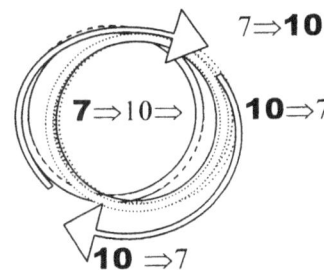

The Titius bode is an interaction between particles in motion through time where time leaves a mark on material ($k^{-1} = T^2 / a^3$) and material leaves a vacancy within time ($k = a^3 / T^2$) and that motion of material or space inter acting with outer space or time is the legacy of gravity by motion $T^2 k$.

Only motion can turn singularity from the spot it has in the sphere to the dot to form a line that takes charge of the independent object, which is in motion and is in relation to another centre. This confirms Kepler's $a^3 = T^2k$.

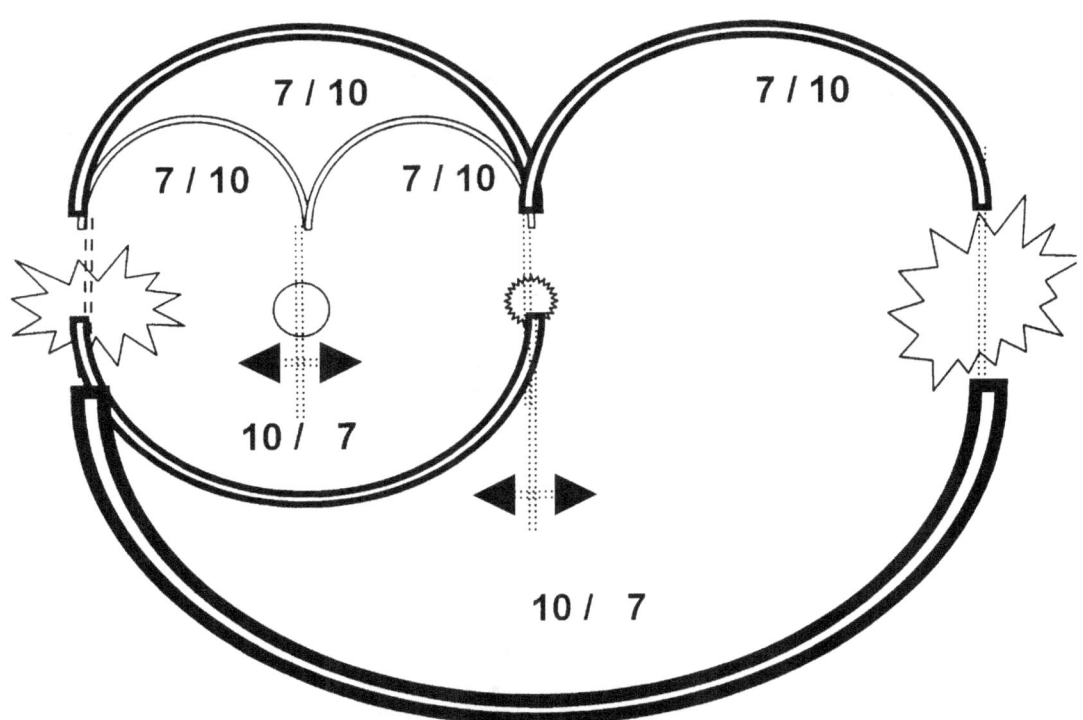

Matter in relation (part of) to the total dimension of space.
(10 / 7) \ (7/ 10) = 2.04

1.4285 / 0.7 = 2.04 Taking from both orbiting influences
SPACE DIVIDED INTO TIME

(7/10) / (10/7) = 0.49
.7 /   1.4285 = 0.49 Taking from both orbiting influences
SPACE MULTIPLIED WITH TIME

7/10 / 7/10 = 1 and   10 / 7 X 7/10 =1 Therefore not influencing change
THE PROCESS PARTED USING THE ROCHE PRINCIPLE

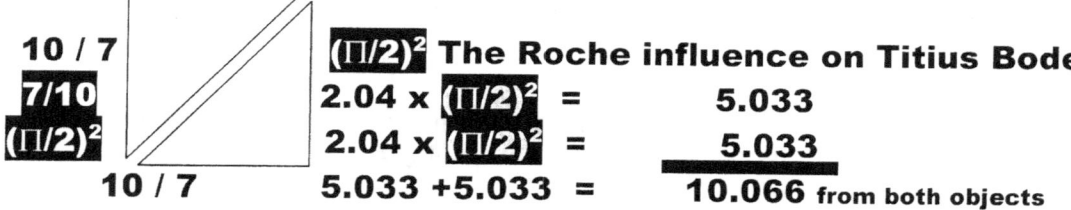

10 / 7
7/10
$(\Pi/2)^2$
10 / 7

$(\Pi/2)^2$ The Roche influence on Titius Bode
2.04 x $(\Pi/2)^2$ =   5.033
2.04 x $(\Pi/2)^2$ =   5.033
5.033 +5.033 =   **10.066** from both objects

SPACE DIVIDE INTO TIME

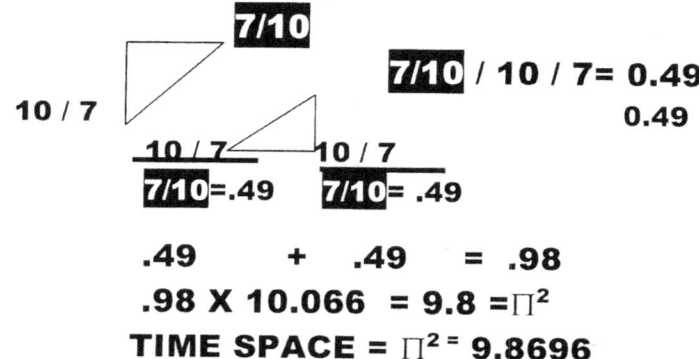

7/10
10 / 7
10 / 7   10 / 7
7/10=.49   7/10= .49

7/10 / 10 / 7= 0.49
          0.49

.49   +   .49   = .98
.98 X 10.066 = 9.8 =Π^2
TIME SPACE = Π^2 = 9.8696

TIME SPACE =Π^2=9.8696= Space and time in a dimensional implication.

The measure of seven standing in relation is all a forming of singularity standing relevant to singularity. In the cosmos only singularity apply. On one side of the sphere there are five plus five which becomes ten and the sphere of material holds six position on every side plus a seventh that stands in the centre and is centralising the six edge point where the point in the centre forms the point that generate gravity through singularity. The other value such as mass is a generated form of spec-time that comes about as time delay and that is what material is. This is rather slight too complicated to divulge in this short profile but is well documented in my books.

The Titius Bode law is one of the phenomena but it is also the phenomena around which gravity is centred. Alongside the Titius Bode law is equally important the other phenomena being the Roche limit wherefrom the Roche lobe derives its value and its measure, the Lagrangian points: and the Coanda effect. At present Newtonians do not even regard the Coanda effect as a cosmological function but sweeps it under the table as something aircrafts implement to fly. Not one of them went as far as linking the Coanda effect and its association with something of value in the cosmos because the Coanda effect does not apply mass as any norm by which it can appreciate the principle of flying. In that Newton is excluded and where Newton is excluded such things just is a hallucination of sorts. Newtonian minds cannot pass off anything that does not support Newton and mass inflicting gravity and anything that shows a lack of supporting the likes of mass they disregard with eagerness. Since there is nothing that actually supports Newtonian thinking everything of value is passed over since nothing collaborates Newton and therefore they have the cosmos being nothing. How desperately wanting and illogically incoherent can an argument get where the prospects of nothing is being offered as a valid commodity to be used by the cosmos in the cosmos where it is supporting the structure to be the ingredient that is forming the cosmos? It was the very nothing Newton introduced when he set his hoax into action and only the nothing he created finds support by the Newtonians since they have nothing other than nothing they used to form cosmic and that can at the same time be implemented to support Newton.

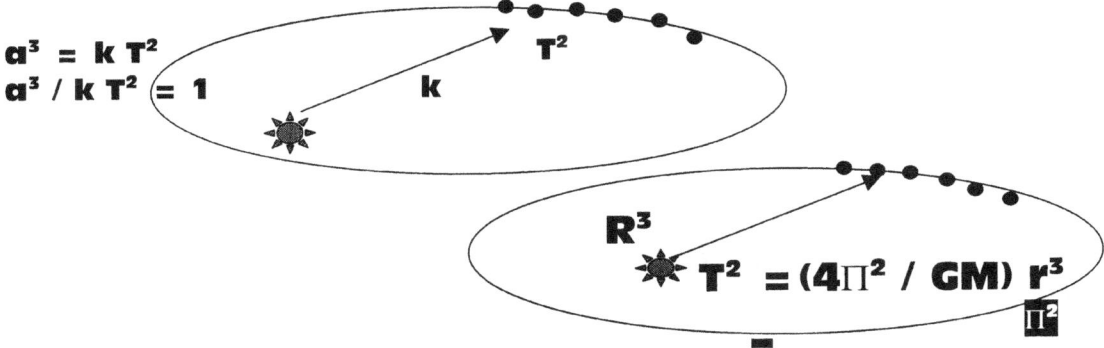

$$\Omega = \frac{d\theta}{dt} \quad \tau = \frac{dJ}{dt} \quad d\theta = \frac{dJ}{J}$$

$$\Omega = \frac{1}{J}\frac{dJ}{dT}$$
$$= \frac{\tau}{J} \qquad J = r \times p$$
$$= \frac{rmg}{J} \qquad = rp\sin\theta$$
$$= \frac{rmg}{I\omega} \qquad = pd$$

$$\frac{d}{dt}(r \times p) = \frac{dJ}{dt}$$
$$\left(\frac{dr}{dt} \times p\right) + \left(r \times \frac{dp}{dt}\right) = \frac{dJ}{dt}$$

$$\frac{dJ}{dt} = 0$$

Taking the argument back to Kepler's law,

a³ = k T²
a³ / k T² = 1

T²

k

R³

T² = (4Π² / GM) r³

If this principle Newton envisage is true then the Titius Bode law of Planetary positioning cannot be true because the **k** as a factor in Kepler's formula of pace-time **a³ = T² k** has a valid position. It has and that is one reason I shout fraud and I shout conspiracy. As one can see there is a distinct relation between the planets where every planet doubles in radius by distance from the Sun. Every planet takes up a position where the planet doubles the distance that all the other planets in total have. This I explain and that I bring in relation to the charging of gravity in that manner and I include an example of my explaining.

Mercury	Venus	Earth	Ceres	Mars
$4 - 4 = 0$ $14 - 4 = 10$	$0 + 7 = 7$ $20 - 4 = 16$	$7 \times 2 = 14$ $32 - 4 = 28$	$10 \times 2 = 20$	$16 \times 2 = 32$

Jupiter	Saturn	Uranus
$28 \times 2 = 56$ $56 - 4 = 52$	$52 \times 2 = 104$ $104 - 4 = 100$	$100 \times 2 = 200$ $200 - 4 = 196$

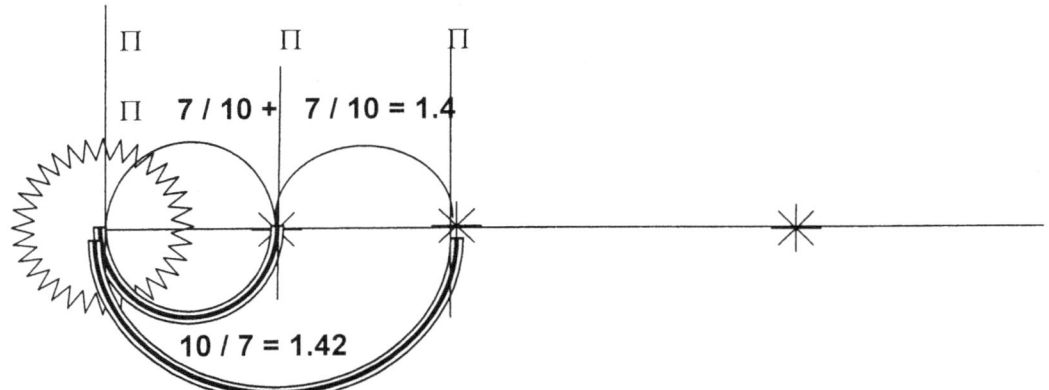

The one inner planet 1^1 next to the inside of the planet taking position forms a marker while the orbiting one positions in relation to the second 1^1 and that forms the dual singularity relation they have with the Sun centre 1^0 singularity.

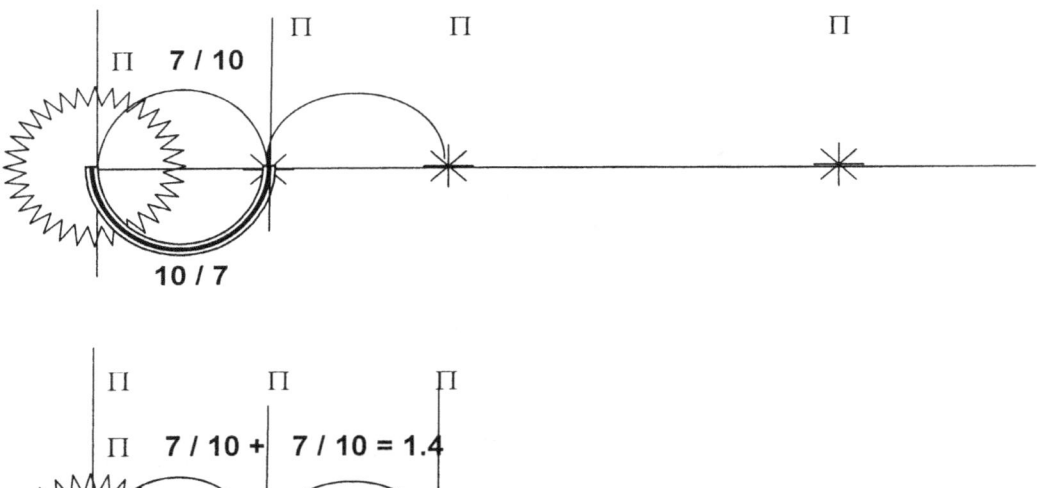

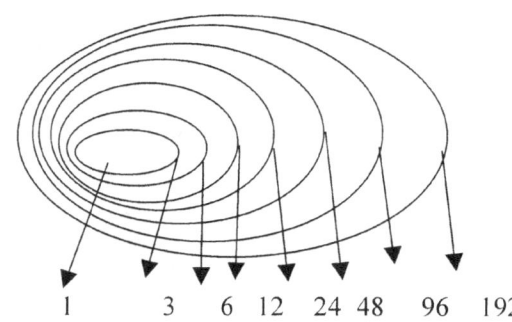

Planet	Mercury	Venus	Earth	Mars	Ceres	Jupiter	Saturn	Uranus
Bode's Law dist.	4	7	10	16	28	52	100	196
Actual dist.	3.9	7.2	10	15.2	28	52	95	192

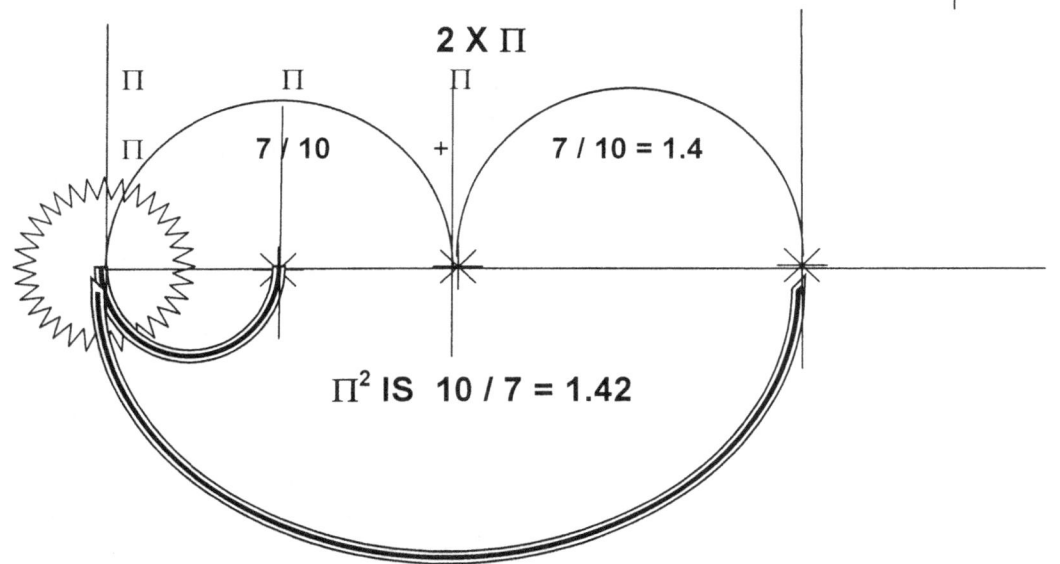

Mercury	Venus	Earth	Ceres	Mars
$4 - 4 = 0$ $14 - 4 = 10$	$0 + 7 = 7$ $20 - 4 = 16$	$7 \times 2 = 14$ $32 - 4 = 28$	$10 \times 2 = 20$	$16 \times 2 = 32$

Jupiter	Saturn	Uranus
$28 \times 2 = 56$ $56 - 4 = 52$	$52 \times 2 = 104$ $104 - 4 = 100$	$100 \times 2 = 200$ $200 - 4 = 196$

With the evidence of the Titius Bode law so much apparent it should be clear to all that mass has no place in cosmology. In the cosmos there is no aspect that works by the principle of mass. The cosmos places all objects on a homogenous footing. The fact that Mercury is equally dispensed as Pluto is or as Jupiter is undeniably clears evidence that mass has no role in the cosmos and therefore mass has no producing of gravity. We have a marker forming singularity by seven and we have a pivot forming a point in seven while the space in between is ten. This is the generating of gravity in the following manner: $7 + 7 = 14$ and $10 / 7 = 1.42$ where $14 / 1.4 = 9.86$ which is gravity Π^2

The spherical positioning layout forming the Titius Bode Principle

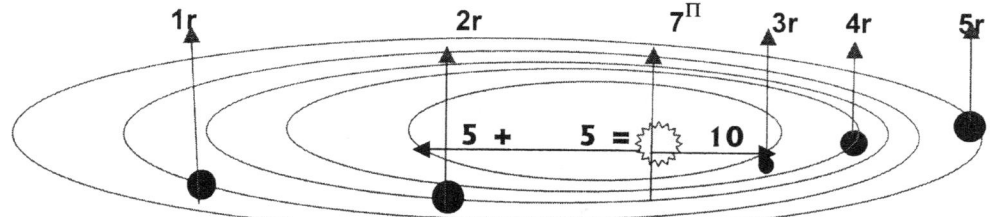

From the matter-to-matter relation in the Titius Bode configuration there are 7 / 10 + 7 / 10 = .7 + .7 = 1.4

From the space-to-matter relation in the Titius Bode configuration there is 10 / 7 = 1.42

$(7 + 7) / 10 = 1.4$

$(7 + 7) / 10 = 1.4$

$(7 + 7) / 10 = 1.4$

$7 + 7 + 7 +$ $7 +$

$10 / 7 = 1.42$

$10 / 7 = 1.42$

$10 / 7 = 1.42$
= .7 /|\ 1.42
= 1.4 /|\ 1.42

The $5 + 5 = 10$ is a position of dimensions as space loses value to singularity. The 7 that matter diverts in points from singularity may seem, as coincidental but is valid. Still in accordance to our perception valuing the number in degrees, it seems coincidental but if it is coincidental, it is nevertheless a figure of diverting proven as accountable in all other calculations and plays a most dynamic role.

The Lagrangian 5 point system results as much from the Curvature of space-time as does the form the Black Hole holds. The Galactica is the opposing equivalent of the Black Hole and has identical but opposing similarities being the five points positioned to singularity. The galactica is generating space and the Black hole is degenerating space.

Because the space-to-matter is in the square at 10 placing the matter-to-matter at a square of .7 + .7 = 1.4 the space-to-matter forces the matter-to-matter to double the distance by number as structures are place father from the mainΠ^0 maintaining singularity.

1 3   6   12   24

Reasons why this does not fully apply to the solar system I give in book # 7.

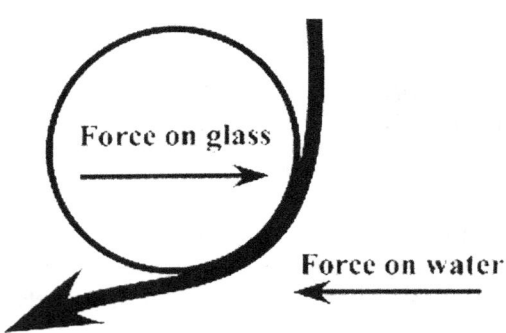

Force on glass

Force on water

The entire truth is that there are no forces of any description but it is a flow of time coming from a liquid or less dense form of heat while it enters a controlled form of heat where spin produces the preserving of time within a sealed environment to contain heat for the accumulation of heat by singularity.

The Roche limit is:

The region surrounding each star in a binary system, within which any material is gravitationally bound to that particular star. The boundary of the Roche lobes is an equipotential surface, and the lobes touch at the inner Lagrangian point, L_1, through which mass transfer may occur if one of the components expands to fill its lobe. It names after the French mathematician Edouard Albert Roche (1820-83).

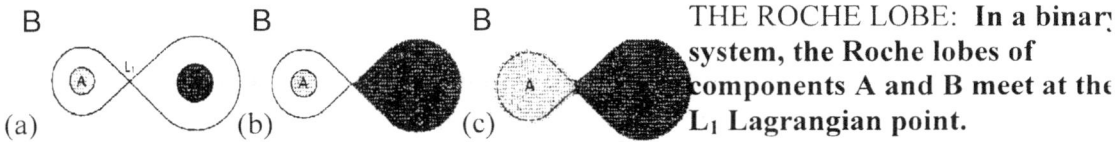

THE ROCHE LOBE: In a binary system, the Roche lobes of components A and B meet at the L_1 Lagrangian point.

(a) In a detached system, neither star fills its Roche lobe. (b) In a semidetached system, one massive component, B, fills its Roche lobe. (c) In a contact binary, both components overfill their Roche lobes and share a common envelope. Lets explain the importance of this Roche limit and how the Universe used the Roche factor to produce the Big Bang. That is where it all started…

Gravity is generated in much different terms and does not work or rely on mass. It has all to do with the law of Pythagoras working in relation with singularity and in conjunction with the combining value of ten finding a square of hundred and seven by the square that stands in relation with the one singularity presents in the square will form o total of fifty. With fifty on both sides we have the effect that fifty plus fifty totals one hundred and the square root of one hundred is ten Therefore we find that seven is related to ten and on that is the cosmos built. From that we derive gravity by the measure of 9.8 and that is the formula whereby gravity is produced. This might sound slightly overwhelming but if one multiplies the Roche factor on relation to the Titius Bode as the Titius Bode form cross-referencing with 7/10 and 10/7 we get the value of $\Pi^2 = 9.86$ which forms the value of gravity. Mass is excluded as any cosmic reality.

The main thing what the Roche limit proves is that mass does not pull mass until mass devours or destroys mass by collision. The Roche limit clearly is rubbishing that and by the Roche limit that rubbishes Newtonian's claim of $F = \dfrac{r^2}{M_1 M_2}$ it puts everything about mass and the pulling thereof, in the paper basket.

$F = \dfrac{r^2}{M_1 M_2}$ should mathematically represent the statement that mass pulls mass by destroying the radius, which was of course the modified blunder of $F = G\dfrac{M \times m}{r^2}$ where Newton and Newtonians after all the evidence are equally determined to convince the world that mass pulls mass towards mass. It is hogwash if ever there was hogwash because no object of cosmic proportions will come closer to another object than the value of 4.67 or then $\Pi^2/4$ and that is directly related to singularity charging gravity.

The Lagrangian point system should be very clear by now as it forms half of the half of the square that forms space in the square. The ten of space divides another halve and there gravity confirms a position allocated to material one point outside the four positions that secures time by $4\Pi^2$ It is the point where the line meets on equal terms as 180^0 with the half circle which also stands at an equal

180^0 and it relays to the triangle holding an equal 180^0. It is how the cosmos started before space became confused with the lagging of time but that I explain so specifically that no one wishes to read it. Again the Lagrangian points offer no proof of Newton's mass and the order that the Lagrangian point system holds has no indication of any relation of any sorts where mass is used to bring some alignment or manifests as a factor being present in the cosmos in any way whatsoever.

5/2
Five sides divided by two spheres.

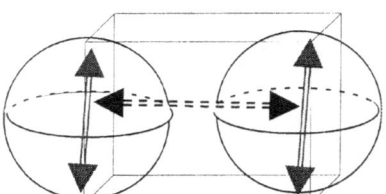

Where Π extends to lock onto the next sphere's extending indicator, Π has to connect to Π forming the square of space and translating that to the half of Π being $(\Pi/2)^2$.

The Roche limit 5/2 becoming = $(\Pi/2 \times \Pi/2)$ = 2.4674 as singularity interferes

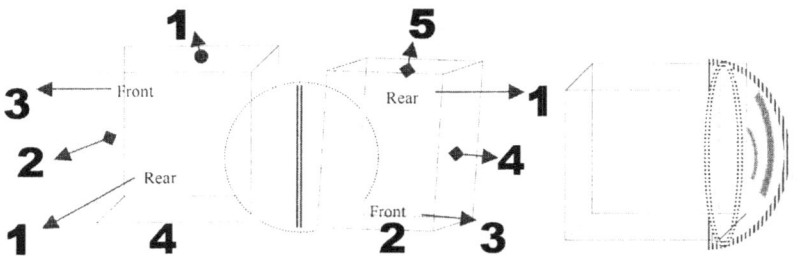

1 Relating to 5

LAGRANGIAN POINT:
*The Lagrangian points
are five equilibrium points
in the orbit of one body
around another, such
as a planet around the Sun*

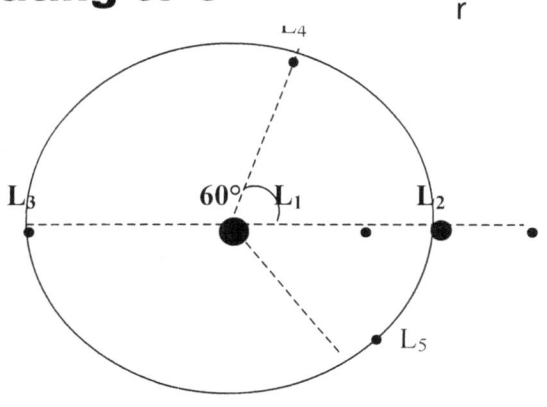

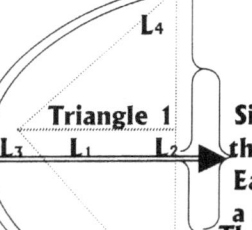

The Lagrangian System implicating the five positions extending from singularity

Singularity dividing the cosmos
Each triangle claiming a side of the universe

1 Half circle $=180^0$ L₃ L₄ L₅
2 Triangle 1 $=180^0$ L₃ L₄ L₅
3 Triangle 2 $=180^0$ L₃ L₄ L₅
4 Straight Line $=180^0$

The half Circle = 180^0 combining as a sphere when comprising Singularity in the matching of the value of the straight line forming the half circle and combining as the triangle and all are equal 180^0

The second one also fits in the singularity influence on the Universe.

The mere fact that no planet is coming closer to the sun or to each other should be enough grounds to launch a serious investigative research into the probability of Newton's theory about mass pulling on mass in order to reduce the radius. That is lacking but moreover my saying that keeps my work unread for six years on the trod. However one can see by the size in pages of the explaining given here and space it takes when considering that the three books cover in total almost 2500 pages. With that in mind then this is nominal in context to the overall information provided in the books given to you for reading. The space it takes is such a small part even of this letter and still this explaining exceeds by far the proof of mass inflicting gravity that Newton established.

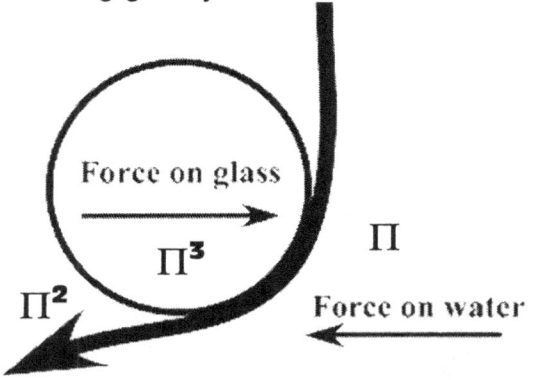

The Coanda effect is the instituting of Kepler's formula where space provides the limit of liquid and liquid confirms the limit of space. Kepler's formula states that space is equal to the flow or moving thereof $a^3 = T^2 k$ where then gravity is the extending of space which includes the liquid forming the total space that forms $k = a^3/T^2$ and liquid is connecting to space by gravity forming the result in contraction $k^{-1} = T^2 / a^3$

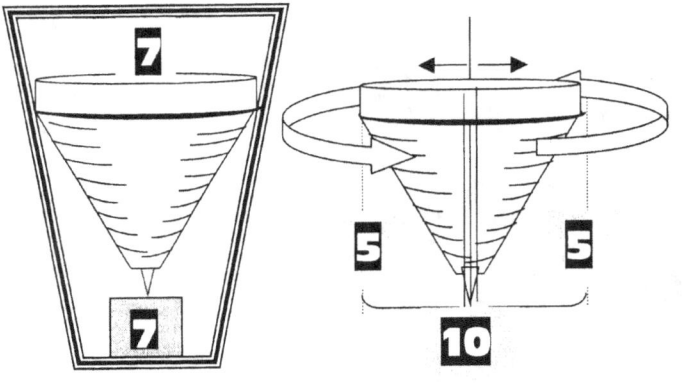

The four phenomina are inseperably one and that forms motion where motio is gravity. It is the crossong of the point that holds singularity where one the has a circle that attaches time to the cirecle ant the other is time attaching to the circle.

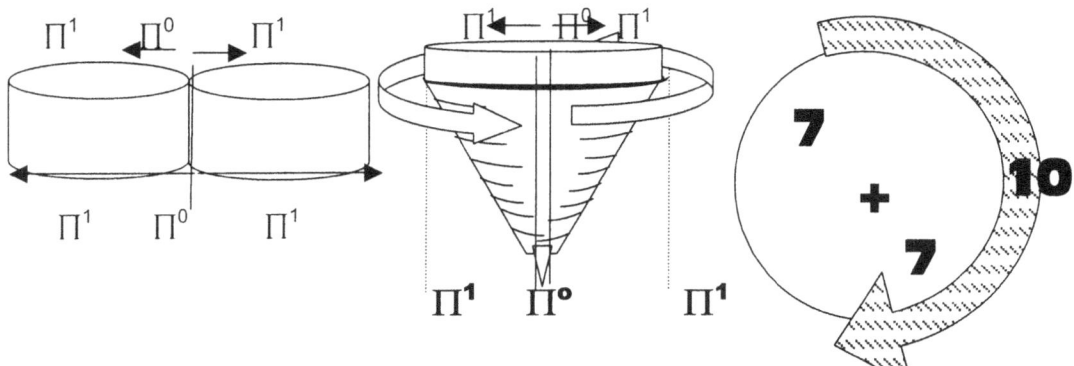

The Roche limit is to my mind the most significant evidence of proving Newton's failure. In the cosmos we have two systems that approached and became too close for comfort. The one is related to the other as a lesser relevant singularity while the other forms the conserving singularity. The two does not engage as Newton would suggest and plummet into each

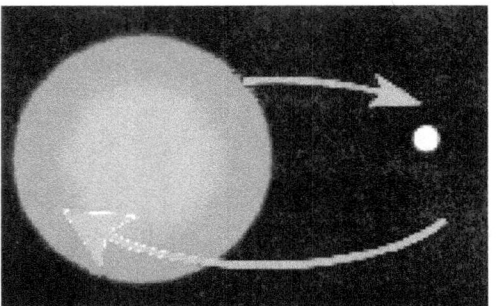

other's space where the mass draws the mass until the radius is no longer parting the structures. At a distance calculated to be $\Pi^2/4$ the two engage into forming a lobe that involve both spheres. The lesser star candidate overheats within this combining sphere that charges the envelope to cover both structures. Both structures losses independent identification as the Roche lobe cover both object and all the material in both objects elect a centre which provides acknowledgement to both stars. Then the one star turns the other star to liquid where the heat forms a cloud that covers both stars. The process that develops turns all the material into heat and it is evident that a process of severe overheating takes charge of the proceeding that follows.

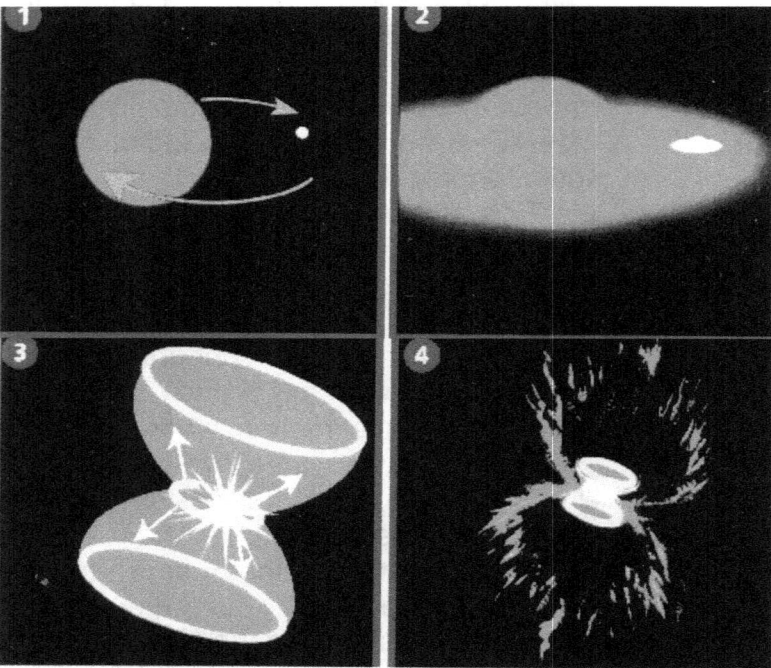

From the centre there is evidence of uncontrolled overheating that deforms the star in a circular patter where the true overheating is found right in the centre of the star. The object that supposedly retains mass holds the mass apart while both remains at a distance of $\Pi^2/4$ times the radius apart but both become liquid and both experience centres that destruct in violent heat outbursts.

Newton is completely absent in this whole picture. No evidence of mass is forth coming but what is very clear is that where the control subsides, heat in an uncontrolled fashion takes over. The collision that was to come about should Newton's mass that is pulling mass close be forthcoming, was not that much forthcoming after all. Again it slurs at Newton and at Newtonian misconceptions while Newtonians hide this under a blanket of corruption.

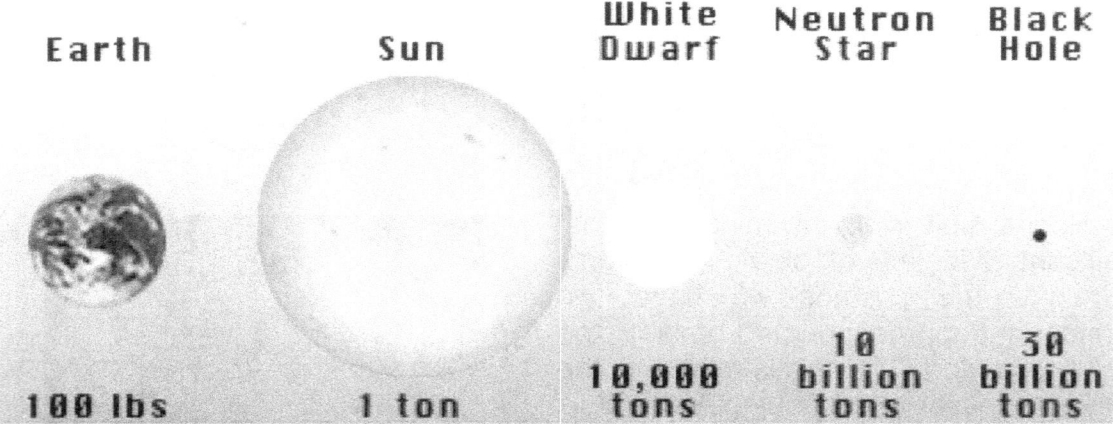

Earth	Sun	White Dwarf	Neutron Star	Black Hole
100 lbs	1 ton	10,000 tons	10 billion tons	30 billion tons

Should Newtonians insist that their constant referring to mass is the space that material holds then the following proves just the opposite. Mass increases in stars in ratio to and where space reduces where it reaches a point where mass almost grows eternal while the space that the mass claims becomes infinitely small. Nowhere is mass a factor except when the motion of material is relinquished and mass smothers the individual purpose of the material by confining the material to a domineering structure. Why that would be is somewhat tedious to explain in this letter but since you never read my work you will never be able to see that part of the explaining. You are fixated with Newton and with mass while Newton's mass in outer space is a pipe dream. Outer space work on singularity $k^0 = a^3 / T^2k$ and space – time $a^3 = T^2k$ and mass plays no part.

You are a Master and a Professor and you see the young minds enter their future Alma Mater in anticipation and excitement every year. The young minds are waiting for their tutorship on the condition they are accepted into the institution of learning. They come in anticipation while waiting to receive their wisdom where they would carry the flame of brilliant knowledge through their generation in the era in time where they will be entrusted to have the opportunity in doing that. They are ready to learn and accept. Some might challenge some authority about thinking but that never involves the basic arguments concerning the founding principles. Notwithstanding their seriousness it is always about matters in a superficial capacity. In reality that is a manner to channel all rebelliousness because of their eagerness to improve on that what they inherit. Before they can improve, they first have to accept what their elders confirm as the basic and the foundation values. In the case of Physics it is accepting Newton to be better than God and my saying that is not intended as blasphemy but it is to break open a gangrenous sore. To manage the introducing of Newton as God's equal, Pavlov is introduced and not the man or his principle but his conditioning. Then comes the mind conditioning that has to be undergone by all students harbouring any wish to become a full member of the future physics paternity. The mind control is automatic in nature. The thought conditioning is a thought association where the student stands on Earth with the experiencing of mass. It uses Pavlov's theory that was instituted by the Physics department at all Universities centuries before Pavlov was instituted. The students are told that mass is the sole contributor to gravity and gravity keeps everyone solid on the Earth. They are on the Earth because they have mass that forms the gravity keeping them on Earth. In their minds no proof is required when they are experiencing the gravity about which they are taught by their tutors. They are programmed in the manner that physics is presented to their young mind to accept what they cannot refute and their standing on Earth is one fact that they cannot confute. They link their standing on Earth to the lesson about gravity and mass and immediately this presents irrefutable proof that they have mass which forms the gravity that keeps them down onto the Earth. Now their tutors have one more task to complete and although it is small, it is crucial in confirming their acceptance in their thinking patterns for the rest of their lifetime to come. They have to associate their standing on Earth while having mass which forms the result of their gravity that they are experiencing with the name Newton. Newton said this is the process where mass pulls mass to remove any distance between objects. They have mass. They can feel their mass pulling the Earth as the mass of the Earth is pulling them and that is stopping them to jump over the Moon. They fight the affect of gravity every second they are on Earth. The gravity coming from the mass is eroding their life power until they will finally perish and go to their graves.

They are in gravity until they are in their grave. Gravity is pulling them to their grave. Sir, madam in your personal lives you too, were victims of this very process and your accepting of Newtonian mass that is generating gravity does not stem from your receiving irrefutable and undeniable proof when you were first introduced to Newton where you then were shown proof beyond your wildest quest for facts during your brittle youth but your sureness about the matter came from the measure of mind conditioning you were resigned to accept. If you do not believe that statement I just made, then sit back and find the day the proof about Newton was presented to you as a student. In all this, one man going by the name of Isaac Newton coined the process when Newton was also a young mind eager to prove what he was capable of. Newton formulised everything he was experiencing as mass and gravity. This concept is so easy to associate with as his frustration of having gravity, which is the process that is sapping his strength by $F = \dfrac{r^2}{M_1 M_2}$. The formula insists that anyone has mass while the Earth has mass and the Earth mass is pulling on all other mass which the two combine to fight the distance that might come about when any person should jump. It erodes any distance that might be in place between the mass of the Earth and the mass anyone is experiencing. What can ever require less proof than $F = \dfrac{r^2}{M_1 M_2}$ for it involves anyone's entire existence. It is a truth the student can never contradict and because of its familiarity and always being present throughout all person's life era. Therefore any and all further proof is unnecessary and redundant. The student has all the proof he or she can ever muster just by standing upright. Is it really that simple or does nature hold more meat that is covering the bone of contention? Galileo said no mass applies when objects fall because all objects fall equal. By Newton's very first inspiration he contemplated mass as that which is generating the force producing the energy by which the falling process of objects becomes a reality. He said mass times mass divide to destroy the distance between objects with the suggested formula of $F = \dfrac{r^2}{M_1 M_2}$. The concept that this idea promotes contradicts Galileo just because Newton introduces mass where Galileo excluded any form of differentiations. From this one would gather that Newton was incorrect even before he introduced his idea to the world, and yet not one of the bright Academics could spot this in three hundred years. If you are as convinced about Newton then explain why would Galileo have all objects fall equal without mass intervening in any way just because all objects fall equal. If mass was a factor charged with the producing of gravity, then all objects with more mass have to fall quicker than the other with lesser mass. If mass is the producing factor of gravity and mass is in place to produce more gravity, it then has more mass which then is in place to be producing more gravity that will encourage more of the falling motion which is gravity, where more mass arrests more space in quicker succession that will allow the object in falling to establish a faster rate of falling. Yet according to Galileo there is no chance that mass comes about to bring any discrepancy amongst falling objects in any way except when the motion is restricted to a stand still where the bodies come to rest against each other. Then mass becomes a factor but not while they are in the process of falling. The absolute uniformity excludes mass being a differentiating factor by elimination because mass clearly has no influence on favouring the more massive while falling. Mass only comes about as a factor when the body is in rest. In my books I explain why mass is a human instituted concept and has no cosmic validity. I prove that anything that reached a point of having mass has also by the same token reached a cosmic death because that object surrendered its cosmic independent existing which is secured in the defining of independent motion where when burdened with mass, the motion of independence becomes annexed by the motion of a larger structure and by capitulating, its motion to mass it accepts the motion that the larger object prescribes. Accepting the motion of the larger object is forming mass. By accepting mass the object becomes estranged from the cosmos and by a marriage to the larger object through instituting mass, the object divorces its independent identity it might have in cosmic motion. Through its marriage in mass it relinquishes independent motion to acquire the motion dictated by the occupying body and being part of the motion of a more developed structure it becomes something with no differentiation in cosmic terms as a distinct object. In cosmic terms any object in mass can never again move out of mass. Can any one imagine a mountain flying about even on another planet? Can any one picture two rocks chasing each other on Venus where life can gain no foothold? It is not possible because the mountain has mass and the mass makes it part of the Earth (or in the case of Venus then part of Venus) or any other planet in question. Once mass forms a part of the object instead of motion, the method of applying mass by removing independent motion is a means to secure the lesser object to the dominating object. The lesser object is connecting by mass to the larger object that captures the smaller one by granting mass and removing independent motion. The giving of mass is a way to enforce the will of the larger motion onto the lesser object by removing the ability of independent gravity from the lesser object. Then all distinguishing motion indicating independence of the lesser object becomes impossible because the motion that the domineering object takes control by measure of mass that dictates the terms of motion. Any object on Earth can only move in relation to the measure in which the Earth is moving except where life produces motion and life is consecrated to one place and that is the Earth alone. Only a paranoid fool having an eidolon would find evidence of the presence of life elsewhere or anywhere in the cosmos in any form other than what life has on Earth when life can only be on Earth as a proven reality because any other idea is merely hallucination that belongs to the insane mind. There is no evidence of life anywhere but on Earth and only rampaging insanity would deliver evidence in support of the contrary perception. Mass is the state in which the object presumes when the object in mass is not any distinct independent cosmic motion any longer. Mass is when the object surrendered its independence and with that it divorced its specific density by becoming part of that which is forming the restriction of the cosmic motion that secures the object having motion. Motion provides independence by alienating mass from forming and mass is anti motion that arrests all cosmic independence.

If a ship is buoyant on the water, it is presumed by science that the ship has mass. That cannot be the case because what then is applying when the ship sinks to the bottom of the water where it then rests without motion on the bed of the Earth? Where the ship finds a location as an individual item laying under the water and not on top of the water, then the sunken ship will have mass because it ceded independent motion where mass then enslaves the motionless object to the larger object. That then prescribes all conditions forming motion. Only when the ship sinks and rests as an individual object on the bottom of the water can the ship qualify to present mass in an individual capacity. While floating the ship takes on all the characteristics one will find when the ship is being part of the water. The ship holds the mass of the water it displaces. The ship or any floating part of the ship is for all purposes just water. The water it displaces while being buoyant is much less than the mass the ship will have when the ship is in dry dock and is fully laden. No Newtonian can ever presume that a ship fully laden while floating holds an equality in mass measured in relation to a ship fully laden in dry docks. The ship does not presume individual mass while floating above the water or even when floating in the water as submarines do. However, when it is at the bottom of the water it presumes the mass of its structure as well as all the water presiding on top of it. But floating on the water the mass of the ship might add to the mass of the water but as such the ship has no mass. If you insist that the ship can float and have mass, then please explain what you would say is the difference between the floating ship and the sunken ship, because there is one Universe of difference between the sunken ship and the floating ship. If you question my logic in this matter then go and look at the difference in the Titanic. In the one the floating ship presumes in the capacity as forming more water and in the case of the sunken ship the ship is more Earth. That is the difference but also in that is all the difference one will find in what there is between gravity and mass. However it does not stop there because relevancies change where motion is concerned. Even if you insist that what you refer to as the material filling the material-filled space of the atom or an assortment of atoms forming a structure that has mass because it is captured in a unified structure in the form it has while it has in the space it holds and being in that position it is mass then that still does not qualify to associate that with mass, because that filled space can go from presenting a few kilograms on earth to where it then becomes billions of tons in a large star. There is a reason to regard form as mass because of the increase in mass and that I explain in very detail in the books. The increase in mass is associated with the loss in form and when mass increase by relinquishing form, then form is not mass. That is not mass that you refer to when you refer to the body that forms the object because the material has the ability to increase the mass by decreasing the space that material occupies. That which you refer to is the space a^3 that is duplicating T^2 k by motion into its next position in time coming from its previous position in time that it holds in the current position in time and that matches all the scenarios that one may contribute to the Coanda effect. That is space a^3 duplicating the position it holds T^2 being relative to an allocated position every time by the measure of k. Gravity is $a^3 = T^2 k$ and that Kepler proved to be gravity long before Newton admitted he had no idea what gravity is. It is the motion of space that is duplicating the space by moving through time in relevance to a measured centre point.

When a body falls into water from an extreme height the fact of having motion such motion brings gravity that changes when the falling body that is falling connects to the water rim where the motion ends by water restriction and the motion of the falling body changes its position from having gravity to become mass. While plummeting k towards the Earth T^2 where the Earth at that point holds water at the surface the water is the solid a^3 notwithstanding it being in a liquid form. The water at that point holds the mass as it forms part of the Earth and is therefore the solid factor representing a^3. While the object is falling it is not the object that falls but all the space that holds the object that moves. Galileo proved this when Galileo proved there is no size differentiation present while bodies fall. In that way the object has no mass since it travels with the space and forms part of the liquid space that moves. The body is a part of a constant waterfall by relevance of motion. The body has to be buoyant because the body has no mass and that Galileo did prove notwithstanding Newtonian denial to the contrary. It is not the body that falls down but the space the body has which is accompanying all the surrounding space as well that is falling and while all that space is falling it is taking the falling body along. That can be the only explaining of not having any distinct differentiations while all objects fall in a similar way without showing differentiation when there is such obvious and great differentiation between such bodies. The object holds buoyancy in the liquid space that carries the object while the lot of space is descending to the Earth in the form of a liquid. The object falling might be a solid but while falling, it moves and therefore it has gravity making the object being part of the liquid T^2k. The ship floating might be a solid but while it is floating it is just more water in the water. That is the only scenario that is able to explain Galileo's finding of equality of different mass in a downward motion. By touching the water the gravity goes on to form mass at the moment of impact and the object turns at the moment of impact from being part of the free falling liquid having gravity, to being blocked by a solid and at the moment of impact ending the motion and aborting gravity to accept mass where this ends the motion or gravity as this results in state of gravity ending that then turns to form mass. The relevancies at that point changes where at the point of impact the falling body will be liquid and the water will be solid. The water holds the mass of a solid and the body being part of the space, which is descending then, at the moment of impact forms the role, which all liquid has. The mass breaks the fall but the body as a factor is swapping from being part of the liquid to being part of the solid, which is the position of the water until at the moment of impact. Just after the moment of impact the water changes the position it has from first being a solid a^3 going onto presume the role of the liquid T^2k and then becomes the liquid T^2k in relation to the solid the Earth holds but only after it first presented the factor of forming a solid a^3 that breaks the falling object. That is why the water is more solid than a human body at the point of contact. The human body forms liquid while the water is solid because the water is without relative motion at first while the body shows absolute gravity and then after the fall the body becomes part of the moving water in relation to the absolutely sturdy Earth. Then, when contact is made the water becomes buoyant since it harbours the object while the object is descending further but this time in the medium of the water and the water this time has six hundred times more density than the space has just above the water. Liquid and solid are allocated to the position of motion that the one factor presents and the position of immobility that the other factor presents. First space was the density factor provider but then after impact the water took on the role as the liquid and the buoyancy moves over from the descending space in air to where it then turns the liquid state over to the more dense water that formed the solid up to such a point. Then at the point where the body comes into contact with the water the roles change where the water forms the liquid that allows the Earth to form solid. The body then will float in the water because the water has six hundred times more density than the liquid air has but is still a liquid in relation to the density the Earth presents. When the body falls on the bottom and remain motionless at the bottom of the water as part of the Earth the body then has mass since the body no longer moves on own accord. Let's investigate Newton's first concept of gravity at the point just before a hunger for personal fame got the better of his senses. Newton stated that $F = \dfrac{r^2}{M_1 M_2}$ which is mathematically expressed as a force that will destroy the square of the radius between two falling objects where the tempo of destroying of the radius by the square holds the mass of the two objects in product to achieve that. The mass factor combines by multiplication with one another in order to diminish the radius from both ends between the Earth and the object at a distance from the Earth, which then forms a square of such a distance. The mass factor makes the bodies conjunct and Newton concludes that it depends on the mass of the bodies in determining how the bodies will fall. The formula suggests this conclusion by eliminating all other possibilities $F = \dfrac{r^2}{M_1 M_2}$. That is Newton's thinking, which he unequivocally expressed in this formula. The formula is no longer in official use but the removal of the formula and the replacing thereof is not as a result of the incorrectness they established when they used the formula in the practical contexts. By using the formula in its original form indicated cosmic destruction and that they had to replace because of the unaccepted that forms part a conclusion coming from the use of the formula. Instead of revising the principle in favour of Galileo they did not revise the principle but went on to revise the formula. By the way it is revised the madness in the concept escalated just because of the protection science found a need to bolster Newtonian ambitions. By implementing Newtonian concept in a manner to determine the forces driving the cosmos it placed science on an equal footing with the abilities and brainpower of God. This is so deeply rooted that after three hundred odd years the error still goes undetected. The formula showed the madness in the matter from the start but science maintained the madness irrespective of any and all indications of such unaccepteble conclusions that stemmed from the use of the formula. They went on the revise the formula by perpetrating an even higher degree of insanity instead of questioning the regularity of the formula. Although this formula, in the way it is presented, is no longer used, it still is the essence in the concept that Newton proposed. It is indicating how he saw that the mass of the apple in conjunction with the mass of the Earth, removed the radius between the Earth and the apple. The Earth was too large to rise to where the apple was and therefore the apple did the falling. The big concern I have with the formula is the lack of concern this formula holds in regard to work that was presented and accepted at the time. At the time and long before the birth of Newton an Italian by the name of Galileo proved undoubtedly that when objects are dropped they all fall at an equal pace because after they fall freely and as they all hit the surface of the Earth the instant of contact is always the same with no regards shown to size differences. The mass amongst the falling objects could be much different but that had no effect on the time the objects travelled in concern with the actual fall while the falling was commencing as long as all prevailing conditions concerning the falling as such was the same. Galileo said $F = \dfrac{r^2}{M_1 M_2}$ was not the case because during the fall all objects fell as if on equal footing whereas mass brings differentiation. Newton came afterwards and put factors of descent on mass and that is no slip of the tongue. That was intentionally contradicting Galileo and that is intentional fraud. When I say fraud I become the dangerously insane with one sole purpose and that is to destroy all civil obedience and wreck civil standards that mankind has gained up to now. My seeing this contradiction leads to all the many times that I was rejected by institutions such as yours and persons with as admirable positions such as you have to such a degree that at present I am suffering from the "being battered yet again syndrome". I say this to indicate that I know fully well that my "repeating of past mistakes" by bringing Newton into the open will yet again bring on all the resentment from all those that are in powerful academic positions. I am fully aware of the resentment that you feel, which I am about to release and what I have already released thus far (should you still be reading this far). I say what I have to say in full realisation what resentment you as the physics administrator at this moment feel towards me. I say this remark to inform you that as much as I am aware about the damage that I am doing to my chances of having my work published and knowing full well I am shooting myself in the foot as to have my work accepted I still am adamant in continuing with such public relation damage that I evoke between yourself and I. While you are denouncing me at this moment please concern yourself with the following. When I fall down a cliff I fall at a steady pace. It is the same pace that I would have, when falling down a waterfall holding a cup in my hand. Should there be water in the cup the water in the cup will not stay behind or spill faster than any other water or myself. The water in the cup will not spill by emptying the content going up or down faster than any of the rest, which is falling. The cup will not fill with water as a result of any motion differentiation there is between the descending velocity of cup in my hand and the water that is falling with the cup located inside or outside the cup. If I had to fill the cup with water while I am falling I will have to supply additional motion contradicting the line in which I fall. I have to establish additional movement in access to the motion with which we fall. The water and I will have the same pace therefore we will fall at the same gravity. My density will not leave me superior. The water mass will not have the water fall more forceful or less forceful. The reason why a body will fall is when it has a denser personal independent density than the atmosphere surrounding the body has. Less density than the atmosphere floats irrespective of mass, which includes all gas substances and in that again it shows form is irrespective of element mass. Argon (Ar_{18}), Krypton (Kr_{36}) and Xenon (Xe_{54}) floats as gas although their equals being Aluminium (Al_{13}), Iron (Fe_{26}) and Silver (Ag_{47}) sinks. In this matter it proves that mass has little to do with falling or not falling. All objects can fall as much as all objects are able to fly. It all depends on the motion of the object in relation to the motion thereof in relation to the motion of the Earth. Floating and falling depends not on the mass but the specific density attached to the mass in question in relation to the space the material holds which forms a factor of the totality of space in motion. That is why aircrafts can fly. The forms the bodies take on as well as the fact that they travel at places where only gas is allowed by nature proves that the density of the aircraft places the flying aircraft in a different category that what it would be when it is stationary. Some has to use a specific form to carry large quantities while others go supersonic and that requires a needle shape. I explain all this when I explain the principle behind the sound barrier. The motion considering all objects is not discriminating on any basic grounds carrying specifics about size, form or weight. That is what Galileo said and Galileo is accepted as truth. There is no reason in heaven or Earth or in-between the dimensions that will establish a coherency between the truth in Galileo where Galileo says that all objects fall equally to the ground and Newton bringing in his mass to have the job of the falling done. Galileo very correctly determined that objects descend to Earth at an equal pace and although admitting to this observation of Galileo by Newtonians, science still declares that Newton was never disproved even on one single occasion. Galileo proved before Newton's birth that size and mass is irrelevant while falling takes place. Then Newton came and insisted that everything about falling while falling depends on their mass pulling the Earth mass and visa versa and in all this confusion Newtonians uphold the opinion that they are of the opinion that Newton said exactly what Galileo said. Newton said it's all down to mass while Galileo said mass has no influence and all along Newtonians see no difference between what was said. That is the result of Pavlov's mind control. In the light of this it is the truth that Galileo disproved Newton even before the birth of Newton because Newton was born at the time of Galileo's death. That makes any ignoring of Galileo by Newton not a matter of pure ignorance but it is then deliberate malice. That makes Newton's statements deliberate callousness of circumventing the truth. And the entire world not only accepts this ridiculous scamming but is also at the moment in agreement with such labyrinthic gestures of inaccuracy. Galileo proved and nature confirms that gravity has not one single common thread with mass. That is because when the object falls it has no mass (Galileo) which means that while it is true that when the object falls it has gravity without having mass. That is what Galileo proved. What is gravity? Gravity must be the part when moving is taking place because when mass is a factor there might be a tendency to move but there is no movement other than the movement the Earth dictates. Newton saw his apple fall and the falling or the movement he named as gravity. The process of falling confirms that mass does not establish or produce gravity since all objects hold gravity alike whereas mass is as individual as the fingerprints of people. If mass does establish the movement Galileo was wrong and that is not the case because modern television confirms Galileo almost on a weekly basis. Yet only Newton and Newtonian disillusions insist on mass, They maintain the factor that is producing gravity because that gives those in physics an unfair advantage in realising what others find to be senselessness. That makes Newtonians clever and the rest of us as stupid as doornails. They can see mass establishing gravity and the rest of us are too stupid to get acquainted with the total concept that the picture holds. The concept involves too much in relation to what our little minds may absorb. Now in the light of logic who seems to be stupid after all? It is shown on television on many occasions where army battle tanks are dropped from aircrafts and the tanks travel the same velocity towards the Earth as human soldiers do. The tanks do need bigger parachutes to restrain the momentum of the fall…yes that is true but the velocity of travel is the same as a bicycle would have when the bicycle is also dropped from the same aircraft as the tank is under equal terms. There was a show on television where some young persons were dropped from an aircraft while they were sitting in the car. There was no perceivable difference between the humans descending and the dropping velocity of the car. The persons got out of the car in mid air and into the car in mid air while descending at the normal rate and at one point they even pretended to get out and be next to the car where they rein acted as to push the car as if the car would not start while they were all falling with the car as a unit included in the total structure of the car or as single components falling on the way. There was no difference in the posture while they all were equally descending to the ground. The persons did not fall slower than the car did neither do the car fall faster than the people. Let your religious belief in Newton show how that confirms the correctness of $F = \dfrac{r^2}{M_1 M_2}$ while the picture I painted in the falling persons totally vindicate all the Galileo correctness. Gravity might remove the radius but it does not employ mass to accomplish the destroying of any parting distance between the Earth and any falling object and that has been proven in no uncertain terms. The whole principle rests on the duplicating of the material in the space the material holds in relation to the time it takes to duplicate the material by motion of the material in relation to the time which confirms Kepler $a^3 = T^2 k$. Newton's mass has no validity in cosmic physics and falling to the Earth forms part of cosmic physics. Mass is just a man made devise or a tool with a purpose to give meaning to ideas that is inspired by man. Just as degrees measuring heat and distance measuring space, in the same sense mass too, is a method of measure. It is a scale that helps man which man uses to establish when trying to connect a sense of worth to some concepts man wishes to establish, but mass as a cosmic function is without proof. In the same sense the other mentioned measuring tools are equally meaningless in cosmic terms. **All other physics form what is a part of man-made motion and life inspired motion. In that field of physics concerning the motion that life produces, Newton was never incorrect on any point anywhere and every aspect of his work is perfect**. But it only involves the life inspired motion that is a contribution of man, produced as the labour of life where life is in control of the motion that comes about. No rock can fly and yet a manmade aircraft can. No rock can cycle in perfect balance as a bicycle can with a human in control. To qualify having motion man has to be involved in the motion, which I mention and that excludes cosmic motion from the man made motion. There is a big difference about what is life inspired and produced by life as an entity and what applies to cosmic standards because the cosmos is fatally alien to life. The cosmos can bring motion to the Earth while man with life cannot. Man with life can have structures with much smaller volume move under precise control that the cosmos is unable to achieve. The cosmos cannot take a rocket into space while man can, but man cannot influence the entire sea while the cosmos can. The cosmos take charge of a large satellite such as the Moon and never require any post- maintenance in order to secure the orbit as the Moon maintains the orbit to a precise control by naturally keeping a time component valid and ignoring the human concept of distance, which is as alien to the cosmos as the notions of temperature and mass are. Manmade satellites require frequent control or face devastation as Mir proved a while ago. What belongs to the cosmos does not fit man and what man can achieve is not present anywhere but where life is in control of the situation. Life is motion and life is the manipulation of motion that can be covered by the range of any form of motion from blowing gently on a feather to the unleashing of a rocket going to the Moon. But that is a part of life where life is confined to the Earth and life as such is alien to the cosmos. Newton motion concerns with life where Newton is alien to the cosmos. Notwithstanding my correctness, my work has been rejected because of this statement and only on the grounds of this statement and to the degree where I now have become punch drunk. Should you decide to reject my work one more time, then be my guest but before you do please take a challenge that I direct in your direction; make sure about your reasons and the accuracy of your reasons when you decide to again reject my work this time. This is no threat because I am far to small to threaten any institution as yours. Rather see it as a challenge of your spirit of fairness but please do see it in the manner that I am taking on the establishment in challenging them to prove the correctness about Newton.

Science underwrites the Big Bang, which is a science of promoting the view where the cosmos is expending and becoming bigger. They officially declare their acceptance that the distance between objects is increasing. Gravity had the best opportunity to remove the radii between all the numerous objects altogether the moment the Big Bang commenced. The influence of the mass subsides as the radius increases. The mass that is producing the force will decline its potency as the cosmos is becoming bigger from a point where the cosmos was so small then. When the cosmos was the smallest the gravity at that moment, had to be the biggest it ever could be at any point or that it can become in the future. The best chance that gravity ever had to produce a force strong enough for gravity to remove all space between particles was when being at the smallest distance apart and that then was at that moment where the radius almost did not exist, because the force will fade afterwards while the separating distance increases where it never again will find a better opportunity to endorse a stronger force. A still better moment of enforcing Newton's contraction was the chance to act by compressing the Universe that came even before the Big Bang came into progress. It should then rather be the story of the Big Crunch as a Universe imploded; that is if Newton did apply. When was mass ever that close in proximity again and when was the radius ever again that small? Take into consideration that mass times mass is supposed to remove the radius and where there was almost no radius the force was the strongest it could ever get. In the formula science officially accept as the cosmic Universal basis that reads $F = G\dfrac{M \times m}{r^2}$,

mass has the purpose to the reduce the radius by unleashing a force equal to the product of the mass in relation to the compliment of the gravity constant and therefore when the radius was almost not in evidence, the mass had the best opportunity to dissolve the radius. Using the formula is a mathematical manner to express verbally that the force sets out destroying the radius by measure of the mass drawing on the mass in conjunction and with the aid of the gravity constant, which combines to form a force. Using this formula puts the mass in relevancy with the radius and the smaller the radius is the bigger in relevancy the mass product will be and in the same event the stronger will the force committing the gravity be. By reducing the radius the mass instantaneously increases, as the one stands relevant to the other as a mathematical response. If there is a partition the size of a neutron separating all objects within the cosmos such as is suggested was the case at the event of the Big Bang, then the force this state of affairs will unleash goes beyond calculations. Yet if Newton is correct his force went on to wither away because with the expanding came the increasing of the radii that is allocated between all cosmic structures. With that the opportunity of mass to increase gravity and destroy the radius altogether simply tarnished. This means the expanding went on while Newton's contraction should either have stopped this expansion or science should have gone silent about their previous notion that was promoted by Newton. In contrast to Newton's prediction they now apply the

Hubble constant as the accepted norm by which science age the Universe. The Hubble constant now forms the norm whereby the progress of cosmic expansion is determined. In using the Hubble constant there is absolute proof of expanding. The results are so accurate that that any person blessed with doubt can study the expanding through a high power telescope and see the expansion personally. The Hubble constant is the confirmation of cosmic expansion. The Hubble constant has evidence, which is without dispute where the evidence insists on accepting expansion of the entirety of the cosmos. This runs directly against Newton's contracting and yet both are accepted because Newton was never contradicted. The Physics Paternity stands by its view that Newton requires no proof in spite of everything in science going on continuously to disprove Newton. Can any one find more contradiction of Newton's validity or that more blatant facts that proves more reasons in finding grounds to dispute Newton, which also at the same time can be more devastating to a ground rule of any nature or have a bigger impact that proves changes has to be implemented or that can deliver a higher degree in disproving of Newton's claims than these two principles I mention above? Yet according to the Physics Paternity that is in charge of all matters concerning physics and acts as the guardians thereof on a global scale, that body controlling academic physics maintains the opinion that Newton teaching was never yet brought in doubt. Following the evidence Hubble brought about, the response of the Physics Paternity was to create more fraud. The way the challenge of proving Newton was met was to conceive more mayhem by establishing more fabrications. In spite of such damning evidence that is contradicting Newton to its foundations, Einstein was ordered by the Physics Paternity to find how much mass was available in the cosmos in order to justify the correctness about Newton. They did not consider the connubiality of Newton or consequently proving Newton's failings that came to the open. They consecrated Newton yet again by finding a need to determine when Newton accuracy would come about. They had to find the measure of the gravity that all the cosmic mass would produce in order to determine when the mass would turn the cosmos around and vindicate Newton. In other words there is an acceptance of the incorrectness of Newton by trying to determine when Newton eventually would be right. By not admitting to Newton being wrong they found the cosmos out of order and gave the cosmos a chance to stand in and prove Newton correct. The spotlight was cast on the cosmos falling out of line by the Paternity avoiding to not admit to Newton's inaccurateness. At that point Mainstream science should have put Newton on ice holding Newton under academic arrest while waiting for his termination from the record books of physics. That is not the case because the biggest criminal act ever launched by any group of persons was set in motion on a scale no other can copy.

Einstein was ordered to find the mass deficiency. Einstein was ordered to see what in the cosmos had the audacity to not confirm Newton! The fact that this order to Einstein came about puts the Universe in dispute because Newton has never been in dispute before. The name they gave to the process used to cover such fraud and inconsistency is the Critical Density, which is the biggest smoke screen to cover fraud ever, enlisted anywhere. In comparison this cover up puts every politician and all criminals in the preschool class. Yet this is the highly respected Physics Paternity, which forms the highest level of brainpower, the world has ever seen. Of course Einstein did not find the critical density to be in order and now they use billions of dollars in tax money to defraud the unwitting public as the Physics Paternity takes the crime one step further and blow the cover on what they established in order to save Newton completely out of previous proportions by trying to find so called dark matter. The suggestion of dark matter throws Newton even further into doubt because it still suggests that Newton is incorrect. The purpose of such a suggestion is intended to cover up Newtonian inadequacy by a declaration that Newton was never yet brought into dispute. By trying to find where the cosmos went wrong (because Newton never can go wrong) they could cover the wrongness created by Newton and leave all suspicions in the basket of the cosmos which then will cover up the deceit by establishing even more deceit. All this is brought into use to cover the brainwashing and mind control they use on physics students throughout the world. Talk about shady spy institutions going unscrupulous by methods too dark to mention and at the end of the line one finds all University Physics departments located where the most extreme serves their mission of treachery. The world will never learn that the Universe is contradicting Newton because the world must only know that Newton was never in dispute. Therefore the fraud is carried on by implementing more of Pavlov's mind control about mass bringing about gravity. However this time the process is unleashed on human children showing their naive trust and is used by unscrupulous academics on unwitting students. The best is that the parents still have to pay for the malice done to their children. Because of such forceful conditioning of the mind and enforcing acceptance it shows that Pavlov's theory most of all applies to Humans with highly developed intellectual abilities. The result of this forced accepting of Newtonian beliefs is the direct result of the fact that Newton statements were never proven. Take the comet for one and find Newton in the gravity that the comet evokes.

Picture what happens when a comet is coming towards the Sun. As Newton would suggest we find the comet is heading in the direction of the Sun because we suppose Newton is correct and the Sun is pulling the comet by strength of mass towards the Sun. The gravity of the Sun is pulling the comet towards the Sun while the Sun is just too big to be pulled around by the comet so the comet is doing all the coming together on behalf of both. The gravity of the comet is focused squarely on the centre of the Sun, which is the focus point of all gravity that is producing such an enormous pulling force. This is happening while the gravity of the Sun has the comet hooked by its centre and is dragging the comet through outer space as fast as the combining mass will produce a force. This proves Newton is a genius. Then when things get critical the comet elude the Sun by missing the centre of the Sun. This proves Newton is suspect. We find the comet eluding the centre point of gravity of the Sun in spite of the enormity of the mass of the sun, which is then able to create a devastating gravity force with which the Sun finds the strength to unleash an inescapable force on the insignificant comet. In spite of the enormity of the mass of the Sun that creates a devastating gravity from which no escaping is possible the comet is defying the mass of the Sun to a certain point alongside the Sun to which it travels. When it reaches the point where the comet misses the Sun centre the comet breaks free and misses the gravity of the Sun altogether. Then the inexplicable happens where the comet defies the pulling of all the mass of the entire sun as it spins around the entire Sun by heading away from the Sun. The comet escaped from its destiny with mass forming in the Sun. The comet avoided a sure death sentence but what brought about the pardon? That small comet evaded capture by finding a means to avoid the gravity pulling force by which the Sun caught the comet. The Comet is going north and it is leaving the Sun in a position South of its Northern direction. This proves Newton is a fraud. Newton's gravity fails to explain this inexplicable action where the Sun allows the comet to defy Newton! The comet is going free in spite of all the pulling of all the mass of the entire Sun. Not once does any Newtonian care to discuss this betrayal of Newtonian principles. No mention of this is anywhere in any class where Newton is taught. That Newtonians allow to be put into the category of the wonders of the Universe and we humans find no answers to any in that category. What went wrong to the position Newton has where mass is drawing mass and there is no escaping clause permitted. The best way to go foreword is to go back to where Newton started science by applying fraud to find answers that he concocted.

Being the Professor with the status that such a position accompanies and with $F = G\dfrac{M \times m}{r^2}$ applying as the force pulling cosmic objects into a collision, can you tell anyone how many comets crash into the Sun every time the Sun's mass is pulling the comet into the Sun? Can you as a Master show how many comets destroy in relation to the number of comets escaping successfully? The Sun is pulling the comet while the comet is pulling the Sun and gravity is reducing the distance between the two. While the gravity should be pointing the centre of the comet directly into the centre of the Sun it does not happen. Every time the comet misses the Sun by a country mile where the comet makes a circle around the Sun and then heads into the darkness of outer space. Why would the pulling gravity have the comet escape every time in the face of the enormous gravity that the Sun produces? How does a comet pass the Sun unscathed while the enormous gravity of the Sun should give the tiny gravity of the comet no chance to escape? By using mass on mass as the force of gravity can you explain how comets get to escape from their definite encounter and avoid the pending collision they have to endure? Can you explain why the comet breaks free and travel all the way into outer space, only to return in a predictable cyclic manner? No…I know you can't explain such an obvious scenario. That is one of the numerous unexplained Newtonian misinterpretations you constantly fail to neither admit nor mention in polite conversation and is another never addressed Newtonian shortcoming in Newtonian science, which makes the pulling of mass onto mass a hoax. It is fully compatible to the likes of middle age witchcraft proposals and you keep it best wrapped where no one ever mention it to the public at large.

The gravity that Newtonians insist on where something is pulling something is a pipe dream. There is nothing that is pulling anything to bring about gravity. Gravity is charged by the combining interaction between the four cosmic principles by the manner they interact with singularity and that I prove! Gravity is the value of singularity moving in formation of space filled with material in relation with time. Gravity is the manner in which eternity parts from infinity and I prove both factors of time without a doubt. It is something that Newtonian science cannot even contemplate because the mass that is producing gravity has as much validation as a witch flying on a broomstick. There are no differences in the forces of gravity that Newton introduced and what mass unleashes by pulling power and a witch bewitching (or whatever witches do best). All four so-called forces are the very same thing except in the motion they apply and what generates to establish the forces. Gravity and electricity is the very same thing but for the scale in which they are generated. Other than that, there is no difference between gravity and electricity. How many times did I send your department these information and how many books did I send where I detailed it? Yet you choose to go the way of the criminal by proposing a concept that relies on fictive science, falsified mathematics, blatant crooked suggestions and highly suspicious arguments. I challenge you Academics in physics too show how much the Moon, is coming closer to the Earth. I challenge you to show that the product of the mass of the two objects namely the Earth and the Moon is having a mass that is pulling by force the Moon towards the Earth, and the Earth towards the Moon bilaterally. I challenge you to show how mass influences orbiting planets by contributing or increasing / decreasing or to what degree the mass is influencing the orbit velocity of any planet in any way or manner. The orbits are in ratio but not like Newton suggested being $a^3 = T^2$ but there is a clear ratio equalising all the planetary motion. However, this means mass is not responsible for the orbiting motion or for any other form of influence there may be on the planets rotating the Sun. The mass of the Sun has no influence on the mass of the orbit of planets because neither distance nor size nor location prescribes any of the orbiting conditions applying. Something else apart from mass is dictating the terms to secure the evenness of the rotating. Notwithstanding, whatever mass differentiation each one has, the lot are still rotating very alike and in fact they are the very same. Notwithstanding the enormous mass discrepancies between the planets where Jupiter has 318 times the mass of the Earth and Pluto has only 0.0025 the mass of the Earth, they all orbit at a shared equality in ratio of about 300. The response to explain the evenness, Newtonians put the equality down to the gravitational constant that is producing the commensurate behaviour. With this in mind I challenge you to mathematically prove that there is a gravitational constant. I challenge you to prove that the gravitational constant is responsible for evening out the mass differences between orbiting planets and that this is the cause why planets all orbit in a compatible fashion using the same velocity notwithstanding mass differentiations. Remember this evidence I mention opposes Newton in no uncertain way by contradicting Newton at the very heart of physics and yet with all the contradicting evidence coming from the cosmos as such, Mainstream science remains to uphold the opinion that Newton was never yet contradicted! In my books that I present to you for your evaluation this time round as it did in the past, this repeat is being an ongoing process about my explanations that is covering this aspect. However I did explain it in the previous books where this time it also is included and is merely a better innovated repeat of that which I presented in the past. However and nevertheless you found grounds to ignore my mentioning this inconclusive assessment in science as you chose to ignore everything I said in all my work. I show that in your manner of calculations you put the gravitational constant as a contributing factor to the mass product that produces the force. The gravitational constant is increasing the force produced by the mass multiplying.

This Newton invented when the formula $F = \dfrac{r^2}{M_1 M_2}$ proved to be hogwash. The force is the measure whereby the radius in the square will vanish and this argument becomes the cornerstone on which all physics hold its foundation. With the original formula in shambles Newton invented more deception by introducing one more unproven factor. He invented the gravitational constant. The gravitational constant is used as the complement of the mass multiplying the mass. In conjunction with the gravitational constant the mass multiply to produce such a force value. This then supposedly produces a force able to destroy the radius between the objects. Then Newton went even further at the time and gave the concept a name: he called it gravity. In the way the formula is presented the gravitational constant increases the effect of the mass of all the planets, which is forming the force of contraction between the various objects and the Sun. By expressing the statement in a mathematical equation it reads as follows $F = G\dfrac{M \times m}{r^2}$ F=G (M X m) / r². The evidence that this is a fable and that mass is not pulling the objects closer was already established with Kepler introducing his work. Newton knew that the planets orbit in equilibrium by the measure if $a^3 / T^2 = 300$ and all of that is in a ratio of about 300. That disputes Newton's claim on mass forming influences because it clearly shows equanimity and the proof was delivered before Newton was born.

Then when all the planets orbit at an even pace and all being the same ratio, where not one is found to come any closer to the Sun or to each other, your officially accepted reason for explaining that, is that the gravitational constant is evening out the mass differences and yet, according to the official formula you use, the gravitational constant is not evening out the mass but is complimenting the mass by improving the so called force. If the gravitational constant is evening out the mass the constant firstly has to divide into the mass to wither the effect that the mass produce and by nullifying the mass to the extent that Kepler's work proves, there is no mass factor present left to use.

The formula then should read $F = \dfrac{M \times m}{G \times r^2} = 1$ F=(M X m) / (GX r²)=1 which in normal verbal English states that the product of both masses is in equilibrium when divided by the conjunction of the gravitational constant and of the square of the distance parting the objects where the product of the gravitational constant and the square of the radius between the planets is in division with the product of the mass which then will produce the force that cancels any influence of the mass factor that both of the planet's will have on the orbit of every planet orbiting evenly around the centre structure. That is very much not what your formula you use $F = G\dfrac{M \times m}{r^2}$ suggests. Your formula shows that the gravitational constant is helping the mass product to destroy the square of the radius. In the cosmos however and in reality that destroying of the radius is the last thing that is happening. To explain this ridiculous deception Mainstream physics go on to explain that the gravitational constant which is supposedly the gravity that is keeping outer space in a unit, where that gravitational constant nullifies the mass influence on the force. Newton said mass is all-important but this explanation now suggests that the mass is nullified by the intervention of the gravitational constant. This is a statement that is not in line with the mathematical equations Mainstream science present to serve as truth. Making such claims that is so far away from the actual mathematical statements, which came into place serving as yet another attempt to cover up Newton fraud with more deceit and becomes compatible with mafia like behaviour. This criminality is the behaviour of all physics paternity in control of all physics worldwide! This is the group of men that is in control of the thoughts in the minds of the brightest developing brains in the world. They harvest the best intellectual group of all students this Earth can offer! To the students questioning or rejecting the Newton claims is not an option. The students has the choice to accept what is taught or face examination failure and total rejection of the institute where they are educated. If that which is presented that the statement would suggest holds water about their explaining how there is such planet mass apathy that will not influence orbit details of planets making them go faster or slower, then the entire Newton presumption is based on yet more Newtonian misinformation. If the gravitational constant were evening out the mass then it would have the gravitational constant in multiplication with the square of the radius in dividing the product of the mass. This will then be evening out the product that the multiplying of the mass will put in place. If that was said and translated in mathematical terms, then it would mathematically equate to show where the force will be in equilibrium because the radius is supported by the addition of the gravitational constant which increases the radius in the square and the total of that is in equilibrium with the product of both object's mass. If that was true then the gravitational constant must be placed where it increases the radius in every event as well as canceling all influences mass may present and it would be stating that $(G \times r^2) = (M \times m)$ **(G X r²) = (M X m)** is applying in all the cases so that the mass has no longer an influence on the outcome of the equation because out there in the real cosmos Kepler proved that the mass has no implication on the orbit of any planet **a³ / T² = 300.** Kepler never even thought of a mass of any description and I can assure you that his not mentioning any mass was not the result of his retarded mind or not because he was mentally inferior to one Isaac Newton. With the corrected implementation of Kepler I can and I do support such statement of equality but then in that event the equation then should read that the gravitational constant is complimenting and supporting the square of the radius and mathematically such a statement is $(G \times r^2) = (M \times m)$ **(G X r²) = (M X m)** That removes mass as a cosmic factor. You know very well that in the planet orbit ratio the space factor (**space a³**) divided by time **T²** leaves a space **(a³) − time (T²)** displacement factor ratio between space and time of round about three hundred in **all cases,** $\left(\dfrac{T^2}{a^3} = \pm 300 \right)$ **T² /a³= ± 300)** which

disqualifies mass discrepancies altogether. With that the cosmos puts Newton in dispute! So why then bring in the use of mass in any event if it does not change the result but has only one purpose and that is to cover Newtonian's misconception and to hide Newton's fraud? However, the formula used by mainstream science clearly shows that the gravitational constant is in effect increasing the mass by multiplication of the mass. The formula undeniably put the gravitational constant in a multiplying position with that which the mass of both structures hold. $F = G \dfrac{M \times m}{r^2}$ **F= G (M X m) / r²** which then in ratio is decreasing the square of the radius. That is pure rubbish and I challenge you to prove me wrong by proving your mathematical interpretation. Then all the while, while promoting such rubbish you still maintain that Newton has never been proven incorrect. When Newton is taken into the cosmos Newton physics goes bizarre. There is no evidence of mass influencing the motion of planets in any manner. Mass doesn't bring the least appreciation the orbits of planets. One can see the lack of any effect mass has on putting differentiation between planets as this table shows and it was available at the time of Newton's plotting. One can see equable motion where not one stands in disadvantage.

PLANET	SEMIMAJOR AXIS $A(10^{10}m)$	PERIOD T (y)	T^2/a^3 $(10^{-34} y^2/m^3)$
Mercury	5.79	0.241	2.99
Venus	10.8	0.615	3.00
Earth	15.0	1.00	2.96
Mars	22.8	1.88	2.98
Jupiter	77.8	11.9	3.01
Saturn	143	29.5	2.98
Uranus	287	84.0	2.98
Neptune	450	165	2.99
Pluto	590	248	2.99

Newton is not incorrect as it is used in normal applied physics but in cosmology there are no grounds in supporting of mass or gravity in the manner Newton saw or in the way Newton suggested mass to apply. On Earth and on the ground Newton's work is impeccably correct and that I admit without any reservations on any aspect of physics in whatever manner. However taking Newton's ideas into space becomes fraud. Mass only serves as a human concept to allocate differentiation when required when gravity is obstructed and in the cosmos gravity is not obstructed. The reason for that I have well documented in An Open Letter To Selected Academics.

In my books I show that mass is a resisting or obstruction by a material occupying space a^3 that prevents the flow of time T^2 to continue by extending the value of k. Mass is the nullifying of the directional progress k to obstruct time T^2 by blocking flow which otherwise would include mass a^3 to participate in such motion. When two particles are in water and are totally submerged, yet they are buoyant, the objects will flow with the water in the direction that the water flows. The flowing of the objects which are in the form using the cube a^3 will pass from a point to another point forming the square of time T^2 and the distance they flow substantiates the relevancy which k renders to the duration of time T^2 and the change in allocated position of a^3 lasting from point to point over the distance of k. Forget at this moment that water is flowing because of gravity. I only wish to present my idea. Kepler stated that $a^3 = T^2k$ where a^3 can only be space in the cube because space is only in the cube since the square is too flat to form space leaving T^2 as the motion in the flat but square motion and k is the relevancy of such motion completed. Also forget Newton's insane arguing about k being zero because of some invalid concept he proclaimed using a gyroscope he does not understand and of which leaves his entire arguments suspect since the argument has no validity. If his argument holds any water the graph has no validity and no generator can charge electricity because it is the current flowing that becomes k and k is that which he claims has a value of null, while in truth k is inspired by the rotation motion T^2 that is generating the electric current to a precise and pre calculated graph. His presenting of some stupid argument that one factor can divide into another factor and from such dividing it will result in zero is mathematical madness.

You as a mathematician should know that the smallest outcome of the dividing of any equation could be an infinitely small number but never can it be zero. Yet still because it is Newton you are prepared to accept what is mathematically absurd since the very idea was all part of his fraud. If **k** is zero then the entire formula becomes nullifies and we would find ourselves living in a Universe comprising of nothing because if **k = 0** then **a³ = T²k** is **a³ = T² X 0** and the entire formula Kepler presented goes away leaving the cosmos without a purpose. Therefore in order to let mathematical sanity prevail, we have to get rid of Newton's incoherency and that is as incoherent as he introduced the idea where having time, (which is the flow of material in relation to each other), find a position that would allow time to stand still. Time in the square cannot stand still because time in the square cannot be a square since only then will it be able to not show progress between two points measuring time. Time is coming from being at while going to. Time is motion from the past through the present onto the future as space moves by time **a³ = T²k**. The only thing that time will find impossible is to stand still because then time becomes eternity in the moment of infinity. When time stands still, time forms eternity in the moment of infinity. Time is the displacing of space in a given relative period where **k** forms the relevancy of the period taking place.

Lets return to the stream and the floating but buoyant objects. When the submerged objects float in the water somewhere in the centre of the water not being at the top or otherwise at the bottom but also not restricted in any way, then we find that the flowing body will go along with the water flowing and will go in the same relation to the water as what the flow of the water demands. Then the movement of the space **a³** would be equal to the relevancy **k** of the distance achieved of the flowing water **T²**. The space **a³** in motion will be equal to the flow of the water **T²k.** The motion of the space secures a free and unhindered location as the water takes the space of material as if the space of material is just more water. The space filled with material can just as well be space filled with water since the material is no different from the liquid water. When there is a retaining by blocking the flow of the space in the water as part of the water of the space changes from being part of the liquid and then becomes a separate solid that helps to restrict the flowing water. By forming restricting part of the motion the space no longer presides with the water as water but becomes a restriction of the water by forming a solid that blocks the water flow. Then only will mass form part of the particle because the resistance will be equal to the obstruction that the obstacle forms in relation the motion.

The obstruction will present the volume of the space that obstructs the flow and that volume in relation to the countering of the flow by restricting the flow in the measure of the size the object retains such restriction becomes the measure of the value the mass represents. In that case size matters because the blocking is in the full volume of the obstruction where the water gives the blocking the measure of resisting that comes about. Then that will be when the space **a³** is standing still by being retained in relation to the flow of the liquid **T²k**. When objects fall they all fall alike (Galileo) and the flow of them is the same notwithstanding size or structure. One can only contribute such behaviour to the indication of the presence of specific density forming buoyancy where the buoyancy derives a value from the liquid T^2k that allows the space a^3 a medium to flow in $a^3 = T^2k$. When the retaining of motion comes in place by preventing the space to further motion then mass becomes a factor but it is not presenting the reason for the flowing of the water but is a result of that which is hindering motion and not enabling the buoyancy. The mass has no part in the motion but is a compliment in the retarding of the motion. That means instead of gravity being $a^3 = T^2k$ the application of mass would result in $a^3 \neq T^2k$. The space is not moving and the not moving while time is still moving without the corroboration of space the result then is that space is preventing a steady flow of time in relation to the object being restricted to flow with the water. What is time is rather more complicated and I leave such explaining to the books. There is a flow of liquid being time that is converting towards the Earth (or any other cosmic structure) where outer space has to be of substance to validate any distance forming between planets and the Sun. Something has to fill the space that is validating the distance between the Sun and the rest of the cosmos and that something cannot be a substitute for nothing or be substituted by nothing.

The substance filling outer space is holding the material in a state of motion and that motion of space can only be when a state of buoyancy prevails. Outer space cannot be nothing or filled with nothing and that idea is as corrupt as any idea connecting with Newtonian dogma. If outer space was filled with nothing then what would that nothing comprise of because nothing actually means it is not. It literally means there is not. It does not mean vacancy because even with vacancy there is a shortage of one part and that will leave an abundance of another. Calling outer space nothing, is not referring to outer space, as outer space is unfilled with particles but whatever should present it in composition when outer space is tagged and branded as nothing, because being nothing then in that event is a case where everything including the unfilled space has removed leaving nothing in its place and that, where I can assure you that is not what outer space is because that is idiocy and that does not represent the substance of what the Universe comprises of. The only reason why mass or size or location does not present any difference in outer space is when these features are absent. It is a sign of buoyancy. Size in cosmic terms is irrelevant and proof of that we find in the fact that a few brittle fragments that once formed a planetary structure rotates with the giant Jupiter and in the same orbit band than Jupiter. In the cosmos mass size and volumetric occupation serves little purpose but to put the unit it holds in a time development. However in cosmic location and in relation to others, the size is meaningless and the Titius Bode law serves as proof of that. The size of Jupiter is to the Sun as equal as the size that Pluto has.

Distance in between structures are a manmade concept that is meaningless in the cosmic context because all planets are at a ratio $T^2 / a^3 = 300$. Further more that is proof that time by measure of k is diminishing in density, because there are motion of k^{-1} going negative, which means in a mathematical equation $T^2 / a^3 = k^{-1}$ and in that sense time equates to $a^3 / T^2 = k$ and from that we can clearly see that all planets are at an even distance from the Sun. We might give their location purpose but that purpose we allocate for human use does not serve the Sun in the least. The very same argument goes for temperature but that is somewhat cumbersome to explain in this letter, however I go into much detail about that in the books. But if you read my books it would all be old news to you by now and most of all, I bring much more proof than the shambles Newton served science to blindfold reality. The cosmos is in two aspects that serve each other by occasion relevancy and that is why gravity is the Coanda effect. Every aspect of the cosmos is in terms of liquids $k^{-1} = T^2 / a^3$ and solids $k = a^3 / T^2$ that forms motion of space $a^3 = T^2 k$.

When that space which is holding what we perceive to be where the solid is in, is in motion then the independence of the motion that is taking the moving space along, secures the object as a cosmic independent structure. Kepler gave us $a^3 = T^2 k$ and that proves that when space is moving $T^2 = a^3/ k$ it proves that such space filled with material is allowed growth $k = a^3 / T^2$ and only penetrable substances like gas and liquid allows motion to continue by distinguishing partition through density differentiations $k^{-1} = T^2 / a^3$. We have to surrender our pretext to what is truly present in the cosmos. The question to what is liquid and what is solid is answered by what is hot and what is cold. Liquid serves as being in a higher state of heat and solid represents being in a lower state of heat. Then we should find what is heat and where it is hot.

 If we fill an air balloon with air it expands the balloon by expanding the inside size of the balloon. There is a ratio in the air between the inside of the balloon and the outside of the balloon and by adding heat to the air inside the balloon the size of the balloon grows. We can vividly see from that action, that the hot air that is pumped into the balloon, the balloon becomes larger in a volumetric measure capacity. From this one can conclude and it is a logically conclusion that by introducing more heat into the balloon the level of heat to material rises in favour of the heat. The heat containing that increases puts the particles in the balloon in lesser quantities in relation to the increase in space by volumetric quantity that increased in space that the heat produce in the balloon By increasing the heat in the balloon the balloon got bigger by volumetric measure but also got less dense by atmospheric measure. The reduction in density is a result of space increasing without having any particles added to the inside of the balloon and the increase was in heat that became space where the heat was air without particles nonetheless. If one fills it with only air particles the growth in size inside the balloon would be far less substantial and the density would not produce a lifting of the balloon. The size of the balloon actually increases by receiving more heat inside the balloon and if we find that heat increases the balloon in measured volume then the air in the balloon is heat forming more of what already is present which is giving the particles of material also inside the balloon less density. If bringing in more heat allows a bigger area to fill inside the balloon then the balloon is filled with heat from the onset.

By adding more heat we promote more space and a decrease in density and that combination allows more motion and therefore the motion is equal to the heat inside the balloon that reduced the relevant density of the material inside the balloon. On the same argument we have to take the conclusion one step further. When we fill what is in the balloon with more heat then that which is in the balloon expands to entertain more of what already is present. The expanding accommodates heat by accumulating what we perceive as heat into an area that makes it after receiving the heat to be a larger area. If we allow such expanding to continue until no further expanding is possible or permitted then we have reached the maximum position heat can expand. We then are where there is no more expanding because all the heat brought about all the expanding that will ever be entertained. Then we are where there can be no more heat added that will effect any more expanding. The heat cannot grow because the expanding nullifies any increase in heat. If no more adding of heat can increase the heat level by enforcing the expanding we know that at that point we reached the position where no more adding of heat can effect the status of the heat level or the increase in volumetric space by adding heat to space. Then we are at the hottest anything can ever go. Then we are in outer space where no more adding of heat can bring any more expanding since it accommodates all the expanding all heat added can deliver. We went and added heat to the balloon to increase the space of the balloon and decrease the ratio density of material and heat inside the balloon and from that the balloon expanded. It did not expand by receiving more particles but it grew more in space and it reduced in density by expanding with accumulation of heat inside the balloon. If that is the case and that is true science then that puts outer space in the hottest position anywhere can ever get and we have to adapt our senses to accommodate reality. Outer space is not the coldest but the hottest place in the cosmos and the cosmos is the judge of that. Forget Kelvin's scale because heat as we humans would like to think of it is irrelevant and non-existing in the cosmos. It is so Newtonian to tell outer space and the cosmos what the cosmos must be because of what we think is true about the cosmos in stead of allowing the cosmos to tell us what the cosmos is. Heat and cold has no measure and the same applies to mass and distance. The entire Universe is all served where every aspect is in ratio of being in a state between what is in motion and what is deprived of motion. If one fills the balloon with hot air the heat will expand the space and put the space inside the balloon, which is in a higher level of expansion in line with a higher state of expansion in cosmic space.

By increasing the heat level inside the balloon the space inside expands and the density factor drops in relation to the conditions on the outside of the balloon where that differentiation drives the balloon to move and that moving is gravity. Reducing the heat will drive the balloon to a lower level of heat with a higher value in density. That is gravity. The adding of heat or the removing of heat, which increases or deprives the density ratio in relation to the cosmos elsewhere is what produces the motion of gravity. By adding or removing heat from the balloon that action puts gravity to the balloon where from that the balloon moves to a higher status of space that is sporting a lesser density ratio in terms of particles between particles and space. The adding of heat produces motion and that removes mass, which will expand the space. If heat expands the space then at the location where the adding of more heat cannot bring more expanding we have to accept that at that point the heat stretched the space to the limit it can go notwithstanding what Kelvin's temperature scale of any other measurement may dictate. The air outside the balloon allows the balloon to flow and such flowing of the objects in motion are unrestricted as they allow the free motion of space filled with material in space not filled with particles but will take space filled with particles along while the lot is in that space which allows the flowing object to orbit the Sun and in that the motion is secured to have gravity $a^3 = T^2 k$ or motion of particles in time to take place. That same space forms part of that space not filled with particles but that is receding to Earth nevertheless. It is that the space not filled with particles, which is receding to Earth in any event and while the space is receding it is also bringing not buoyant objects along in any case. The purpose of the gravity the Sun unleashes is to remove heat from space and not to go futile and put more heat into an area where the adding of more heat has no purpose what so ever. The balloon clearly shows why the space is floating in the air. By floating or coming to the Earth can only be if such material holding the filled space is the object that is in a state of buoyancy in the space it penetrated. Why they have buoyancy is a matter I explain when and where I explain the sound barrier. If Jupiter is "swimming " in the same "puddle" as Pluto the "puddle" will affect both planets in the same manner and Kepler did prove that statement with $\left(\dfrac{T^2}{a^3} = \pm 300\right)$ $T^2/a^3 = \pm 300)$ being the measured flow (k^{-1}). If the "floating" was caused by a density not of the planets but of the substance in which the objects floated by buoyancy, then condition of similarity would prevail where all are effected alike and again Kepler proved that as is stated by $\left(\dfrac{T^2}{a^3} = \pm 300\right)$ $T^2/a^3 = \pm 300)$. The fact that the planets do not show mass as an influence proves the motion is in a state of buoyancy $a^3 = T^2 k$.

The fact that Galileo saw objects fall equally notwithstanding size differences proves the objects falling equally are in buoyancy. The fact that the neutron shows no signs of mass puts the neutron in buoyancy while the electron as well as the proton shows the characteristics connected to forming restriction of a flow or depriving something its ability to flow. It is being indicative of restraining the flow of space, which provides both with mass or with space restriction. I have explained it in every book I sent you thus far in the attempt to get you to read my work. By this time I am anticipating that you are not going to read my work again because it happened for so many times in a consecutive sequence of events before on previous occasions. I have found solace in the fact that you don't care to read it. I mention this on purpose because you have the idea that you are served with a position putting you beyond any attack from other quarters and that there is no person that can rattle your cage since you have Newton to defend you. Make sure about that! Although I do realise by this time that everything I say should be old news to you since you heard it before when you evaluated my work in the past I honestly think that there is no academic that even got this far in reading this letter. But if by any chance in the slimmest of possibilities there might be one still reading and in that event if it seems all new to you it proves that in the past I am correct that no one cares to see another view except for the fraud of Newton. I challenge you to show how Newton's mass allows motion to the balloon where Newton's mass forcefully dictate mass on all that is willing and not willing to participate. If by chance you get this far in reading this letter, you do reached this point in the letter because the work interests you in some way. If you did read my work in the past you would have seen it was written in my work that you previously rejected / dismissed / ignored.

But I would be highly surprised if you are still reading at this point where I accuse the entire physics paternity and not you in person as such that you are deliberately, with malice in mind, entertaining a misconception while you are innovating mass that is responsible for establishing gravity and by that you are applauding the misgiving which is in association with the detestable corroborating of Newtonian deliberate fraud. By not investigating my accusations you then are contributing to fraud that hides any possibility of finding scientific truth. While you place mass where it does not belong and where it clearly shows having no influence at all on the working principles that guides the cosmos, you deliberately circumvent the truth by committing a heinous act of calculating deception on the part of Master Newton. Please think of my statement that there is no mass when objects fall because there is buoyancy present and see that it rings far truer than putting mass where no mass belongs just for the sake of protecting Newtonian deception. There is no mass in planets orbiting because the process works on buoyancy. While you know very well there is no mass in any of the places I mention you still ignore my work and commit to the fraud of placing mass where mass does not belong.

I challenge you to show the glorified graviton and I challenge you to prove that mass is pulling planets closer by destroying the radius distance there is between cosmic objects. You claim that this which is out there for us to see and that which we now have which is serving us as a Universe had grown from where it once had the size of a neutron as it expanded to what is now present. Just go outside at night and look at what came about since the lot out there was the size of a Neutron. On the other side there is Newton's deflating or contracting Universe. You maintain that this one principle of Newtonian mass inspired contracting is applied while at the very same time you then are in an official capacity also promoting the Big Bang development and the Hubble constant, which absolutely contradicts the contraction idea.

The Big Bang contradicts Newton in all aspects as it clearly shows beyond doubt that there is expanding going on in the cosmos. The two proven phenomena that promote expansion is acutely indicating an increase in cosmic structure expanding while Newton is projecting a contraction whereby the final conclusion will come about as all structures unite. There is no manner in which to marry Newton's contraction ideas with the Big Bang expanding and Hubble's cosmic shifting of matter notwithstanding that the latter is proven beyond doubt. There is a cosmic parting of material moving further away and that is the proven concept forming the norm in the cosmos. Mainstream Science has to cover the fact that Newton is deceptive by producing more improbity. To try and con the public in realising Newton's theoretical failure and deception Mainstream Science do not directly admit to Newton not panning out but Mainstream Science distort any such realising as they invent more misleading and scamming. Mainstream Science diverts from the truth even further with more falsified facts as they come up with more theories by which Mainstream Science puts the attention not on Newton failing but the mystery why the expanding is in progress without finding it necessary to reflect on Newton. If Newton is not, then when will Newton start? Mainstream Science do not ask the question why there is this anomaly and total contravention of the rule but would rather find no need to reflect on Newton altogether…no they do not admit to Newton's failure at all but they respond by diverting his failure as they put his success into the future. They don't admit Newton is wrong because they establish when Newton will be correct. Newton is correct, but they just have to find when he is going to become correct because they must find why the cosmos would behave in such a disgraceful manner by contradicting Master Newton! They divert all attention from Newton's failure and of course also from the proposition that science has no idea about the facts supporting physics to finding the flaw that the cosmos would present to have.

This hiding of the truth is to coward from reality and to prove the illusion Physics wished to create and substantiate. If science doesn't have Newton science in all its splendour they have nothing and science promotes the idea with atheism that science knows more about the cosmos than what a God might know. Therefore and in that light Newton may never be wrong because that would leave science as the laughing idiots that can't even explain their faculty.

With the latest admitting of The Big Bang and the accepting of the Hubble constant there now is much proof of Newton's failings because disproving Newton directly comes via the proof of the Big Bang and that is also the proof of Newton's failure. The proof that embraces the Big Bang is also the very proof of Newton failing and how much Newton was being incorrect all along. The whole concoction of the Critical Density that should prove mass deficiency in the cosmos as a whole is just a criminal smoke screen to cover up fraud. Mainstream Science puts Einstein to task to measure all material throughout the entire Universe. This idea in its individual capacity of finding the ability to calculate what there is in the Universe by own merit is outrageous and I am prepared to take on any person disregarding his or her status, to task to prove the feasibility of such an action to take place. To measure the Universe Einstein has to be in the centre of the Universe, which will enable him to have a panoramic view about what there is to measure throughout the entire Universe.

If he were slightly off centre then that would place more material on one of his sides and less on another side. How would he then be able to gauge the entire spectre when there was more material on his one side than there was then on his other side, which would then fall outside his parameter of viewing. It is madness and deception that is thought up by fools and is purposely hoaxed to serve as hogwash to all the idiots they think the public at large are. When the evidence Einstein could gather did not underwrite the fraud, another even bigger scam was suggested. Then to further exaggerate the criminality and to achieve more deception with the sole motive to cover the failing of physics, then a plan was set in action where Science was put to task with an order to start looking for some ridiculous non-detectable dark matter. If the matter is not detectable it is not contradictable which makes it automatically acceptable and overall plausible because it cannot be disproved as much as it cannot be proved and by not disproving it, the deception may just stand in for the truth. It is what Newton invented the first time round and the plan will merely follow in the Master's original guidelines because that is what Newtonians got away with for three hundred odd years. If no one can prove that mass is responsible for gravity then no one can prove mass is not responsible for gravity.

This debacle can again play into the hands of the fraudsters because if dark matter can never be proved it therefore then can never be disproved either and is very much similar to what Newton established working precisely on the lines of Newton's gravity by mass deception. If no one can detect the dark matter because it is undetectable dark, matter everyone will be hesitant to disprove the validity because it can be proven just as little as it can be disproved. The only logical reason as to why this order was given is to cover the Newtonian misconception and the hallucination of mass that is producing gravity by contracting material. Newtonian mindset dictated that after all there has to be a presence of mass should there be gravity because if mass is the producer of gravity then in that is the truth that there cannot be gravity without mass...and all along there are still those that insist I do not understand Newton! They persist that there has to be enough mass to produce enough gravity for there cannot be too little mass to charge enough gravity to pull the Universe into confinement. While there is the expanding going about, their motive is to find gravity producing mass to stop the Universe of making a fool of Newton! Therefore to protect Newton and with that to protect mainstream Science from being covered in mud and hogwash they had to find enough gravity-producing-mass, either proven or otherwise, notwithstanding whether the mass is detected or undetected.

Locating mass became a smoke screen not even Lenin and Joseph Stalin could produce which would prevent every idiot in the general public to realise the fatality suffered by science as Newton is drowned in contradiction of Hubble's Universe that insists on the moving apart of the cosmos. No method was too unscrupulous to save Newton and proof had to be found, which counteracts Hubble's disproving Newton and stands in as truth for the gravity that should start contracting imminently or otherwise science will face what was coming their way all along... In this matter as to why mainstream science committed so much treachery one can only have one conclusion and the only sensible argument is that they worked feverishly to contain the proof of Newton's irregularities by keeping the truth from the general public whereby the entire world will remain deceived as to the sorry state Newtonian science finds its position. Spelled out in more fashionable colours it would be that there has to be enough magic everywhere to establish the presence of a force that will conceal the truth and establish another populous blind spot. The deception once again became acceptable although it was undermining the truth and it was created to help with introducing more of in despicable fraud Newton started on a Global world wide academic scale. I just do not accept that Newton could mislead the world's brightest brainpower and all that brainpower was blinded by their shortsightedness and naivety. That is very unaccepteble. Not with such intelligence being in demand and in command of the subject. It was done with intent to defraud.

By duplicating Newton's first act of fraud and repeating the manner again, Physics can still pretend they are the Master in control of their faculty while no one realises the degree of deceit that is prevailing in the corridors of all the physics paternity and all is done just to give Newtonian delusions legality. In doing that the academics are defrauding the taxpayer in their falsifying facts in order to use a racketeering process whereby they steal the money they defraud the public with by promoting a falsified claim to cover Newtonian fraud. Every one concerned with physics has to realise this shortfall of theoretical consistency, which includes the solution attempt that is so detestably untrue. To falsify and defraud facts by misrepresentation in as far as presenting falsified claims as if being the truth is a way to criminally illicit money from an innocent and unsuspected public, which then is a criminal offence. It is racketeering. If there were any other convincing argument that can support honesty prevailing in the midst of madness I would love to hear about it.

The truth of the matter is that planets rotating as celestial bodies show no mass influences. The two phenomena mentioned being the Big Bang and Hubble's expanding is proof of the expanding going on in the cosmos, which by expanding and not contracting is putting distance between cosmic objects while Newton's principle promotes contraction and the reducing of radii. The two phenomena mentioned are based on highly tested and well - proven merits about facts and are not accepted on cultural merit, as is the case with Newton's claims. There was never proof about mass forming gravity. Yet while it was never proven, it still went on to become the foundation of Newtonian physics. The idea of mass that is producing gravity forms the corner stone on which science is founded and every person on earth (excluding me) accepts unconditionally that Newtonian views are accepted truth but in so many years not one shadow of evidence was once introduced to substantiate any Newtonian claims.

No one could ever produce even on just one occasion some proof that mass instigates gravity. The accepting of Newton's cosmic views comes from implementing a principle conceived through the culture of what was thought to be correct through the ages and accepting these facts as proven truths is founded by preconditioning the human brain of young students. It is founded in the culture to accept the past unconditionally and is not founded on unwavering and undisputable factual correctness born from reputable proof. The Newtonian thinking about mass forming gravity is religiosity enforced on students by conditioning their accepting the correctness thereof and linking such unconditional accepting to their failure or success. By the repeating of what is reflected on, that is coming as accepted from the past about what was taught in the past without conditioning such information to newly required proof and with no wavering of the truth about the content, is that what allows their maintaining on the course of learning about physics.

It is definitely in the case of Newton not depending on the accurateness of facts. It is instated by accepting without questions that they either gain a pass allowing them to stay on as an accepted student, where they then become part of the system or when not accepting without reservations and because of that to be shackled as a failure and then be removed from the institution. This programming of the human mind to accept facts unconditionally is called examinations and it forms the pivotal part of a system that runs through every institution of learning generation after generation. The young mind is taken in the prime time of potential learning and by teaching the culture that comes from the past generations about Newton being correct and that Newton is accepted with no questions, forms the required foundation of Newtonian science. The system allows only room for students that unconditionally accept Newton. Others not willing or not able to just accept Newton on a say so basis (be it those that require some conciliation of proof by facts being presented to fill their investigative needs to back such claims of whom I am one in that class) are implicated as being somewhat retarded and short supplied in brain matter.

Not accepting Newton without any substantiating proof classifies any person with such mentality as to be presumed being incompetent in brainpower to "understand Newton!" Those that become accepted have to take on the role of acting mindless by "understanding Newton" in their accepting what they are taught regardless of the sensibility such teachings would otherwise dictate. They stand highly apart from the crowd that wishes to be served with evidence before accepting Newton as the unqualified truth. Yet the Masters as well as the institution towards those showing total and unconditional mind control by accepting Newton regardless of any condition attached to the accepting irrespective of questioning reflect the total opposite. They are the ones tolerated to become future members because they are intellectually superior in their "understanding of Newton". In that matter of finding the acceptable candidates a climate is established wherein there are no margins of doubt allowed about any part of Newtonian issues. It is a mentality that is created by the Superiors suggesting that only those with a superior mental capacity and clarity of mind can become the chosen ones because they are intellectually much more advanced making them better suited to fill a position where they are able to see and accept Newton because they show "understanding of Newton". The others "that do not understand Newton" are detested as those who are mostly rather slow witted. Those being incapacitated by doubt because of the lack of proof and therefore are not able to see what Newton claims are the ones that is lacking the faculty to understand Newton.

They are portrayed as the dim ones and outcasts as they are slow in grasping the situation by identifying their shortcomings with having questions and doubts. It is presumed that the bright ones can clearly see what Newton claims and that makes them the cream of the harvest. That is the institution the student is exposed to with the full knowledge that the student also is compromised when "not understanding Newton". This is part of the student's life when it is also well understood that at such a vulnerable age in the era when the students are dogmatised, that in their young minds questions are not yet part of the student repertoire and their conditioning of accepting is well rehearsed, constantly repeating the idea that Newton is the unqualified and truth unconditionally. To solidify the foundation this repeating is based further founded by all the testing that goes along with the process of dogmatising, the final acceptance are entailed through examinations.

The questionnaire that is called examination papers sets the standard of measuring their conditioning. By the level of the thought conditioning relating to those standards of accepting without preconditioned wavering in doubt or insisting on proof. With the admittance accordingly come such standards of testing where on those standards which the acceptability rests to enter the academy. Admittance or rejection is subject to the level of accepting Newton notwithstanding and regardless of whatever other requirements should prevail becomes the focus point of them going further with their studies. Professor, be honest, and admit to your person in the privacy of your mind how you will react to a student that desires to carry on with physics studies while he reject Newton due to the lack of academic proof on the matter of mass producing gravity. Explain in personal honesty to your conscience what you would do to a paper wherein a student refutes Newton as I do just because there is any evidence in support of

Newton's claims. How will you respond to a paper that rubbishes Newton on the grounds that Newton never brought significant proof to the investigating table? What will you do to a student that declares his rejecting Newton is because to his mind he was never presented with acceptable proof about the merits of mass forming gravity and the student therefore rubbishes such statements on those grounds? If he should dare not to complete the paper on the grounds that he has not witnessed any proof about the planets revolving around the sun by measure of mass or that he finds proof lacking to prove that objects are falling by means of mass applying because it contradicts Galileo and where he then will demand such proof before he would complete an examination paper, then I ask you what would your response be as the Headmaster of your faculty? If you are honest to yourself you will admit to your personal integrity that the student has the choice of either to accept Newton and surrender all doubt or to surrender his scholarship at your institution and in that be rejected and banished.

Yet all the doubts are present because I ask you for proof and you know there is no such proof. Students are never given proof about the matter of gravity and mass but are expected to learn Newton by over and over repeating the repetition and rehearsal of the absolute correctness one has to show which then will serve as proof to their commitment of trusting Newton. This they have to do to prove the degree of their personal ability in accepting Newton unconditionally or to use their more commonly used phrase as they put it "to understand Newton". This is so lasting even after years later no proof on the matter is ever demanded by any student of Newtonian dogma. It proves that those Masters that is in control of the future paternity in the past chose the future candidates well. When those chosen candidates are by then part of Newtonian culture, they never waver on the idea that mass provides gravity and the accepting of that fact goes without any questions asked. Any future questions are unheard of and doubt on the matter is never tolerated. I should know this well since any of this was never tolerated when I presented my case in person in previous meetings where such meetings were tolerated and such meetings were agreed on. Most Academics of importance told me to leave my books at the door of security at the gate being on the desk of a security officer at the gate because they can find no reason for us to meet in person. By instituting such forceful accepting of the very much accepted but also never yet proven, teachings is going on through generation upon generation. It is carried from the past into the future by tutors repeating their past and projecting their shortcomings onto the students which represents the position they now have somewhere into the future. It is a culture of a process where the dogma of science is unleashed onto students and without ever raising thoughts the matter becomes undisputed without any measure of proof ever served or any question raised in the line of proof.

Let any professor say why does he or she unequivocally accept that gravity is the process where material is drawn closer and that mass is responsible for such behaviour. A better name to use in describing this conditioning process will be if one uses the term "brain washing". The conditioning of the mind in accepting or face rejection by physics is the most outrageous form of mind control ever instituted by man. It is what this conditioning of student minds is boiling down to and it is echoing Pavlov's theory by confirming the total accuracy in Pavlov's conditioning theory. You may wonder what made me come to such a conclusion as the one I portrayed in the paragraph I just mentioned...well I am still at this moment subjected to just such treatment by those I accuse of deliberately practising what I accuse them of. I do not admit to Newton therefore I am subjected to rejection.

To present falsified facts as proven veracity that is equal to God given verity set without limits on the scale it is done and with the intent in which it is done is nothing less than unscrupulous plunder of the truth in unspeakable dishonesty and is an unpronounceable violation of truth with criminal intent to defraud which is comparable to delusion. It is a hoax to defraud the unsuspected by ruse and trickery in order to defraud money by means of establishing the hoax as the indisputable truth. One may not serve with chevalier in whatever capacity and no matter who does so or whom it will benefit, no matter for what reason or to what goal it eventually will achieve because nothing can merit the right or no person has the right to present falsified statements as honest truth and by doing so will be expecting to gain financially notwithstanding that the gain might merely take on the scope of claiming a monthly salary. When doing so is not only equal to swindling for it is treachery.

My work offers the solution. My work explains what applies and why there is a Big Bang while contraction is continuing. For the first time ever I am the one that is able to explain the Bode law, The Roche limit, The Lagrangian points system and I prove that the Coanda effect is gravity. The Coanda effect marries the other three principles to form gravity. Newtonian science views these phenomena in a range going from not existing to not understood to being coincidental, to being just a tad beyond explanation. Those four phenomena are what forms gravity and yet no Newtonian once had the mind to look at my work and see how the four phenomena form gravity. All investigation into my work is suspended the second I denounce Newton. Those phenomena are there and all the Newtonian denial about them or where the Newtonian science finds the explaining thereof impossible or where by using normal known methods they are not being able to explain the existence of the phenomena is casting the attention on their doubtful abilities in understanding cosmic science. While marking the phenomena off as coincidental, puts a question mark on science and not on the phenomena being or not being. But in reality the gravity generated by the four phenomena has nothing to do with mass and while that is true, such truth comes just because I can show how I am able to dispute Newton's claim on mass being the producing factor of gravity because I can conclude from what I can prove by using the four phenomena and what that idea represents where the four phenomena produces gravity, that the idea of having gravity which is founded on the basis of mass producing gravity, such a concept is utter nonsense. Mass in every purpose has nothing to do with the science of cosmology or astronomy. Therefore my work departs from that of Newton like a rocket flying away from the Earth. There is no way to console my work with the work of Newton because the work of Newton does not apply in the presence of cosmology. Because I reject the work of Newton and incriminate Newton as a falsifier of facts where he presents conditions that do not apply I get no one to evaluate my work in sincerity. I reject Newton because mass is a fabricated lie of the Newtonian imagination. Yet because of my mentioning that no one even read what I offer although I offer the solution. In spite of what I offer all in physics are fixated on the untruths that

Newtonian astro physics present. The truth is that Pluto is equal to 0.0025 times the mass that the Earth has and Jupiter is 317. 8 times the mass of the Earth and all the while all three are spinning merrily at the same pace carrying on at a ratio as Kepler stated. Incidentally Kepler was dead before mass was invented and that should also serve as a strong suggestion about all "Kepler's laws" that Newton invented. Kepler never mentioned mass but Newton defrauded the work of Kepler to fit in mass through much cunning. Now fit mass into that explanation about the mass differentiation in planets while they orbit at equal space-time ($\mathbf{T^2/a^3}$) and see what I mean by misrepresentation through open fraud. My explanation does away with mass but it introduces so much more in its place…and yet you reject it because it is not in line with Newtonian thinking. Or as one Academic once put it: "it is not in line with acceptable Mainstream Physics" Newtonians do not know what time is because Newton went along and killed time by putting time equal to zero. Anything equal to zero does not exist and therefore putting time at a pace of zero kills off time. When time stands still there is no time and with Newton allowing time to stand still he removed all evidence about time from the records. He stopped time and time cannot stop because time flows. But he had to put time in at zero to find a reason for mass to apply. By taking away the value of **k** it gives validity to mass pulling mass. By saying that $\dfrac{dJ}{dt} = 0$ such a claim can give rise to the idea that **k** will diminish because the influence that **k** represents is zero. This fraud he brought about because he had to find a manner in which to incorporate the mass he was fixated about with such misleading evidence. You Newtonians talk about space-time but you know neither what space is nor do you know what time is and I mathematically prove both space as well as time.

Should I not find a publisher In South Africa willing to publish my work and one that offers me positive results this time round I am going to publish the books by way of "Print – On – Demand". Again I repeat: Please do not see this statement as a threat because from my poor academic status and low position I am far too small to be a threat to any one. But if in later development someone should find merit in my work being correct, then it will be an embarrassment to your institution when it comes to everyone's knowledge that the work has been presented once again which then will also again include those times I sent work previously which is now forming part of so many other times in the past where my work was again rejected. Then somewhere there will be some academic that would learn about the merits that was used in your rejecting / ignoring / dismissing of my work. When my book goes on sale on an international scale, there then must be some academics that will put an asserted effort in scrutinizing my work and when they do investigate my claims, they then have to learn that my presentation of cosmology is proven to the most and finest detail unlike the Newtonian principles that is furthered by the Pavlov methodology. In the books I now present I have gone into the most ridiculous small detail to present my case and the format in which they now are is down to a standard that a child can understand.

If I do not find any joy from South African institutions I am going to self publish my work by the way of another method of publishing that has much limitations but I hope this very letter addressed to your office might circumvent some of those limitations I face. The correctness of my work is beyond dispute by the uncompromising proof that I deliver. This stands as a challenge to the physics paternity while their use of mass in Newton formulae boils down to despicable fraud. Somewhere along the road someone of the right stature will read my work and will then recognise my correctness concerning this matter. It is what Max Planck did to the work of Einstein. It took one person with vision to recognise the correctness and award the claim. I do not compare my work or myself with Einstein but on the other hand I am way past being modest or being boastful and you can bet on that!

It would be to my benefit to find assessment and recognition through a South African institution, but if that cannot be done I am going to turn to the international press by way of using the system called "Print-on-Demand". Although I agree that this will not give you any sleepless nights I still am trying to secure another window to the world through which others may consider my work. I plan that in the hope that there someone will see that my arguments are correct and bring the acceptance I hope that will follow. But when that comes about, it might just bring embarrassment to those that rejected / dismissed / ignored my work in their official capacity. That is only a suggested warning of a possibility of the tiniest proportions but ask yourself if you think it is worth the chance of embarrassment when all you have to do is to consider my work without being pre-judgemental about it? In other words just go on to read my work and then try to find valid arguments when you go about dismissing / rejecting / ignoring my work. Just read it first! All I ask is when you again reject my work the following time in doing so then see to it that the reasons why you reject / dismiss/ ignore my work is based on water tight and sound principles. I show and explain the sound barrier, which is totally unexplainable when using Newtonian science. I show where one can find and physically see singularity. I show where one can see singularity by looking at the spot where singularity is not. I challenge anyone to prove that that spot is the spot that is not holding singularity. I show where time has an infinitive value and where time has an eternal value. Any one can look at the positions I allocate to where time holds eternity and time is in infinity. I show where one can see the place they both are located. Anyone can see where it is, should the person make an effort to look. I show what time is and where time is. One just has to follow my instructions to physically see it. I show by examples why time is there. I show where and how time influences the cosmos by generating gravity. That I do by mathematically proving how the Titius Bode law generates gravity. I show what gravity is and I prove that gravity is not a force in the manner that Newton saw the force.

I show physically as well as mathematically where and what space-time is. It is there in an allocated position where one may physically look at it. The information that I now present is the most basic there is. In my previous work I was under the impression that everyone could see where singularity is, or should I more correctly say where singularity is not, and why singularity has the value it has. Even a tiny child can see why singularity is allocated in the sphere because of the simplicity in the argument. I took it for granted that everyone should be aware of what time is and how time influences the cosmos. Now I do not take your realising of such matters for granted. I take the reader by the hand and guide the reader as I show where every aspect of the cosmos is. I show where eternity is. I show where infinity is. I show that gravity is motion that partitions eternity and infinity. The best of all is that I prove that gravity is the parting of infinity from eternity, which then results in the Universe we have. With using that information I then could understand how the cosmos advanced from singularity into space-time. There is no need to be a genius to understand it because even I can understand it. I explain that process how the cosmos came about in the finest detail as I take you through a mathematical tour how the cosmos came in place. All you have to do when looking for reasons to again reject / dismiss/ ignore my work is find one argument out of line or one mathematical conclusion that is irregular or one reason I give that is not conclusively proven. You can condemn my work by finding just one reason to dispute my work on. You don't need to hide behind Newton's institution status because my work offers page after page of other reasons than hiding behind the purity of Newton and if it is with reasons based on sincerity it will bring merits when you denounce my arguments. I hand you that challenge. Don't just dismiss / ignore / reject my work because I have proven Newton wrong but I challenge you to find serious grounds about the correctness in my work in order to do so. I do not have to prove Newton incorrect because nature, Galileo and Kepler was amongst the first to prove Newton was wrong and that happened even before Newton did not know what gravity is. Newton knew what Galileo said while Newton knew what Kepler said and there are a lot of others Newton also sidelined when he forced his fraud onto the world...that Newton did that intentionally is what you have to admit to that took place or otherwise prove it is not deliberate fraud!

I admit that there might have been some chance where in the past where I was guilty of cramping too much information into too small a printing space, but I am not guilty of that in the presentation I propose this time round. There are three books dealing with three different aspects of cosmology. If I could, I would suggest that the reader might start with **"an open letter to Selected Academics"** but that is not really important. This book I name, deals almost exclusively with the matter of gravity. It will be the book that will naturally flow out from this letter. In the event of you choosing again to reject my work, I ask you to make double sure about the foundations and the grounds on which you base your rejection. Make sure you can prove that anything on Earth can have mass when that something is displaced in a location being inside the Sun because to have mass when inside the Sun it has to be there by filling space and not be a mere suggestion. To have mass inside a Pulsar whatever is there has to be there and not be a few photons the pulsar rejects. To have mass inside a Black Hole it has to be an atom first and not a thought. When you connect a meaning to mass then define what it is that will have that mass you surrender to the object because whatever is in a massive star is very different from what it might be on Earth. Make sure about the argument and what such an argument produce in sensibility.

If my work has merit and such merit is recognised by other acclaimed academics then put yourself in a position that is ensuring that their recognition does not become an academic slap in the face for your institution. Be sure your rejecting my work again is on valid grounds that you can defend the integrity thereof about your reasons and your reasons are not just to defend the integrity of Newton. With that I then again must sincerely advise you to find grounds for your rejecting / dismissing / ignoring as it just might even by the smallest of chances otherwise lead to your institution becoming the laughing stock of the entire academic world.

Make sure that when you once again reject / dismiss / ignore my work this time, that you have reasons of integrity in doing so. I do not make any presumptions on your behalf about my work, but all the reasons for rejection my work on previous occasions in the past by publishing institutions was no more than an insult to me because it proved that those responsible for assessing my work never came round to even reading my work. The comments they made had no implication on my work but only were merited to a few pages of reading, normally the first ten pages of the book with seven hundred pages to offer. Be sure this time the insult is not on your own account.

I have reasons to press the point that in the event of you deciding to reject the work, I ask you politely to make sure about the sensible validity you apply on which you base such a rejection and that your doing so is about all the aspects in my work which you took into consideration and not only the information you gathered from glancing over the first ten pages or so which was done by some pre graduate when that person had no real incentive to do so or felt any ambition to understand what I say but was doing so because the person was ordered to read my work and because the superior academics in charge of the department had neither the will to do more than the bear minimum nor did they find the academic drive to do extra work. Again I acknowledge as I repeat that this can be no threat because I am far too insignificant to be a threat to any one at this moment in time but should my work be correct and should I find such recognition internationally in the future which there is a slim chance of I admit, but in such an event ensure that your rejection could not lead to an academic embarrassment to your institution and also to you. The work is of such proportions that it either could lead to nowhere because it is never been read by any person of status that agree with my views or it is going to develop into the biggest thing to hit physics in many years. There is no middle road.

When you reject my work again then all I ask is to do so on defendable grounds that can support you in doing so later on. By your rejecting this time then make sure that your rejection is valid and that you can defend your reasons in the presence of other academics. Make sure that it will not later come to be an avoidable academic embarrassment because again you have made no effort to read when assessing the work as you did on other previous occasions when judging the validity of my views in the past. There is a slightly less than absolute chance that somewhere in the past your institution opted to use one or more of the following reasons for ignoring / dismissing / rejecting my work when they did so in the past. Choose any one or more of the reasons I supply and it will most probably reflect on a choice your institution made in the past when they chose to dismiss me. The rejections by now have almost become countless and that includes almost all institution in South Africa as well as numerous international institutions with highly admirable reputations that has as much credibility and yours have and are amongst those I implicate. Notwithstanding your institution's name you may have, it does not matter because your institutions supplied one or more of the reasons. When I inquired about why such rejection was done at the time it happened, I was given the following reasons. I was given these reasons by administrative personal about what merited the rejections in the first place and on what grounds the rejections were formed. Academics in principle motivated such rejection at the time because they did use one or more of the following reasons in the past to reject / dismiss / ignore my work. Should it please you to use any of the past reasons for rejecting my work again this time round, then please do not think of it as very original because by using it, someone already jumped you and you can therefore only repeat again what was done before. Such are the following examples. When using it, then be sure to motivate those reasons by finding reasons in my work as such, which you can state to be incorrect since that part was never yet done. These are some reasons on which you rejected my work:

1) My English is of a poor standard to use. I am Afrikaans in every atom my body holds, which makes English not my strongest point but to put it that my English comes across as being unintelligible is defamatory. I see this rejection resulting because everyone in South Africa would immediately see that Peet Schutte is as Afrikaans as the person with the name Jan Botha. Make sure that you do find my English to be below the level of comprehension. I am Afrikaans and I would never care to disguise or try to hide the fact. In my books however the part that I present for your attention and request, your evaluation is written in mathematics and as a Mathematician Master you Academics that I have approached on this matter, have failed miserably to read what I communicate when I am in conversation using mathematics in the past.

2) My arguments about the outer space not being nothing but is filled with substance is incoherent. If you use this line of argument again by insisting outer space does comprise of nothing between cosmic bodies then also make sure to multiply the distance between the Earth and the Sun with zero and prove the coherency about such an answer on your part because $150 \times 10^6 \times 0 \neq 150 \times 10^6$ km but is zero. If you are unable to prove how nothing can be anything else than a factor that introduce zero as a value in a calculation then you have to admit that there must be some form of substance to give meaning to the distance that is between cosmic objects.

3) My arguments on gravity hold no merit. Prove the merits of Newtonian gravity and then compare that to the proof I introduce in my presentation on gravity where the Coanda effect is gravity.

4) The publishing of my work will be too expensive to justify such a process.

5) Newton cannot be wrong because Newton was never proven as being incorrect thus far because time proved Newton correct and therefore Newton is without dispute.

6) My work does not fit the required institution criteria while no one ever says what their criteria then involves or what about my work falls outside the criteria of astronomical science.

7) The Acclaimed academic responsible for scrutinizing my work distances himself from my views as he stands by the principles Newton holds.

8) Newton requires no justification from any person.

9) The academic simply sides with Newton with no reason given for doing so.

10) My academic history and qualifications does not merit their attention or investigation into the matter.

11) A response in silence that is rather more befitting the mortuary.

In the light of all the rejection all I say is that mass forms the restriction of gravity and there is either free gravity which is unrestricted motion, which is what gravity presents without mass or there restricted gravity with the tendency to move which then is mass. However my saying that falls on incompetent ears all the time because of a Newtonian fixation on the ridiculous. When a body is in outer space the body floats above the Earth as a body would float on water. The body can only float when the body gravity can sustain a position it holds in relation to the velocity of duplication by motion that the Earth holds. In my book I present the facts why this is in the form of density that is applying to form gravity. The body maintains independent gravity but the gravity totally relies on motion that has to be in line and the rotating motion has to be superior to the orbit motion of the Earth. If the orbit motion is inferior the motion ratio will put the body in a smaller orbit and such orbit reducing will lead to the body entering the atmosphere of the Earth. It has nothing to do with mass since all bodies require a specific velocity irrespective of the size that it poses. The tarnishing of the orbit will lead to the body falling and that too has no indication of mass. When a body enters the Earth space the body will sink just like the ship sinks as the body loses buoyancy and that buoyancy stands related to the speed differentiation there is between the Earth and the independent object that enters the atmosphere at such a point. When the body falls it loses its independence in gravity since the Earth then takes command of the body's motion. The floating or not floating depends on the motion and velocity that the body can maintain when it is in orbit at a distance above the Earth in outer space and that would establish the required density attached to that. I explain this principle in very much specific detail but in short it is exactly following the rules that Kepler's prescribed that gravity should have even before gravity had a concept and that is being $a^3 = T^2k$. As is the case with the floating ship it is the case with the falling object. The only difference is when a body is no longer buoyant in the water on Earth the loss of density is because a difference in personal atomic density and the density the Earth puts down and a condition. But even that too is a case of velocity that provides the density through motion and that motion is the gravity factor.

Supplying an adding of heat gets all mass stricken objects moving in relation to the adding of the heat and in the time that the adding of heat took place. We call it either drive or exploding but in essence it is just adding heat to the relative density that the particle in focus has to endure in relation to the time the endurance is taking place. As is the case with the ship at see, once the ship sinks it becomes part of the space the Earth claims where it can no longer move except by moving with the Earth at the velocity the Earth dictates. It then has mass because it is the Earth by the extension of space that the Earth has because it then is part of the Earth. On the other hand when the body is in the water while it floats it is part of the space that the water reserves while the water presumes in the capacity of a liquid and as a fluid claims the position liquids claim because it can move independently in a variation of alternating its relevancy from the dedicated precise motion the Earth demands. The displacement is added to the value of the water making the floating ship extend the measured value of the water. The ship then takes on the mass the water provides and relinquishes any claim to independent mass it might otherwise have. It exchanges the gravity or the motion it would otherwise have for the mass when accepting all motion that connects to the Earth moving.

We that sank to the bottom of the Earth-soil have mass because the Earth sustains our motion or gravity. While we are swimming we bob as water and that motion substantiate our position as a liquid that is independent from the solidity of the soil. While in outer space we provide our independence from having mass because we hold a position where we are unattached to the Earth because we sustain our individual gravity or motion. Science call this micro gravity but it should much rather be micro mass and macro gravity. Being part of the Earth puts us in the class the Titanic at present has but being in outer space makes us part of outer space as we have independence on the precondition the independence can be sustained by a motion that match the required gravity. Then we provide our independent gravity as we supply the motion and with that establish independent mass, which Newtonians gave a name: they call it micro gravity. Being on Earth, the Earth provides the motion or gravity and sustains us in forcing us to hold the gravity of the Earth by supplying us with mass. We deprived ourselves of having independent gravity by accepting the Earth gravity on the condition that we receive mass which would make us a directly linked extension of the Earth. Now I ask you: Is what I say in my books when I say what I said above in a mathematical context so hard to grasp especially when those whom I address should be masters in mathematics? When a body takes on mass it becomes a part of the structure of the Earth and therefore loses its cosmic independence in motion as it presumes the motion of the Earth while being a part of the Earth. It then is the Earth by extending and only life can establish motion when the object in motion still has mass as it is attached to the Earth where it is the Earth that still provides independent motion but with such independence we measure that performing in the manmade measuring tool of mass. In the cosmic reality objects however, can either move as well as having gravity or it becomes part of the Earth and has mass as it is being part of the Earth in total. Being part of the Earth, the object can find mass as much as it forsakes independence and gravity. In true cosmic reality there is no chance of having both. Once an object surrenders independent gravity it will have mass because it will never have gravity afterwards except that, which the Earth provides.

The mass is the manner in which an object arrests material and claims it as confiscated possessions where by the arresting object provides the gravity the object arresting will subscribe to which the arrested material surrenders space by receiving mass. The apple that Newton gave a value of mass would remain part of the Earth in the location it fell if not for the intervention of life. Then again the reason why the apple fell was because it had life to start with. Rocks just don't drop from trees and roll around afterwards just because they can find motion independent from the Earth. Rocks have mass because rocks do not run around and relocate in new positions. Life creates gravity because life is about motion and motion provides independence by giving gravity-motion above and beyond the gravity motion the Earth provides. Life can command mass whereas in the absence of life mass destroys independent gravity.

Again you have the choice to prove to yourself and see where does mass have any prognostic influence in establishing gravity and show your honesty in the matter. I am of the opinion, which I set out and prove in my work, that mass is a man made unit of measure, just as temperature and distance forms a tool with which to measure to a specific standard applying only on Earth and has no real significance in the cosmos at large. Mass is a human invention to measure what in the cosmic context turns out to be truly beyond the measure of man. Go on and show yourself how much the Earth / Moon mass is drawing the Moon closer or how much is mass having any object fall faster or slower than the one with more or lesser mass or how does mass influence the planet Jupiter with its enormous mass to orbit faster than the individual debris in the clusters that is sharing an orbit with Jupiter. Show how mass introduces gravity other than by means of misinterpreted culture. You now have the option to side with those instating the criminal intent or you can admit to yourself what is motivating the need to find evidence that will confirm any critical density factor. With that information then bring proof that mass is producing gravity at the moment throughout the entirety of the Universe.

Make your consideration is made in a sober environment where you are detached from other influences and by standing away from performing as one that would create a screen to cover fraud when you reconsider what the critical density theory entails. Think what influences have to be subject to such a turn around where each atom is at present on route in a direction of expanding from every other particle and particles that ever formed. The idea that what is there forming what has no end can stop in its tracks is childish. Think what an effort it takes to stop a ship such as an oil tanker, which is fully laden and at full power while speeding at full capacity into a harbour. Think of what it will involve to stop the Earth from moving away from the Sun. Then stop the entire solar system, which would include every speck of material throughout in one instance. Stop the massive atoms at the same moment as one would stop the less massive atoms. Stop a star expanding while at the very same time the force will control the hydrogen on the outside with equal precision than it controls the iron in the core. Explain how the force will find a way to discriminate and apply more fiercely to the heavier atom and have them stop at the same rate as the force will apply and stop the lesser massive elements such as hydrogen and helium. Find a way to explain how the core will stay intact to the star centre and not dislodge from the star when such force of total equality enforces a will on all that is totally in inequality. How would such control on different elements with different mass deployment work? The thought must fill one with soberness. Think what it will take to stop something as gigantic as the Milky Way. Think about the impact required to turn around the expanding of every atom that there is in a body that forms something as large as the Milky Way and then multiply that with a hundred billion times the Milky Way. If that doesn't shake your foundations, you have no clear mind left. Just the consequences of what this suggests indicates the contemplating of the farce this presents being without equal of having a parallel. Think what the impact that just the thought alone embraces and the amplitude of what this entitles and what the con would suggest that such action would accomplish. What will it take to stop the entire Universe in direction and force a turnaround of the entirety that is out there? See what I mean by criminal minds blowing facts out of proportions because it is not motivated by clear minds but the motivation is to cover up deception by producing more deceit. That suggestion is that all of us (the mindless many standing in to represent the thoughtless populous) we that stand outside the academy of physics are stupid enough to swallow such slander. What on Earth would guide Academics to regard us with such dismay and have so much disrespect for our ability to judge by using sober reasoning and where we, the brainless masses, will lack any ability to also contemplate and find reason by thoughts? That entertains an insult by merely addressing such a gesture.

Prove to yourself that even when finding the so-called critical density how that then will convince the cosmos to introduce Newton's contraction and even how that will establish a start of the implementing of the system that will firstly stop the expanding. Think of it in the manner that every particle has to come to a stop. Then determine the forming of a revised working principle method in the cosmos. Show how the cosmos will stop expanding and come to a standstill by stopping all directional motion throughout the entire Universe, which includes every particle large and small and especially down to the most insignificant of particle detail. To top such madness then explain how the cosmos will then go on to start moving again from a stationary stance as the motion will start in another directional flow and where the most domineering will start to move at the very time the most insignificant is starting to move by applying the same motion velocity equally to all there are. The entire stopping and restarting must be simultaneous and if not, that will bring annihilation to an entire cosmos.

It involves the charging of the most significant as well as the minutest particle in motion as well as involving all of the entire cosmos that will change the directional motion. Go into exact detail how that will that come about where the action will involve every particle without having massive collisions and total destruction? Calculate every one of the actions of the smallest particles in an independent detail as mass will have to discriminate between large and small to equalise all. Such turnabout includes every tiny particle as well as every star in every galactica big or small throughout the entirety of the Universe everywhere. Every atom that was going the one way will be in a position where every one will have to stop at the same time, then turn hundred and eighty degrees in direction and start to follow the particle which was following the particle that then is behind the one that was behind it at first. The follow my leader has to change to lead my follower and it has to be done in a way that all possible atomic collisions are averted simultaneously. It has to stop every atom visible or otherwise, and then after that start every atom atom visible or otherwise not visible subatomic structure again to follow the particle that was in front of it but now finds it to behind, which was following the particle that then is behind but from then on will be in front. The one was at first behind but suddenly finds it in front of the one it followed up to that point and no atom may collide because the collision will bring nuclear destruction as atoms destroy the atoms in an escalating crash which is dooms day by the stretch of any imagination. The smallest unseen subatomic particle may become a missile, which will be destroying every atom that stops before the subatomic particle could stop. Let Newtonian gravity choreograph that one for beginners and let Newton's gravity explain how this will come about. In all of that it shows how much sensibility there is in the Intelligence of the most intellectual group on Earth where they attempt to circumvent reality and introduce total madness and all that is to cover up Newton's fraud and criminality. To those suggesting such a possibility, it is playing a game of equalling God's abilities by finding matching abilities with God on a mentally mathematical level by setting mathematical rules applied in the games in order to avoid reality by creating overbearing deception. I challenge those to come into the world of reality by leaving the world of playing mathematical games with physics. Go on and calculate the process where every particle will destroy the particle that was in front of the one that will be behind of where it was up to that point, if the force of inequality does not bring total equilibrium to everything participating in the process. The contradiction in motion will make redundant all distances achieved up to now wherefrom collisions will come about that will end the Universe then and there.

If that proves how the greatest minds in the Universe form conclusions with the worlds most intellectual group in action, then thank God I am one of those excluded and considered as being merely stupid for I am considered to be one those that is not blessed with the brains to understand Newton! After all I was accused on so many occasions by those in power of the academy with subtle suggestions where they literally suggested quite frank and to my face that I lack the insight to grasp Newton, and this they do, while they display the bravest understanding in absolute sympathy towards my insufficiencies and disabilities especially on the mental level. I was told to my face in numerous conversations when I presented my work for evaluations that I am not mentally capacitated to comprehend what Newton's work imply, but also I must add that it was done in a very sympathetic and gentle manner as not to hurt my feelings too much or degrade me very much.

Be honest and show the sensibility in such an argument and while you are about it, then show to your own view in the privacy of your own company how you envisage how gravity will overcome all the obstacles, as gravity will stop the entire cosmos when it could not even halt the progress of the expanding of the cosmos up to this point. Remember to rush and be very quick about it because with the escalating of the distance of particles the relevant gravity between particles is demising by the square so there cannot be much time wasted to prove Newton correct! The longer you linger, the weaker the force of gravity will get because the further the distance are separating structures, the less effective will the mass be in its task of creating the force. If you do not act with haste to bring a sudden solution, more deception has to follow and Newton will have to be proven yet again with another scheme of deception somewhere in the very near future! You can do that before you read my books and you can do that after you read my books. Then you can decide where you stand in the matter of the criminal defrauding of the truth forming part of my accusing Mainstream Physics of the falseness' of Newton principles. When you read my work I challenge you then also to prove me incorrect on my insisting of the relativity between particles, which is what I suggest in my work as a counter theory to Newtonian incoherency. Realising what stupidity prevail in the formulating of the critical density puts everyone of that era at the time when the formulating of the Critical Density theory was suggested, which is also including Einstein that was the one that was blessed with the task, in either a state of pitiful stupidity, or criminal conspiring. They, who were taxed with the formulating of such a theory, were the most brilliant minds that ever lived! Think in a supposition of sincerity and not a supposedly manner to cover up an hallucination that lasted since the dark ages about what was suggested by those that should know of much better and is thought to have the best minds ever, being in a position where the lot was performing in a mindless manner, acting as if thoughtless, while judging the merits of their deeds. Was that what they tried to accomplish brainless or just conniving. When using the slightest scrutiny one has to come to one of two conclusions, which are that either their intellectual status tarnished or their morality became suspect. There is another option, which is slightly slanderous but serves a better elaboration and that was that the lot that were involved in the act of conniving by creating the Critical Density fiasco intended to rape science and to rape the common courtesy of the publish at large.

I challenge you…no, I dare you to charge me with slander because if you do, that would stir the hornet's nest and bring the hidden into the light of openness. Let us then present our individual cases to the public at large and find support in the unwavering belief we have about or different interpretations that present correctness in this way. If you are of the opinion that my remarks are besmirching those viewed in admiration with intentional deformation on the characters of some of the most outstanding and remarkable persons in the world of the academy of physics then prove to your conscience how mass forces planets to go faster or slower in accordance to the mass they individually have or how planets orbit the Sun faster or slower in relation to the mass they have or how mass would have objects fall faster in relation to the mass they have. Show how I commit deformation when I accuse those deforming the truth in physics of doing just that in an unprecedented case of swindle and betrayal. Show the innocence of those with highly regarded integrity where I accuse them of being unscrupulous fraudsters. I again dare you take me to a court of law where both sides can show the justice in their views on their matter and make it a legal issue. Just go on to show where in true nature does mass play a part in anything other than being a man invented measuring unit devised to be used as a measure of calculations in a method that establishes a means of measuring.

Sir, Madam, the choice is yours to make where you may come to see the criminality in the matter and decide to investigate my work that proposes the truth of what is out there or to side with the criminality of falsifying the truth further in order to substantiate the ongoing fraud and continue to support and validate all such criminal activity. You have to agree that when one introduces falsified facts and intentionally admits to the authenticity and integrity of lies of any nature which you present as absolute unblemished truths that the doing so thereof then is criminal conduct irrespective of motive or the innocence behind such conduct or in order to pardon the bent motivation of what eventually will have a benefit from such conduct. You now have to decide which way you are taking the matter because the matter has come to your attention.

I will admit that this letter does carry another agenda than seeking your approval this time round or that this letter might find another motive than be trying to establish hopes of receiving your acceptance. I have passed that desire of finding academic approval as a motive in my effort to contact you some time ago. This is a warning that when I publish my book, I am going to do my utter most to reveal what I consider to be a case of fraud in physics as a whole. I am going to go as public as what my abilities would allow me and find as much publicity to further my conclusion about a conspiracy going on in the world of science. I will establish as big a cloud of dust as I can muster through as much press exposure as I possibly can generate with one accomplished goal and that is to portrait not only Newton as fraud but the entire Physics paternity in a state to conspire to defraud. I will do my best to show that this state of affairs has been going on for years. I will base as much as I possibly can that this is founded and partly due on their constantly ignoring me. On your part make sure that your position is as rock solid as you believe it to be and that your facts are as proven and as defendable as you might consider it because your security in your faculty rests entirely on the public's idea that are seen as unblemished. There opinions are resting on their surmising that you are being accurate beyond all compromise and in that the public belief is vested in your unshaken correctness. All about physics ride on the public opinion of never having the least bit of doubts about your accuracy.

That idea might tarnish and take with it the ultra white of your reputation. Just make sure that I cannot prove that your facts are not as unblemished as you pretend they are. On your part take the challenge and prove Newton is as defendable as you believe he is. However in all fairness, I also admit that I see this letter as being just a charade because I have to pretend to participate in this façade one more time. I have to pretend that I might still be convinced that you will be in mind to accommodate me this time when in truth you never even acknowledged receiving any of my mail in the past. If you ignored my work in the past there is no reason why you would read it at this point and if you did read it in the past but failed to act on it, then you are guilty of conspiring and in that you will not change. On the other hand I also realise that if you would acknowledge that my views are correct then you must also admit at the same time that you know less than nothing about a subject you pretend to know more about than God in person knows about science. However in the event where you go on pretending there is in science what there is not validation of and that Newton is a God instead of the hoax he is from the very start, then I guess we shall confront one another through the press. I say this as a test because no Academic would ever endure this much insulting directed at Newton and therefore I am one hundred percent sure that no Academic would read this letter this far. I don't envisage any person in academic science that would reach this very line because science never confronts the truth since they believe they and Newton sit on a pedestal from where they and Newton govern us little people.

Since no Academic will be reading this letter I therefore feel free to say that I am going to use this very letter to promote my theory to the broader public through whatever form of media a might muster and with the aid of whatever publicity I might gain on the matter I am going to call on the broader public to assemble their awareness about the matter and find their opinion in the matter. It will also then help me to sell my books without the benefiting from the benediction of science. Lastly just consider the following tomorrow where you stand in front of your classes:

Sir, when you address your students about the wonders of Newton, then tell your students tomorrow how much the Earth drew closer to the Sun since the days of Kepler, and Madam, announce to the world how much the Moon came closer to the Earth since the days of Tycho Brahe while you then remain convinced that Newton is still flawless after three hundred and fifty years.

Sir, Madam, the choice is yours to make where you may come to see the criminality in the matter and decide to investigate my work that propose the truth of what is out there or to side with the criminality of falsifying the truth further in order to substantiate the ongoing fraud and continue to support and validate all such criminal activity. You have to agree that when one introduce falsified facts and intentionally admit to the authenticity and integrity of lies of any nature which you present as absolute unblemished truths that the doing so thereof then is criminal conduct irrespective of motive or the innocence behind such conduct or in order to pardon the bent motivation of what eventually will have a benefit from such conduct. You now have to decide which way you are taking the matter because the matter has come to your attention. I will admit that this letter does carry another agenda than seeking your approval this time round or that this letter might find another motive than be trying to establish your acceptance. I have passed that of finding academic approval as the divide some time ago. This is a warning that when I publish my book, I am going to do my utter most to reveal what I consider as fraud in physics as a whole. I am going to go as public as what my abilities would allow me and find as much publicity to further my conclusion about a conspiracy going on in the world of science.

My aim in writing this letter is to confess my doubt about the manner in which science regard gravity forming. I have serious reservations about the matter in as much as how science assumes that mass is the reason behind whatever is forming gravity. I equally hold strong doubts about gravity being a force of pulling or gathering. This institutionalised concept has formed the basis of science in physics during the past three hundred and fifty years but in the light of evidence that came to the attention of science in the last hundred years and more so during the past fifty years there is reasons that puts those assumptions previously formed in the past about gravity in doubt. The new evidence points away from the previous direction and assuming. Modern evidence calls for re-examining in regards to this matter. The facts that are now available to science to form any prognoses makes the old principles that was used for such assuming much more questionable.

One only has to study aviation principles to form the doubts. Aircraft flight disproves the assumption that mass is instituting gravity and in that context the notion forming the concept of mass and gravity becomes extremely doubtful. I am about to prove in my writing of this letter that mass is motion discrepancy between two bodies and gravity is motion pure and simple not a force. Any object holding any mass of whatever description can become airborne with the correct conditions applying. It only depends on motion coming about achieving a speed over and above the earth speed that must be sustained. Mass of whatever magnitude can fly. There is no limit on the mass becoming airborne. What is required to achieve such a fete is a speed. Only that limits flying. A Jumbo jet can get into flight as easy as a micro light aircraft. The only requirements are a differentiation in velocity of air movement above and below the aircraft wings.

When a water-skier stands on the water the skier represents a certain pace time which is in truth not magic or mystery but in relation to of space used or occupied in a certain time. The slier holds its space in time $a^3 = T^2k$ in relation to the water holding its space-time $a^3 = T^2k$. At such a point where the skier is stationary the skier represents the solid and the water becomes the fluid. However not long after the skier stared moving with the aid of a drawing boat the skier represents the liquid and the water takes on the role of the solid. His becomes a reality since the movement is now on the side of the skier forming the liquid and the Earth is in relevancy solid which represents the stationary factor.

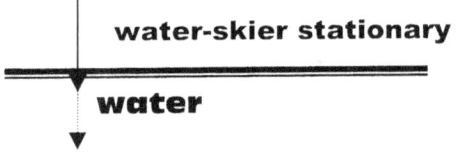

The space-time the skier represents is Π^0 in addition to the Earth holding $7(3\Pi^2)$ of space-time. The skier represents a factor of 1 and not zero.

As soon as the skier moves the skier becomes the liquid as the roles presenting each factor changes. Since the earth motion of $7(3\Pi^2)$ also belongs to the skier as well as the Earth the Earth factor becomes 1 and all motion becomes part of the skier in $7(3\Pi^2)$ $.25\Pi^0$. The earth then holds the factor of the stationary 1 that stands still although it is moving and the skier moves although in comparison the motion difference is minuscule.

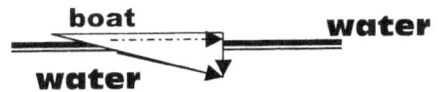

A boat floating on the water has a certain mass allocated in the proportion of the water it displaces. The mass would be considerably more if the boat were on dry land. But with the boat on the water it only represents the mass it displaces while it is stationary. The boat matches the conditions fitting a solid while the water acts as the liquid applying motion.

Then with the least of motion coming about the boat raises from the water and the

depth of the previous water line reduces considerably. This can only be because the boat then represents much less mass as it displaces much less mass in the instant. However in the period of time the boat came into contact with much more water than was the case when the boat was stationary. The water became more and therefore in the same context did the water become less if the relevancy was looked at from the other side. If the water it is in contact during a period becomes more the instant of contact will have the boat reduce in mass. That is why the boat lifts from the water. It is because it shows less body in relation to more water in the time in contact with the water. The triangle in the water reduces as the motion lifts the boat from the water because the duration per time instant became less and the water in contact with the surface of the boat became more. The boat is under all conditions maintaining Kepler's true law that motion is equal to matter which is $a^3 = T^2k$

A situation arrives where the boat hardly makes contact on the surface of the water with the surface of the water. The mass of the boat per moment in the instant is so little that the boat hardly touches the water which proves that the mass of the water reduced the size of the boat in relation to the water to a fraction of what it is when the boat is stationary and bobbing on the water. The same that happens to the hot air balloon rising into the sky by loosing mass and receiving gravity so it also happens to the boat where the boat ultimately rises into air to become air and all is done on the balance of motion.

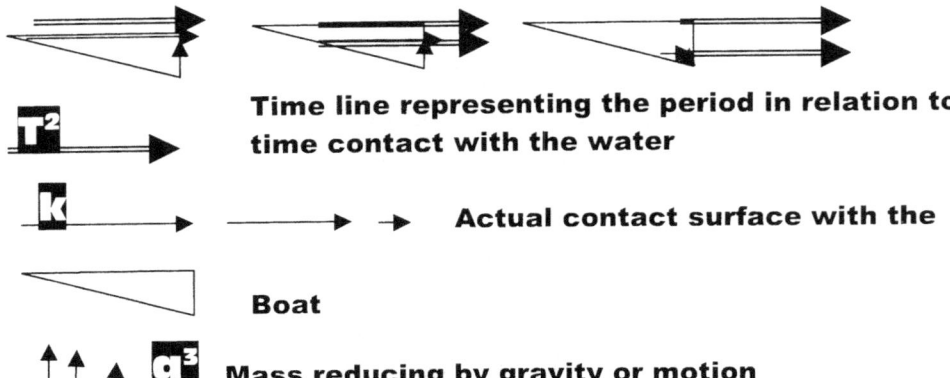

Time line representing the period in relation to time contact with the water

Actual contact surface with the

Boat

Mass reducing by gravity or motion

The relevancy of motion brings on the ultimate triangle of space-time. The more the Earth representing T^2 dominates the factor of motion the more the space is represented by mass a^3 where the relevancy of independent motion k is reduces to a factor of 1. However as the relevance of individual and independent motion gains dominance, the factor that the earth represents as T^2 diminishes and as it diminishes so does the relevance of mass diminish as the factor of space in the instant reduce. The boat lifts from the water, which indicated the mass reduces. The boat displaces less water, which proves the space in the instant is reduced by the water taking proportionate dominance in the period. This proves the solid truth behind Kepler's findings where space a^3 is worth the motion T^2k thereof in relation to the duration T^2 as well as the relevance brought on by the instant k motion create the dismissing of mass and the lack of motion increases the fact of mass. The release of the boat from the water shows clearly a release of immersing of the hull of the moving boat and the faster the motion is by velocity, the higher is the boat releasing from the water. It reaches a point where the boat finally leaves the water in total and no part of the hull is in contact with the surface of the water. If that is not flying there is no flying anywhere possible on Earth, and yet Newtonians fail to recognise this by connecting it to gravity and mass.

The lifting connects directly with the water sliding through underneath the boat in a specific duration or period in time. The more water comes past the hull through motion the less of the hull makes direct contact per time instant with the water. When the boat slows down the contact relevancy reduces as much as it increases again once the velocity increases again. The boat became a gas while the water presumed the part of the rock hard impenetrable solid. The specific density of the otherwise solid boat became that of what a gas is because it floats in the air in a position of being air born just as a gas would be while the water is so solid it will break the boat to bits at any point of contact. No matter how hard Newtonians try to allocate mass to what ever they wish to have mass there are certain limitations that even all the falsifying of facts cannot hide the truth. The electron has mass because the electron restricts the flow of time in relation to space. The proton holds mass because the proton holds a restriction the flow of time in relation to space to represent mass. The neutron has no mass and whatever grossly falsifies if Newtonians try to establish in order to accommodate their hoax in mass the neutron simply has no

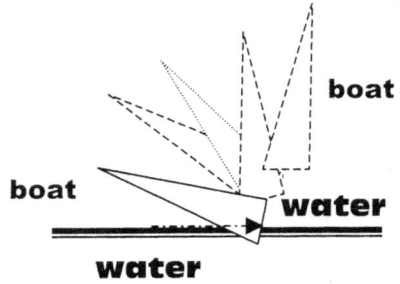

mass. That means the fact that they wish to attain mass to the fact of the presence of the particle is a contamination of the truth because the neutron simply has no mass. That means the neutron is all motion and the neutron is all gravity. The fact that it shows no indication of having mass as a clear fact represented, is not an incorrect connotation that the cosmos connects to mass but shows Newtonian incorrectness about the presumption they and their Master Newton has about mass.

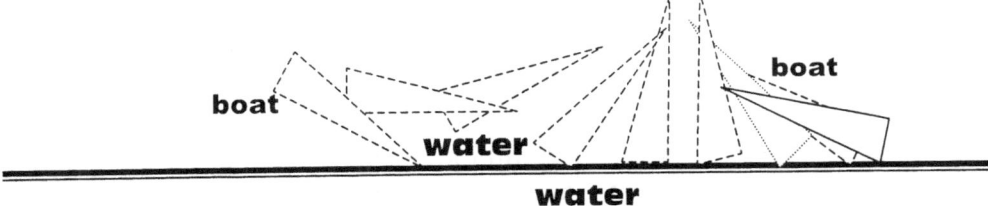

In a frame-by-frame one can see the boat staring to fly but since the drive seizes to commit to the motion we get the idea that the boat flips. It is just a case that the front part of the boat starts to end the duplicating by motion sooner than the driving rear part and the end comes charging past the front. If the drive at the back were consistent the boat would start to fly as airplanes do.

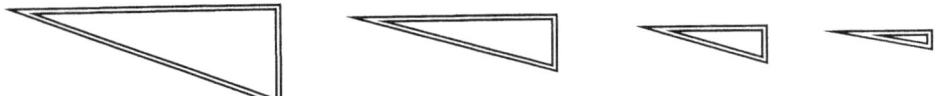

The Earth did not become bigger and the boat did not really become smaller as much as the water did not become really more but in relation to time space duplicating takes more time to produce less space. But there is a relevancy between the motion of the boat and the motion of the Earth, which the water represents in the relation that the boat holds with the Earth. As the motion of the boat accelerates the moment-by-moment contact the boat has with the water allows the boat in space to diminish by the same measure as the ratio in the boat to water increases on the side of the water. The water becomes more because in the flow of time the boat makes more contact with a line of water where the water makes less contact with the space of the boat. That proves that a boat can fly and that proves that man in motion also can fly. That confirms Kepler and by the same margin it denounces Newton and Newtonian mass inducting gravity. Motion secures gravity and mass stops gravity while mass is the restriction of gravity and objects in total unrestrained motion can have all gravity that presents no motion as the neutron proves.

Mass aside, laden or not, aircraft of all description will fly. The wings determine the required velocity while supporting all mass. The aircraft support the same mass it has after becoming airborne as it had before it was airborne and the mass remain in the air because of the speed which enable the wings to have air pushed past the surface of the wing and that is what allows gravity to be countered by flight. Science thinks of movement as momentum but in fact it is antigravity since it counteracts the containing of Earth gravity. Momentum in fact is only an increase of gravity on the object that is taxed by mass, which extends the gravity of the burdened smaller factor beyond the limitation of Earth restraining by mass. An object has mass because it moves slower than the Earth and is therefore dragged along by the Earth. Once the object moves faster than the Earth even within the Earth atmosphere the influence of mass abates. Mass is only the friction caused by speed differentiation whereas the sound barrier is the limit on the other end of the speed differentiation scale. At the moment science differentiate in terminology used to describe what I deem to be the same issue but that puts a huge question mark over the use of terminology and the concept of mass playing any part in forming gravity. That what is considered as momentum is gravity in addition. Notwithstanding the pulling and the tugging brought about by the mass, the ultra light plane finds as much flying capabilities as does the enormous cargo jet aircraft. The pulling of the mass is many times more in the case of the heavy cargo plane than what it is in the micro airplane but still at speed the plane overcomes the pulling of the mass by only establishing motion differences between the top of the wing and the bottom of the wing. When flying having more mass does not implicate or discriminate in any way because with the required motion both craft keeps in the air equally effortlessly. Once the craft is airborne it is a matter of maintaining speed that would secure the flying ability and not the loss in mass that would increase such flying possibilities. In the case of a satellite such flying demonstrates my argument even much better. The space station Mir was in space for many years but it fell when the space station could no longer maintain a specific orbiting speed. While it was in space in an orbit there was no changes in mass required to maintain an orbit. The changing of the mass could not secure Mir a longer orbit life. At the point where it started to plummet to the Earth, the mass did not increase but the orbit motion decreased and that allowed the plummeting. The fuel eventually ran out and by not further being able to accelerating in order to maintain a specific required velocity. Without the speed requirements needed to match the motion equilibrium that was required to maintain a steady orbit between the Earth and Mir, differentiation set in and that caused mass, which came as a fall towards the Earth whereby Mir met its end in a fireball. It had nothing to do with mass. It had everything to do with speed being at an equal pace between the Earth in orbit and Mir's orbit. While the satellite is able to hold a speed in a ratio with the speed the Earth manages, an orbit is secured and such securing of an orbit has no bearing on the mass that is up there flying through the sky. It has everything to do with matching a speed in relation to the distance from the centre of the Earth and in relation to the motion of the Earth that requires a specific velocity at that distance. It has everything to do with the time it takes the space to go through the motion from one point to another point. It is the required heat in the space that the space needs to move at a certain distance in a specific time of motion. Gravity is a relation between space and the heat required to provide motion which is $a^3 = T^2 k$. It is space (a^3)-time ($T^2 k$) and that brings me directly to Kepler. The only way mass becomes an issue is where a bigger object such as the Earth frustrates or hinders the natural motion of any individual particle in its capture (atmosphere). Then the gravity or movement results in a tendency to move but still the tendency is manifesting the motion where mass is the restraining thereof. There always has to be motion but there is not always mass. The tendency of motion, which becomes the mass when the motion is restricted, remains a part of free movement the frustrated object still possesses as motion. There always remains a tendency to move. The Earth set a speed of motion and in order to beat mass the object has to excel by providing amplified motion in excess of the Earth motion granted to the object and gravity extending beats the restricting effect that mass has on any body connected to the centre of the Earth. That is motion and I am about to show that gravity is all about motion.

The value of space that Kepler's indicated as a third dimension a^3 does not depend on indicating a structure forming a^3 that is in rotation T^2 but only needs one position having a constant of some sorts. There has to be a space to point to a space in defining. Any point where k as a line is ending at that point k indicates a position between the start of the line k^0 and the end k^0 of the line k where one will find a spot with the value matching a^3 and the matching location will fit T^2 at that point. The line k is coming from the centre and ends at a point a^3 but such a point establishes another space in which a^3 rotates. The line k that forms by a^3 rotating T^2 is the relation there is in the solar system between all planets and the Sun . The Sun always indicates the centre k^0 and the planets always indicate the rotation. But $a^3 = T^2 k$ is only producing a relevancy of three dimensions that is equal to two plus one dimension. The fact of a^3 being at the end k^0 of the line k is how a^3 secures k and where k then secures a^3. The line k cannot have validity without proving the validity of a^3 just as much as the rotation T^2 of the space a^3 defines k. There is no possible removing of any of the factors without removing all three factors equally. The space a^3 serves to prove T^2 that defines the line k that presents a^3 an existence.

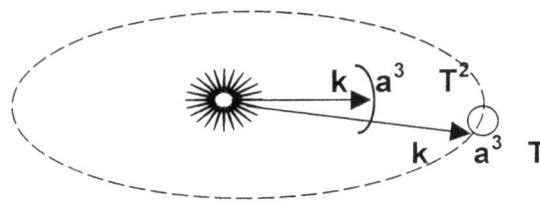

From the Sun there are three points moving between two points from one point to two other points giving the six dimensions we find in space. It is space in time or space converting space through the movement of time. It is a location of a point in the third dimension a^3 that will move according to the second dimension T^2 that will implicate k as a reference in the first dimension. It is about dimensions in reference to one another.

Let us take it from a point where the Sun provides a centre as one starting edge of k then that centre k will provide a line from the centre and the line k will provide three spots in a formation that produces a structure by the square T^2 of the dimension. Not once did Kepler indicate size as a contributing factor to a^3. That means every single point that k indicates there are three positions a^3 implicating sides of a double dimension. In the same manner is k not limited to distance or is T^2 lesser by size $k = a^3 / T^2$. That is what Kepler said. There are three dimensions a^3 between any two points T^2 flowing as time from the centre of the Sun , which is indicated by the line k. The implication of the relevancy produced by the use of the formula $k = a^3 / T^2$ brings about that when dividing T^2 into a^3 there is k left. The fact is that a^3 is a three dimension (^{3}) of single k (^{1}) showing one or T^2 is two dimensions of k being the one dimension it means that k is a part of space a^3 or T^2 which is time. It is the same thing in a double dimension or space being a triple of k then k is one factor and k cannot show a position of zero. If $k = 0$ then there is no possibility of $k = a^3 / T^2$ because $k = 0$ then $0^3 / 0^2 = 0$. That does not make sense. Mathematically space cannot be zero because those being of the opinion of space being zero or nothing must first prove mathematically that space is zero. Moreover they then must prove mathematically how does zero grow through the Hubble constant. By translating Newton's vision of the circle in completing a cycle would become zero through rotation...well that does not count the use of the formula a^3. If k cannot be zero then k could not start from zero. With $k = a^3 / T^2$

no point can be zero because **k** shows space $a^3 = k\ T^2$ is no reference to the volumetric mathematical formula used to calculate $a^3 = 4/3\ \Pi\ r^3$. Nor does it show the use of the circle ion the second dimension being $a^2 = \Pi\ r^2$. In the case of the Newton formula the circle factor becomes the square as indicated by the duration of the time T^2. The factor standing in for the line which normally would be r and then be the square value is in the case of Kepler not the value indicating the square. That means Kepler never indicated a circle of mathematical procedure but said mathematically the distance of the planet from the Sun **k** holds space a^3 in relation to time T^2 **Lines mathematically cannot start at zero because there is no evidence of zero as a factor in mathematics. Should you disagree with my statement** the question in need of answering is this: **What will the length of the shortest hypothetical line imaginable be and moreover, what would the total overall length be in that case?**

Locating zero

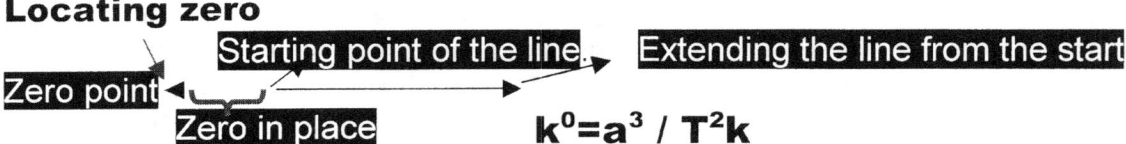

Zero point   Starting point of the line.   Extending the line from the start.

Zero in place $k^0 = a^3 / T^2 k$

The fact of form proves that the sphere captured all sides that can possibly influence the sphere. The sphere therefore holds $k^0 = a^3 / T^2 k$ within the boundaries designated to the sphere. When a body is placed in a location on the outside of such spherical borders that object seem to float in any direction. There is no control one can establish which will secure movement in any specific direction of preference except by releasing heat to counter act the required motion in a specific direction of choice. We all have seen what happens to any object that comes into the border area of a sphere. The object suddenly is motivated by motion to follow a specific designated direction and the motion leads the object to move towards the centre of the sphere. It is as if the support of the six opposing sides has lost one side where the sphere took over the control and movement starts in the direction of the Earth centre. The support of one side is literally removed by the centre of the earth where Einstein claimed the strongest gravity is and the motion of the object starts in that direction. There is no pulling on the object but there is removing of space by the centre of that specific point leading the object and the space it is in as well as the space it carries to move to the centre spot. In the sphere the borders the sphere holds are deliberate and very distinctly placed edges forming a specific distance from the centre. The centre is also proven beyond any debating. The centre of any sphere has to be at the very point where space completely falls away. That will put that space at that point in the single dimension and centre is the single dimension.

The fact in the matter suggests that all three factors hold the same identifiable measure and the differentiation between the factors is in name alone. In what I explain at a later event the true value should stand connected to singularity and in singularity the value is allocated to singularity is Π. The formula should read that $\Pi^0 = \Pi^3 / \Pi^2 \Pi$ and when used with the correct connotation the formula becomes more sensible. But be as it may the factors indicate the same measure carried in different dimensions and should all read the same value. The cube is a loosely connected structure with any form possible but the only precondition is that there must be at least six sides connecting. The six sides hold a relevancy or a responsibility to one another and provide a Universal accepted form maintaining the universe. From the structure one can see gravity is not strongly present. All six sides support what ever are inside evenly form all sides. The sphere is the form securing gravity. In the centre of the sphere there is a point where space vanishes. At that point where space vanishes gravity is the strongest. From the centre point where gravity is the strongest gravity hold the sphere true to form. At the edges of the sphere there are also point lining in 90^0 and 180^0 holding relevancy and responsibility to one another but the centre spot being the gravity point positions all the points in a location that the centre point allocate. In the centre where all lines cross one will locate singularity but I am explaining that fact a little tater on.

Newton changed the symbol of **k** by using the mathematical equated symbols G $(m + m_p)$. This is just a longer and probably a more detailed manner of indicating **k** and better defining of **k** but it symbolises precisely to the point what **k** stands for nonetheless. I wish to draw your attention to the matter of Johannes Kepler's findings that Mainstream science considers as resolved and closed for many a century while it is not. My investigating Kepler helped me to resolve other unresolved matters but it was only possible by using Kepler's work. This changed the aspect of gravity in cosmology fundamentally and as I am about to show most and totally incorrectly.

Let us for one minute leave Newton's surmising about Kepler's failure out of the picture and concern us with what Kepler found long before Newton thought about what Kepler found. Kepler said that the space a^3 is equal to the motion T^2 of the space a^3 distant from a specific centre k. That then is $a^3 = T^2 k$. Reading this mathematically encrypted coded formula of the cosmos given to Kepler and keeping it removed from Newton it reads as the space a^3 is equal to = the motion T^2 of the space a^3 in ratio to a centre k.

What this proves is that gravity is the motion of space provided by time being the liquid.

Please allow me to explain. In the formula $a^3 = T^2 k$ the space forms as the space is in motion. Newton suggested that $\dfrac{dJ}{dt} = 0$ where he stopped time to have the motion of the circle demolish the work that the circle does. That means he got time standing still or being T^1 and the motion **T= 0**. Let us ponder on that thought for a while, while remaining with the formula Kepler suggested it will seem that according to Newton $a^3 = T^2 k$ and in that T^2 then becomes **1**. Should that be the case then we have space going flat because $a^3 = T^2 k$ where $a^3 = T \times k =$ forming a square instead of a cube, and the Universe we have is a three dimensional cube in every aspect there is. If time stands still then time becomes $T^0 = 1$ and **k** being in one dimension all the time it too has to become zero.

$a^3 = T^0 k^0$

$a^3 = T^0 k^0 =$ being $(T^0 \times k^0) = 1 \times 1 = (1)^2$.

That is mathematically incoherent and a complete fallacy. To proclaim that a cube can be a square at the same time it is a cube is ridiculous, even coming from a man that has the stature of Newton! By taking the thought further we find the same blunder in removing **k** in Kepler's formula as a factor.

$a^3 = T^2 k$, then $a^3 \div k = T^2 k \div k$

being $a^2 = T^2$.

That is totally going against all mathematical principles. It would be more correct when saying that $a^3 = T^2k$. Looking at the formula in this way we find that $a^3 = T^2k$ bringing about that $a^3 = T^2k$. That proves the following:

$T^2 = a^3 \div k$ or $T^{-2} = k \div a^3$. ($T^2 = a^3 / k$ or $T^{-2} = k / a^3$)

From the implementing of $a^3 = T^2k$ we can see that:

$$\begin{aligned} k &= k^{3-2} = k^1 \\ a^3 &= a^{2+1} = a^3 \\ T^2 &= T^{3-1=2} \end{aligned}$$

$k = a^3 / T^2$	$a^3 = T^2 k$	$T^2 = a^3 / k$
$k = a^{3-2} (T^2)$	$a^3 = T^2 k^1$	$T^2 = a^3 / k^1$
$k = a^{3-2} = k^1$	$a^3 = T^{2+1} (k^1)$	$T^2 = a^{3-1} = T^2$
$k = k^{3-2} = k^1$	$a^3 = a^{2+1} = a^3$	$T^2 = T^{3-1=2}$
is the same as	**is the same as**	**It is all the same**

$k = k^{3-2} = k^1$ is in direct relation to $a^3 = a^{2+1}$ and that is, is in direct relation to the formula $a^3 = T^2 = T^{3-1=2}$.

With this information staring mainstream science in the face and scream pleading at them to recognise the information they turn around and ask why can man not fly off to other galactica at the speed of light

$k = k^{3-2} = k^1$ ⟶ It takes time for space to fill **k** in the distance. In fact, it takes the distance that **k** developed since the Big Bang $k = k^{3-2} = k^1$ to fill the distance.

It also takes time $T^2 = T^{3-1=2}$ to produce the distance forming k^2

It takes space $a^3 = a^{2+1} = a^3$ to form k^3 since coming from the Big Bang

That would give gravity meaning and that would explain not only gravity but also the Coanda affect which to my mind is gravity by principle.

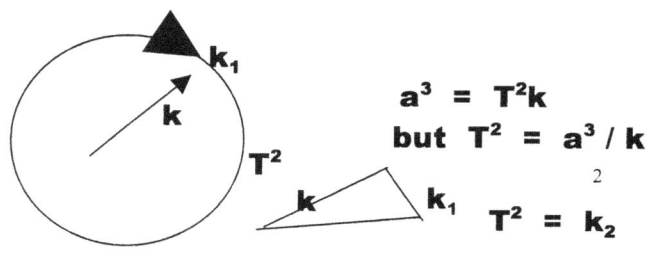

$$a^3 = T^2k$$
$$\text{but } T^2 = a^3 / k$$
$$T^2 = k_2$$

The fact is that although the symbols Kepler used were not the same, the value they measured were very much the same. The values of the symbols were interchangeable.

From the implementing of $a^3 = T^2k$ we can see that:

$$\begin{aligned} k &= k^{3-2} = k^1 \\ a^3 &= a^{2+1} = a^3 \\ T^2 &= T^{3-1=2} \end{aligned}$$

$k = a^3 / T^2$	$a^3 = T^2 k$	$T^2 = a^3 / k$
$k = a^{3-2} (T^2)$	$a^3 = T^2 k^1$	$T^2 = a^3 / k^1$
$k = a^{3-2} = k^1$	$a^3 = T^{2+1} (k^1)$	$T^2 = a^{3-1} = T^2$
$k = k^{3-2} = k^1$	$a^3 = a^{2+1} = a^3$	$T^2 = T^{3-1=2}$
is the same as	**is the same as**	**It is all the same**

$k = k^{3-2} = k^1$ is in direct relation to $a^3 = a^{2+1}$ and that is, is in direct relation to the formula $a^3 = T^2 = T^{3-1=2}$.

With this information staring mainstream science in the face and scream pleading at them to recognise the information they turn around and ask why can man not fly off to other galactica at the speed of light

$k = k^{3-2} = k^1$ ⟶ It takes time for space to fill **k** in the distance. In fact, it takes the distance that **k** developed since the Big Bang $k = k^{3-2} = k^1$ to fill the distance.

It also takes time $\mathbf{T^2 = T^{3-1=2}}$ to produce the distance forming $\mathbf{k^2}$

It takes space $\mathbf{a^3 = a^{2+1} = a^3}$ to form $\mathbf{k^3}$ since coming from the Big Bang

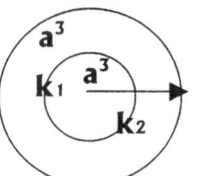

Man could create motion but at first, such motion was far less than that motion which the Earth provides. The motion of man's ability was vested in what his muscle power could provide. But a very short while ago man grew wise to machines and the fact that machines can provide more motion much faster than could animal muscle bring about motion. By supplying machine motion it gave man extra ability whereby man extended the relation between

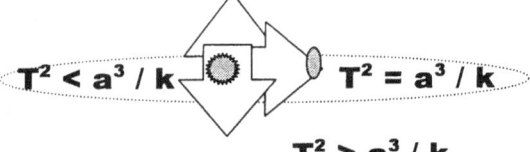

$$\mathbf{T^2 < a^3 / k} \qquad \mathbf{T^2 = a^3 / k}$$

$$\mathbf{T^2 > a^3 / k}$$

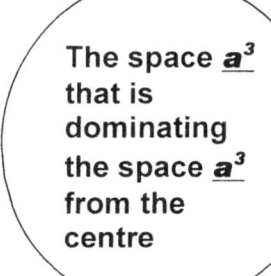

The space $\mathbf{\underline{a}^3}$ that is dominating the space $\mathbf{a^3}$ from the centre

The space $\mathbf{\underline{a}^3}$ that is dominated by the space $\mathbf{\underline{a}^3}$ from the centre

what the object has when in normal contact with space and when extended by extra motion allowing more space to apply to the surface of the object, thus enlarging the object surface in the relevancy brought on by motion.

The smaller space $\mathbf{a^3}$ is distinctly distinguishing the larger space $\mathbf{a^3}$ as the larger space $\mathbf{a^3}$ is housing the smaller space $\mathbf{a^3}$ where the factor $\mathbf{k}$ is as much indicating by length the larger space $\mathbf{a^3}$ as much as it is indicating the end of the length of the larger space $\mathbf{a^3}$ at the location of the smaller space $\mathbf{a^3}$ by directly pointing at the position the smaller space $\mathbf{a^3}$ holds. Where the larger space $\mathbf{a^3}$ ends the smaller space $\mathbf{a^3}$ is. The two remain as an inseparable single unit in double motion where the motion identifies the unit as much as distinguishing the separateness in the unity and always remain in absolute relevancy.

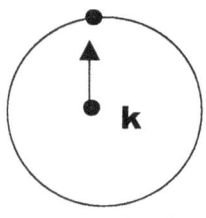

$$k = a^3/T^2$$

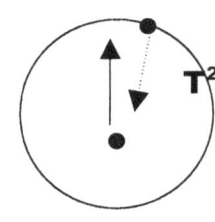

$$T^2 = a^3 / k$$

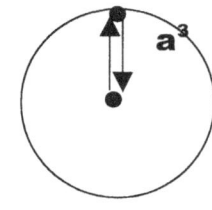

$$a^3 = T^2 k$$

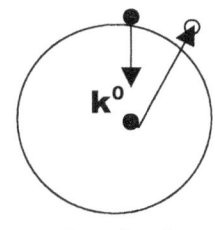

$$k^0 = a^3/T^2 k$$

By duplicating the space of any particle sharing space within a larger cosmos structure such as an atom inside a star or a human inside the Earth there are two relations applying.

The independent object serves as the outer relevancy while the contact the Earth has with its centre forms the contact with the containing singularity. When the independent object starts moving the atmosphere allows flexibility, however this flexibility is only limited to a point. As the speed increase the limits on the space that retains the object in motion starts to stretch and such stretching has definite limits.

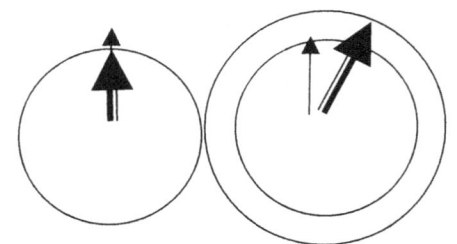

Kepler's formula is the exclusive display of _space-time_ and Newton's version of Kepler's formula is plainly and commonly misleading and a deception of the truth. There is another correct connotation to Kepler's work that divert completely from Newton's vision. In the picture to the left we see the all-familiar Moon that every one blessed with the ability to see has seen at some point in his or her life. Any person not agreeing totally with what I have to say in

the next sentences knows less about mathematical principle than a newborn baby. The Moon is according to Kepler's studies, (and discounting Newton's inconsistency if not totally flagrant rubbish) from our perspective in the space a^3 that is equal to the time T^2 it is in placed at the distance between us by the factor **k.** The Moon consists of the **space a^3** the **time T^2** ad the distance **k in time not space.**

The observing "size" of what we see depends on the time factor **k** relating to the space a^3 we see in

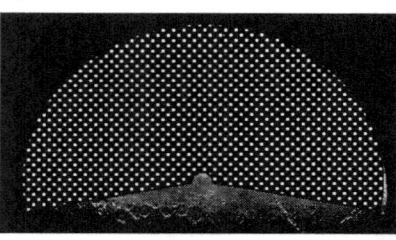 

relation to the time T^2 we see it in. The bigger the space in relation to what we see gets the smaller the time factor is that allows the time frame or the motion that frames the space. From serving my cosmic position allocated to the space I claim in the time I hold Kepler stated space-time is $a^3 = k\ T^2$ and therefore mathematically it is correct to say $k = a^3 / T^2$.

When I observe the Moon the Moon is at a distance **k** from me where I see the space a^3 in relation to the time or the moving or the spinning T^2

The figure to the left I covered the area of the Moon that I use to indicate the space a^3 is served by the sketch to the left. The symbol to the left of this point is what I use to indicate the area time T^2 is in which the space moves or the picture to the left depicts the motion aspect of the space.

The bigger a^3 will seem the less **k** has to be in order to reduce the motion or time T^2.

A large space a^3 will bring about a small time T^2 factor because of a large distance **k.** That means that $10a^3 = T^2 \times 10k$, and $a^3 / 10\ k = 10T^2$ or then the distance **k** is **10** times less. $k = a^3 / T^2$ then $k / 10 = 10a^3 / T^2$ or from another perspective it is $k /10 = 10a^3 / T^2$ that would bring about space that is 10 times bigger in the time that is relating to the space. That is space-time and that forms gravity and that is what I am about to prove in this letter. The qualities Newton attributed to mss and the principle what Newton tried to pin to mass is just the opposite of what can be pinned to mass. That, what Newton regarded to be mass, serves the very opposite to mass because gravity is motion. Gravity is the moving of objects through a container every one considers being space but the truth is that the container is time.

An object coming closer will have the relative factor **k** indicating a negative indicator as a k^{-1}. In such an event the space will increase by decreasing the time aspect and this Kepler's formula most accurately show to be $k^{-1} = T^2 / a^3$. In such an event the time putting distance between the object and the space would reduce allowing the space to relate much stronger in terms of the time as part of the overall picture.

In the overall picture $a^3 = T^2k$, the time factor, which we incorrectly see as a distance **k**, has gone very small and this is putting the space a^3 as an enormous part of the overall view and ity reduced the time T^2 as a small fragment of the entire picture. The very same picture is used but by reducing a^3 we take it that the time factor **k,** which we mistakenly put as a distance and which it is not, as being huge and that large factor increase change our relevant factors in the entire picture completely. According to our use of mathematics by increasing **k**, that reduced a^3 because also T^2 reduces a^3 and by reducing $a^3\ T^2$ gets a bigger share of the entire picture. That means the space a^3 stands related to the overall time being **k** as well as T^2. That is unmistakably exactly, precisely and to the iota correctly what Kepler said when Kepler said one is observing space (a^3) –time (T^2k). However from that we can see time increases and when time increases it is the **k** factor in the relation of space-time (a^3)= (T^2k), that reduce the space (a^3) in relation to the allocated position the onlooker has. Reducing the value of **k** allows the object (a^3) to get closer by the reduction of the time **k** between the object and the onlooker reduces making the relevancy falling to the space a^3 larger and by the same measure does the time factor in the overall picture take a smaller part. By the same token does the time factor T^2 reduce to allow the space more prominence in the entire picture. The second picture shows where the time factor **k** reduces giving the space factor a^3 more prominence. This will indicate a much reduced time factor **k,** which reduced the time factor T^2 and this gives the space factor a^3 much more prominance. The time factor is connected to both aspects of time being (T^2k) and that effects the space factor.

The expanding is a process where time develops space. It should be obvious to anyone concerned that as space became developed it was in line with the process of time that progressed and time progressing establish an increase in what ever is parting objects in the Universe. That is what the connecting of space and time should tell. It is space that parts time. It is the task of science to find what space parts time and how space parts time. Kepler thus gave us the answer about what Hubble found what was happening in the Universe centuries ago and centuries before Edwin Hubble's discovery. From Kepler's formula one can see that time and gravity is the same because as gravity weakens so does time reduce and as space expands so does the influence of gravitational reduce because gravity has less time per unit to control; more space per unit. Gravity is $T^2 = a^3 / k$ since the object cannot depart at any further distance between the centre and the object and is captured at that distance. Also gravity is $k = a^3 / T^2$ for the very same reason. The circular bonding T^2 of space a^3 is enforcing an orbit T^2 to gravitationally circle around a specific centre **k**, which indicates the gravity $T^2 = a^3 / k$ in relation to the other gravity component $k = a^3 / T^2$, and it means T^2 is a circle of gravity and **k** is the straight-line distance of gravity applying motion. Still Mainstream Academics ignore my statements that gravity is space in motion and motion of space is time: precisely as Kepler said. Any area to the cube is space a^3. However since the Big Bang conception produced the concept that time brings development, and we know it took time to develop the moon at the position the moon is now in, that means the moon is at time **k** from the Earth. In the next 10 billion years the Moon would be much "further" and therefore we know much more time would have developed the Moon in another allocated position.

In our minds and us being those blessed with life this picture shows the Moon closer or further away. That is the image life holds. To the cosmos there is no misplacing of displacing and every aspect comes with the development of time. Travel is not a concept in the Universe and only Newtonians see that as an aoption but as all Newtonians do that also is a pipe dreams of fools. It takes time to get the Moon that much further away from the Earth. The Moon was much closer when the cosmos was younger because it started when everything was much closer. It time that is pushing the Moon and the Earth apart. In that there is proof of what time is, what space is and what the concept is dividing the two cosmic differences. Space is defined while time is eternal. That black stuff there is filling the night sky is time and not space. Since that is time and not space, that time not being space is eternal without ending. By Newton throwing the time aspect relevancy away by declaring it as a value of zero, Newton threw the baby away and kept the bathwater instead.

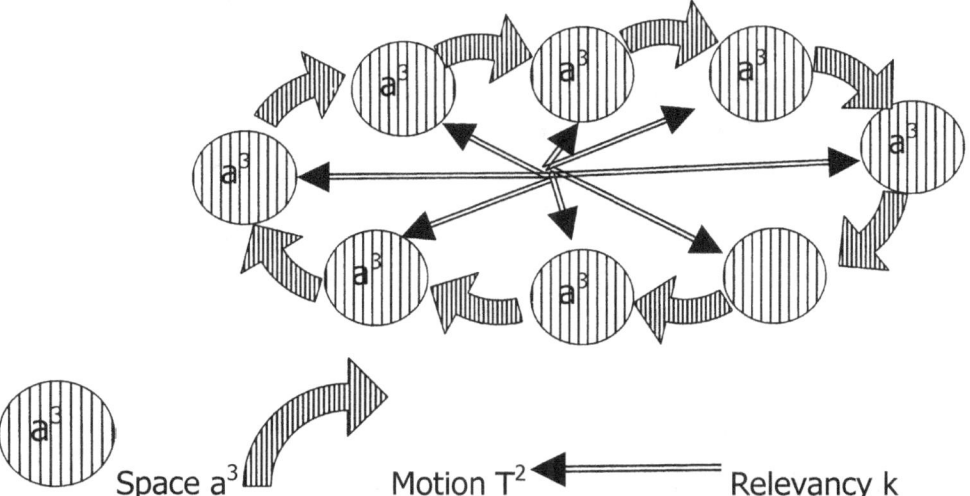

Space a^3 Motion T^2 ◀━━━ Relevancy k

That Sir, Madam, is mathematics on a much higher level than the mathematics that Newton could understand. That is the mathematical interpretation of words put in mathematical terms that the cosmos used to speak to us. That is a language that goes beyond the language of us mortals. That mathematical interpretation is a language that makes sense when one who is schooled in mathematics has a mind to read what is said. That is the language that the cosmos used to teach us what the cosmos wished us to know. That is how the cosmos told Kepler what Newton never came to understand. That is words Newtonians still cannot read because Newtonians wish to interpret into the cosmos what Newtonians want to tell the cosmos what the cosmos has to be and then humans would accept the cosmos the way they want it to be.

If science assert them and truly display a need to know instead of a need to

Time T^2 k

Space a^3

inform the cosmos in the effort they display when studying the cosmos, science would find progress. What is gravity? After three hundred and fifty years of research there is no one that vaguely knows anything about gravity. Someone came up with an idea of having a graviton but that is even more elusive than finding the gravity that the graviton supposedly is producing. The cosmos told us what gravity is but Newton knew more than the cosmos so Newton told the cosmos that the cosmos has mass. Newton told the cosmos that gravity is mass inflicted. The result is that after three hundred and more years we are as close to determining what gravity is than Newton was and Newton admitted he had no idea what so ever what gravity is! Today with all the wisdom there is going around there also still is not one Newtonian that knows more about gravity than the nothing Newton knew. That alone should guide Newtonians out of the dark ages but it does not. It only confirms the Newtonian stubborn nature to tell the cosmos what they want the cosmos to be without listening for one instant what the cosmos said. This letter is about listening to what the cosmos said on the principle Newton named gravity.

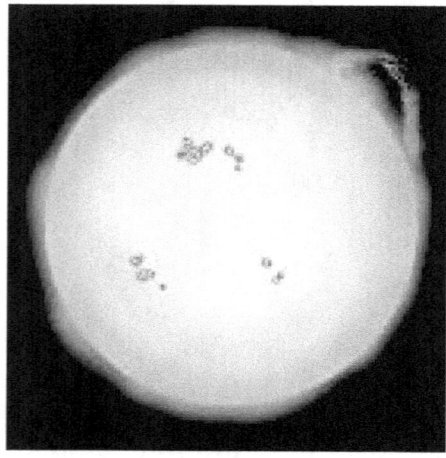

The Big Bang today is the cosmic development profile that is accepted by most and correctly so too, indeed. But in the cosmos development the Hubble constant is represented by the factor Kepler indicated as **k**, the linear time between objects. More important is that once any one accepts that this factor in time **k** grows, and then the factor cannot respond alone but has to influence the other factors it is related to by the same measure. If **k** increased, then so too had space a^3 had too but also included the rotating time T^2. Kepler said that there are three factors in very close relation and the three factors are the space a^3, which is directly equal to the time factor T^2 that is placed relevant to the time factor **k**.

All spinning matter has the point where the spin is still there but the radius is to small to measure by any means. That point is standing still in relation to the rest of the spin. In relation to that logic I do not except Newtonian science holding the radius of s spinning object unaccountable in the spin, whether the spin is applying or not.
Applying Newton's second law F=ma

One arrive at the formula
$GMm / r^2 = m (\omega^2 r)$

By replacing $(\omega^2 r)$ with $2\Pi / T$ we obtain Kepler's third law
This law predicts that $T^2 = a^3$

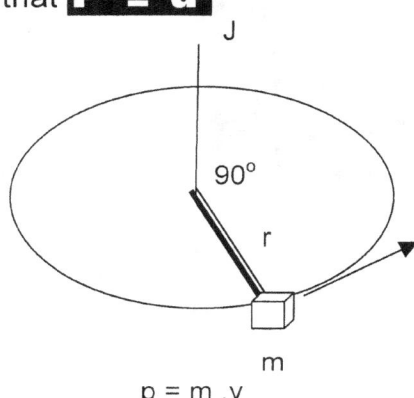

$p = m .v$

The mass (m) multiplying the speed (v) forms a new value J AND THEREFORE j CONTINUOUS TO IMPLY J = I ω
$J = r \times p$ where $p = (v = r \times \omega)$
$J = r.m.v = m.r^2 .\omega = I. \omega$ and becomes interpreted as $J = I \omega$

This establishes that r = dJ / dt

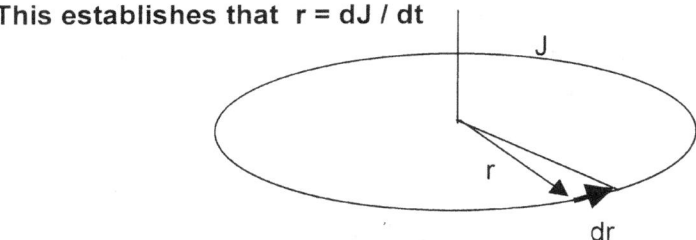

r = dJ / dt In the case of planets in orbit around the sun r forms a value of zero because dJ / dt = 0.

What this statement implies is that r does not exist. When anything has a value of zero it is for all purposes non-existent. Only when an object is following s straight line can the radius be non-existent because the radius alters value through time development.

The logic behind giving the reason that substantiates this claim this argument seems almost an insult to the intellect of others. How does a Sun the size we now have fit into a Universe the size the Universe was at the event of the Big Gang when the Big Bang was the size of a Neutron?

Every claim any person ever made when suggesting the Big Bang theory was that the "distance" between the objects in the Universe grew and it is still growing to

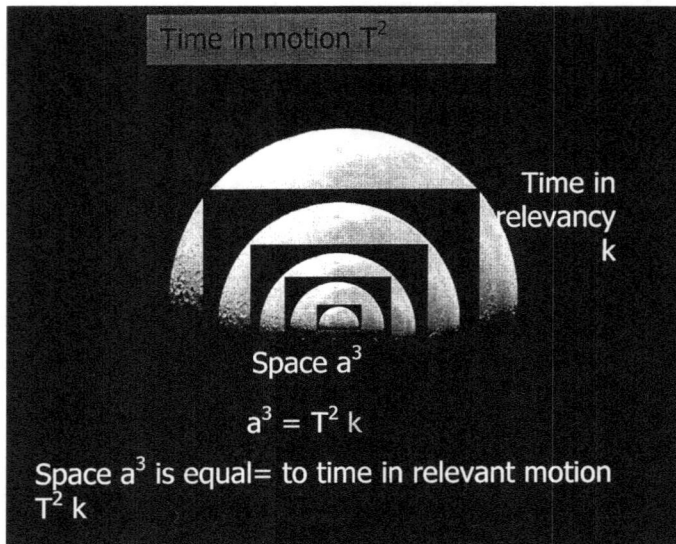

Time in motion T^2

Time in relevancy k

Space a^3

$$a^3 = T^2 k$$

Space a^3 is equal= to time in relevant motion $T^2 k$

this day. It is accepted as the Hubble constant. If **k** in $a^3 = T^2k$ grows, then the material a^3 must grow but also to the same measure must the circle forming time also increase to fit into the total relevancy of space-time or then $a^3 = T^2k$. It is clear that the time factor **k,** which we connect to the distance between objects are as important in the formula Kepler handed down as any other part of space-time. That completely goes against the grain of Newtonian interpretation of this significant formula. The more **k** would find significant the more k will increase T^2 and the more **k** will reduce a^3. This conclusion is part of Kepler's formula $a^3 = T^2 k$ which is declaring that $k= a^3 / T^2$. Time decreases space when time increases relative motion.

If we have a close look at Kepler's formula we find the space is in motion centred by a specific distance, which forms a conclusive part of the formula. We can see that the space as well as the time effecting, the space varies in accordance to the distance **k** and it is the distance **k** that produces the time, which produces the space. How is that possible you might ask when Newton has k down to zero?

We find that the distribution of planets relating to the Sun is all in precise sequence and the name of such a sequence is the Titius Bode law, and a law it surely is! The gravity extending from the Titius Bode law forms the entirety of the building of the Universe by constructing the Universe in the using of the atoms to form the Universe in the entirety thereof. Bode's Law:

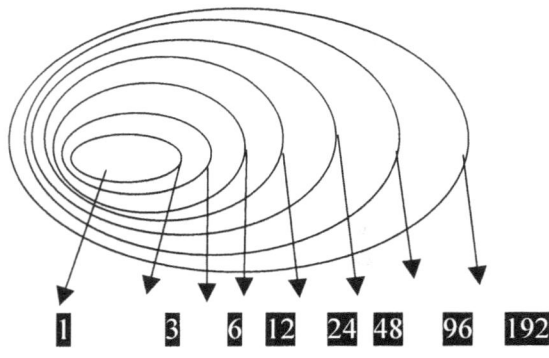

1   3   6   12   24   48   96   192

Outer space cannot be a vacuum because if it were a vacuum it would implode unless it was secured with very stringent outer walls. There are no such retaining walls and therefore there is no vacuum. It is again another Human myth transported into outer space that has no validity. The absence of material proves the abundance of structural composition and if matter is reduced then something else is absolutely overbearing to fill the absence of material. That what fills outer space is what keeps outer space rigid.

Planet	Mercury	Venus	Earth	Mars	Ceres	Jupiter	Saturn	Uranus
Bode's law distance	4	7	10	16	28	52	100	196
Actual distance	3.9	7.2	10	15.2	28	52	95	192

A numerical sequence announced by J.E. Bode in 1772, which matches the distances from the Sun of the six planets then known. It is also known as the Titus-Bode law, as it was first pointed out by the German mathematician Johann Daniel Titius (1729-96) in 1766. It is formed from the sequence 0,3,6,12,24,48,96, and 192 by adding 4 to each number. The planets were seen to fit this sequence quite well – as did Uranus, discovered in 1781. However, Neptune and Pluto do not conform to the 'law'. Bode's Law stimulated the search for a planet orbiting between Mars and Jupiter that led to the discovery of the first asteroids. It is often said that the law has no theoretical basis, but it does show how orbital resonance can lead to commensurability. The importance that becomes known is the sequence the Titus Bode law saw in the number arrangement of 3; 6; 12; 24; 48; 96 etc. The incorrect application of the Titus Bode law lies in subtracting the figure of 3 from 10 leaving 7. The other way of reasoning is to add four each time to the first value of three starting with 3 and so on.

This puts the entire assortment of planets in neatly allocated and precise arrangements according the composition and outlay of outer space. The location of the planet does not depend on size neither does it depend on mass. The planets are arranged irrespective of any particular factors in accordance with anomalies or other uniqueness on the side of the planet. The arrangement of the located position runs by way of number and the distance doubles every time the number is one more in position to the Sun . I know I am on record of saying distance has no value in the cosmos and it has no merit other than to be a man made device formulated by man to be used in some insignificant calculation concerning some needs man may have. This arrangement in the layout I do explain later on but at this point I wish to show that there is absolute prominence in the composition of outer space and the planets has no particular preference accept to be where they are in the cue they take from the centre of the Sun . Let's

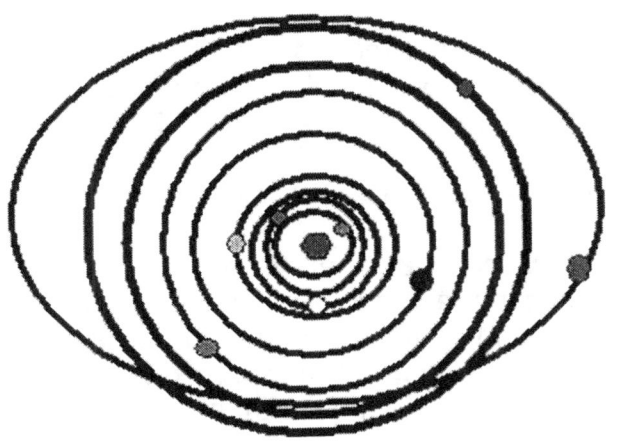

for one minute forget the childish Newtonian argument of outer space being zero because I cannot afford to waste time in this book in disputing the validity of such an argument. I concentrate in another book where on this issue of explaining. In order to have a precise arrangement or pattern of positioning the fact of outer space must present some merit in as much as its composition. The planets as such can even be fragments because a structure forming a unit indicates that it does not participate in the issue. There is an arrangement of positioning in outer space, which shows that outer space holds the measure and gives the relevance.

Outer space is not just another vacuum of no importance but has absolute say when put into context with the layout is has related to the position of the Sun . The proof is there in much more undeniable prominence than the unproven mass mesmerising nonsense Newton presents. The Titius Bode layout shows clearly no mass has any validity in the position allocated to any specific planet and therefore mass as attraction or force can be ruled out totally. It is not there while the Titius Bode law is there.

Furthermore we can find that the Kepler research proves beyond denying that **k** as a factor goes beyond dispute in contradiction to Newton's claims.

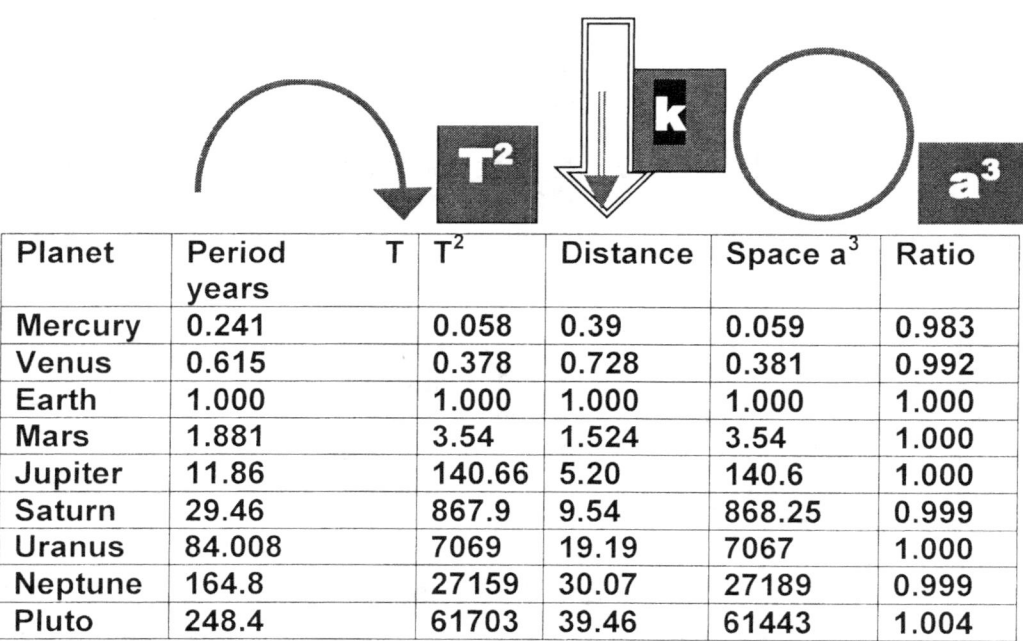

Planet	Period years	T	T^2	Distance	Space a^3	Ratio
Mercury	0.241		0.058	0.39	0.059	0.983
Venus	0.615		0.378	0.728	0.381	0.992
Earth	1.000		1.000	1.000	1.000	1.000
Mars	1.881		3.54	1.524	3.54	1.000
Jupiter	11.86		140.66	5.20	140.6	1.000
Saturn	29.46		867.9	9.54	868.25	0.999
Uranus	84.008		7069	19.19	7067	1.000
Neptune	164.8		27159	30.07	27189	0.999
Pluto	248.4		61703	39.46	61443	1.004

The factor **k** has a value of a^3 / T^2 and not even Newton can take such an argument away. Studying the evidence that was in hand even before Newton's birth where we find that the arrangement is such that it clearly defines the tempo in relation to the space of orbit at the location where such orbit can take place. The space has the dynamics of the other two time factors and the time factors indicate that the three factors are inter reliant on one another. The one gives the other value and the one takes value from the other. The space $a^3 = T^2$ **k** to the time in motion. This means the space receives its value in relation to the value of the motion, which is in relation to the centre of the Sun . Again mass is a pipe dream.

The **space $a^3 = T^2$ k** to the time in **motion,** which would then have **k $= a^3/T^2$** and in another ratio it is **$T^2 = a^3 /$ k**. That makes mathematical sense. Compare this mathematically to what Newton suggested and compare the mathematical sanity that Newton portrays when Newton declare that $a^3 = T^2$ **k** $= a^3 = T^2$ **X 0 (because according to Newton k=0) $a^3 = T^2$ X 0 $= a^3 = T^2$**. It is blatant corruption of mathematical law and any child that would put this down, as a valid mathematical argument in any maths class will find that child has failed the examination. Yet Newton is allowed to go insane. $a^3=T^2 k$ can only be happening if there were two relating factors in the entire issue. It can only have meaning if outer space stands in relation to the Sun and no planet has any meaning in this partnership.

	Distance (AU)	Radius (Earth's)	Mass (Earth's)	Rotation (Earth's)	# Moons	Orbital Inclination	Orbital Eccentricity	Obliquity	Density (g/cm³)
Sun	0	109	332,800	25-36*	9	--	--	--	1.410
Mercury	0.39	0.38	0.05	58.8	0	7	0.2056	0.1°	5.43
Venus	0.72	0.95	0.89	244	0	3.394	0.0068	177.4°	5.25
Earth	1.0	1.00	1.00	1.00	1	0.000	0.0167	23.45°	5.52
Mars	1.5	0.53	0.11	1.029	2	1.850	0.0934	25.19°	3.95
Jupiter	5.2	11	318	0.411	16	1.308	0.0483	3.12°	1.33
Saturn	9.5	9	95	0.428	18	2.488	0.0560	26.73°	0.69
Uranus	19.2	4	17	0.748	15	0.774	0.0461	97.86°	1.29
Neptune	30.1	4	17	0.802	8	1.774	0.0097	29.56°	1.64
Pluto	39.5	0.18	0.002	0.267	1	17.15	0.2482	119.6°	2.03

Again there is no trace of mass of individual objects having any significance in the outlay or in influencing the time of rotation to go faster or slower on the value of the mass or that the orbit is bigger or slower in relation to the size of the planet. If one scrutinizes the whole issue as a unit, it seems as if there is no significance in the fact that there are planets participating in the orbit process. The allocated position shows order but the order concerns the planet and its location in relation to the Sun and one inner marker. I explain that later on.

The Sun it seems could care less about any planets being in the formation of outer space. If they were all in a specific allocated position that proves a given formula as to where they find such a position and the allocated position is in no way served by the influence of mass or size, then that proves buoyancy. Only when buoyancy enters the equation will size and mass not reflect any influence. If they rotate in precise equal fashion and the orbits in which they orbit shows identical relation to the space and the time we can surely claim that it is more the soup that matters that the size of the crumbs in the soup.

In relation to the Sun every planet is not only at the same timing but also at an equal relevant timing of $k = a^3/T^2$. That means in ratio to the Sun all planets are at an even distance and that could only be if the Sun regards the outer space arena as liquid where the lot is floating in a bowl of liquid with a specific density applying. That puts the arguments about gravity in a totally different principal. Where liquid stand in affect to a solid we find the Coanda effect in place.

Newton stated that $a^3 = T^2$. That too is corrupting Kepler's fact as far as possible only to give veracity to the falsified claims Newton presented as trueness.
If that were the case then the space divided by the time would leave on $T^2/a^3 =1$. That is not the case because we have the tables to prove that it is a fallacy. Neither is $k = 0$ or is $T^2/a^3 =1$ because $k^{-1} = T^2/a^3 \neq 1$.

PLANET	SEMIMAJOR AXIS $A(10^{10}m)$	PERIOD T (y)	T^2/a^3 $(10^{-34} y^2/m^3)$
Mercury	5.79	0.241	2.99
Venus	10.8	0.615	3.00
Earth	15.0	1.00	2.96
Mars	22.8	1.88	2.98
Jupiter	77.8	11.9	3.01
Saturn	143	29.5	2.98
Uranus	287	84.0	2.98
Neptune	450	165	2.99
Pluto	590	248	2.99

From Kepler's space-time $a^3 = T^2k$ formula we find that the relevancy of all planets $k^{-1} = T^2/a^3$ in relation to the Sun is alike. This is only possible when the planets are floating in buoyancy because when in buoyancy all objects are equal in relation to the water holding them.

There is no big or small but just those having specific density in relation. If there was no buoyancy mass or size would form some sort of resistance that would allow more and less restriction in some or other form to be present. In the sea and to the sea in that case there is no big fish or small fish but there is only fish. To us humans we think in perception of distance but in the cosmos space in the form of distance is the measure of time developed. The same goes for temperature and mass. Mass is good and mass is a product of the Human mind to put perception when it is needed but mass is dysfunctional in relation to the cosmos.

All the planets are more or less 299 in ratio from the Sun, which makes the time affecting all the planets $T^{-2} = k / a^3$ which is then $T^2 = a^3 X 299$ and in relation to space $a^3 = T^2 / 299$. What this does is it puts all the space in ratio at an equal distance to the center of the Sun and it puts the time in motion rotating at an even period around the Sun. From where we stand the planets are assorted but from the Sun it is liquid of outer space spinning in relation to the Sun that is serving as the solid. Being liquid and solid is having motion or not.

The Sun is at a level with all the planets parading past the Sun in the given ratio of 299. All the planets are precisely the same "distance " $299 = T^2/a^3$ from the Sun and has precisely the same "mass" $a^3 = T^2 / 299$ in relation to the Sun and the rest also floating about the Sun while they all travel at the same velocity $T^2 = a^3 / 299$ around the Sun.

The lot is in a bowl of liquid and it is the liquid that keeps a regard to the Sun where the Sun forms the solid as the regard to the liquid. In that regard gravity loses all it's mystery as the Coanda effect starts to explain gravity and the purpose of gravity. In accordance with and depending on what Kepler's law indicate, we find space-time is $a^3 = k T^2$. I wish to remind you where Newton came up with the idea that $a^3 = T^2$ when he disclaimed Kepler's findings to validate his position on mass and the way he interpreted gravity. Let us again go back to investigate....

Applying Newton's second law F=ma
One arrive at the formula
$GMm / r^2 = m (\omega^2 r)$
By replacing $(\omega^2 r)$ with $2\Pi / T$ we obtain Kepler's third law
This law predicts that $T^2 = a^3$

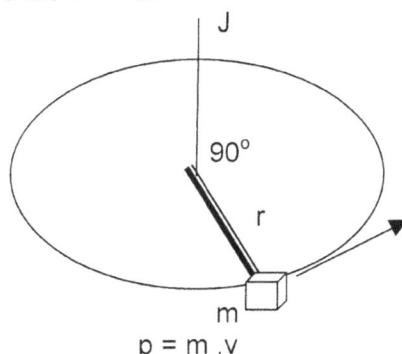

The mass (m) multiplying the speed (v) forms a new value J AND
THEREFORE j CONTINUOUS TO IMPLY $J = I \omega$
$J = r \times p$ where $p = (v = r \times \omega)$
$J = r.m.v = m.r^2.\omega = I.\omega$ and becomes interpreted as $J = I \omega$

This establishes that $r = dJ / dt$

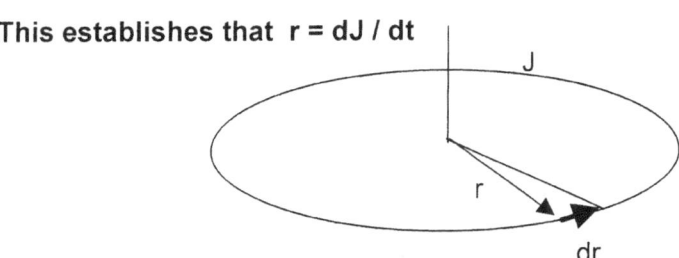

$r = dJ / dt$ In the case of planets in orbit around the sun r forms a value of zero because $dJ / dt = 0$. The big incorrectness is that this presumption works on stopping time. That is impossible to achieve.
This is totally going against the grain of what Kepler said because Kepler said the motion of the space is equal to the space. On this rests the entire Universe because as soon as time stops, space collapse. The motion provides time a space to move and that is the Coanda effect.

$r = dJ /$

$r = \underline{\qquad\qquad\qquad\qquad\qquad\qquad}$

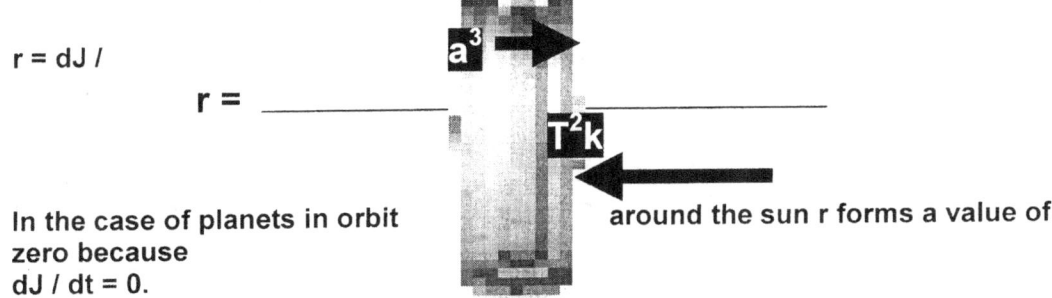

In the case of planets in orbit
zero because
$dJ / dt = 0$.

around the sun r forms a value of

Then Mainstream science still has the audacity to say Newton was never proven to be incorrect...it should rather be said that Newton was never investigated and Kepler was left undiscovered because Newton raped everything about Kepler. It is as if Newton stole a vision Kepler got by studying but never saw. Let us for once seriously reflect on the matter while forgetting about Newton and Kepler.

If it is true that $a^3 = T^2 k$ and we dismiss Newton's obscenity while going back to basic mathematical principles we find that:

Taking the argument back to Kepler's law,

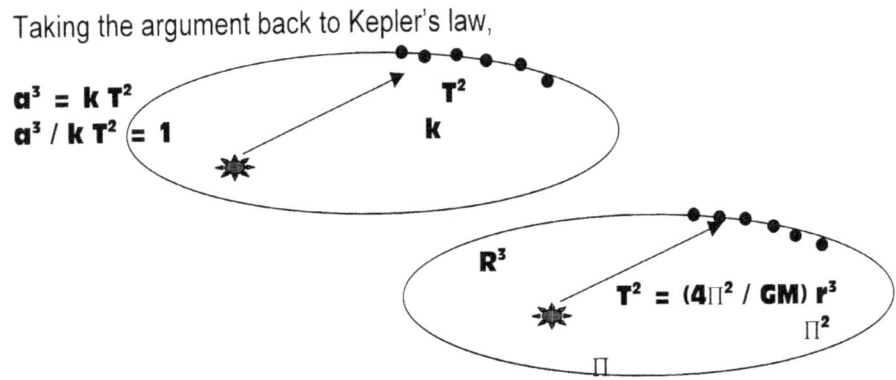

$a^3 = k T^2$
$a^3 / k T^2 = 1$

$a^3 = T^2 k$
That means the space will move in a straight line while it circles. That can only indicate a controlled expanding because $a^3 = T^2 k$ (also is); that the line between the centre k^0 and a^3 / T^2 is increasing at a rate of $k = a^3 / T^2$ which indicates that expanding is happening at this very moment. The line k is not zero as Newtonian madness would suggest or the line is not 1 ($a^3=T^2$) but the line is gaining by one dimension every rotation that is completed.

$T^2 = a^3 / k$ which means the time it takes to move is in ratio with the distance the space will move.

$k = a^3 / T^2$ which means the distance the space would move depends on the time allowed for moving. However if that is true and it is a mathematical statement then it also must be true that:

$k^0 = a^3 / T^2 k$ there is a appropriated centre

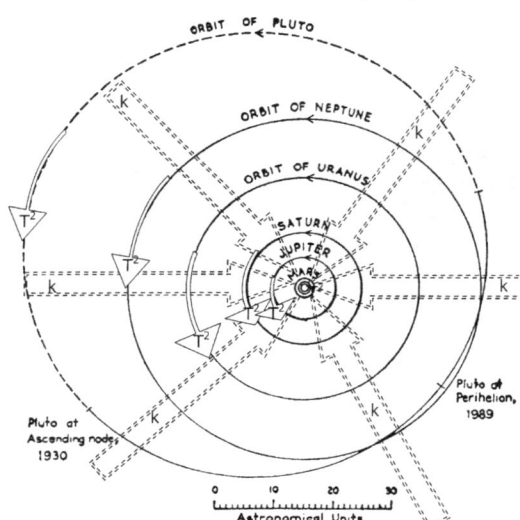

$T^{-2} = k / a^3$ Time can increase by manipulation as well as time that can reduce by manipulation. One can travel by increasing or decreasing the valid time.

$k^{-1} = T^2 / a^3$ The distance is moving between the objects. If the objects are not coming closer it must be the substance holding the objects that is coming closer. When we have a table indicating a shift towards the centre $k^{-1} = T^2 / a^3$ that does not involve any of the planets coming towards the Sun it would then show the space holding the planets are moving inwards because it shows definitely that something about the radius is decreasing and that means something is shifting towards the centre. The space a^3 is rotating T^2 and the only other counter action could be when the space holding the rotating space is reducing $T^2 / a^3 = 299$ by the margin space is expanding $a^3 = T^2 k = k = a^3 / T^2$ that proves the Big Bang is in progress. **Kepler told so much about so many by using so few syllabifications that a mathematician such as Newton was unable to comprehend the full implication.** Allow mw to explain with the aid of a manmade machine.

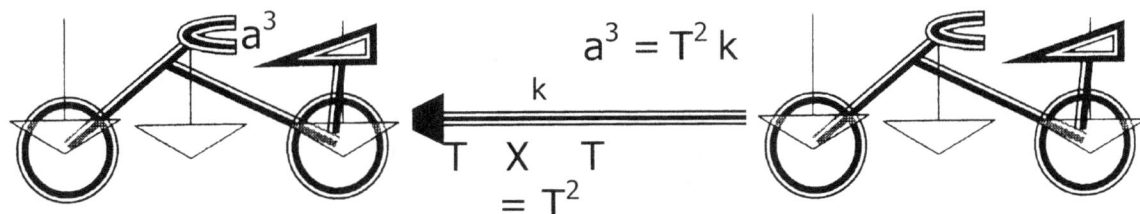

We know time connects to the moving bicycle where the position changes from where the position allocated is putting the bicycle from the one instance to the next instance. The distance the bicycle moves is associated with time because the distance than **k** has is the duration in time that it took the bicycle to replace a^3 during T^2 all across the length of **k.** We know the bicycle cannot jump from the one position to the next position notwithstanding whatever drive we connect to the bicycle. The bicycle constitutes of a massive number of atoms forming the

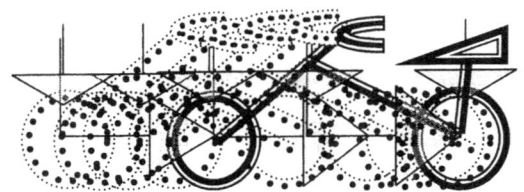

structure that forms the bicycle.

The get the bicycle from one point to another point we have to take the one lot of atoms forming the bicycle structure that is the bicycle unit to the next lot that forms the same bicycle at another location.

That means to every atom there is a replacing of the one atom to the next location of the same atom at a distance of **k**, which is time during a period, which is time. Time therefore has two components being **k** and being T^2 and all the while **k** multiplied by T^2 forms space a^3. If Newton had no regard for Kepler then at least he should have had some regard for mathematics.

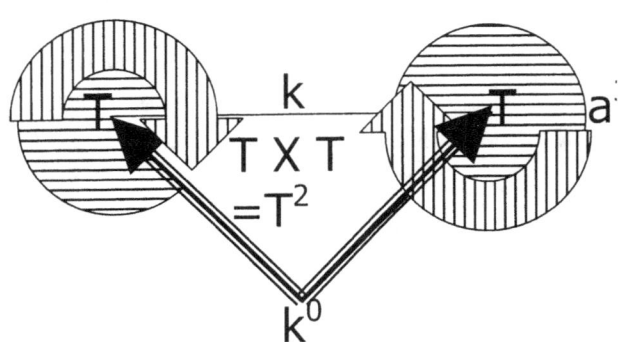

It takes the time that **k** allows to move a^3 from **T** to (X) **T** during T^2 and that is what Kepler said about space-time. Kepler mathematically said that $a^3 = T^2k$. Time is the motion that time allows space to move from and towards $a^3 = (T$ to (X) $T = T^2)$ **k** It proves that the factor **k** shares all the dynamics which the factor T^2 presents in allocating the position in time of space a^3

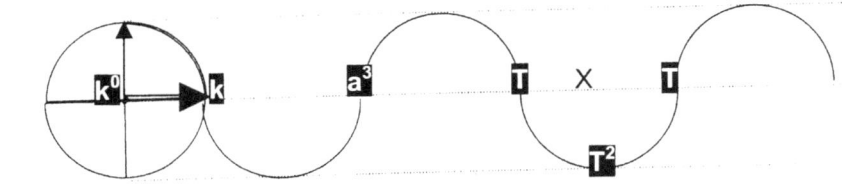

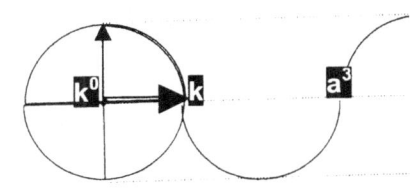

The motion of the atom follows the exact curves of the graph. Because there is motion the motion is in the square and since the square is present we can refer the result to the triangle. The motion is always 90^0 to line of connecting the centre to the motion and where 90^0 become present the law of Pythagoras becomes the dominant mathematical factor.

The factor **k** is always an extending line from where **k** as a straight line started to 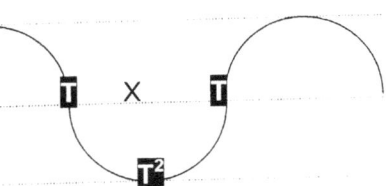 where **k** then will end. The line that forms the basis for the triangle is k^0 going on to **k**. If **k** is implicated as zero as Newton said then the entire Universe loses all its validity in being. That throws the entire cosmos into not being and that is the only thing that cannot ever be a factor in the cosmos since the cosmos is everything that is between infinity and eternity. The motion of a^3 is from where k^0 starts the line to where **k** ends the line and in between the start and the end there is a beginning and an end **T** to (X) **T**. That is also forming the other end of what totals as time T^2, but it holds time in duration T^2.

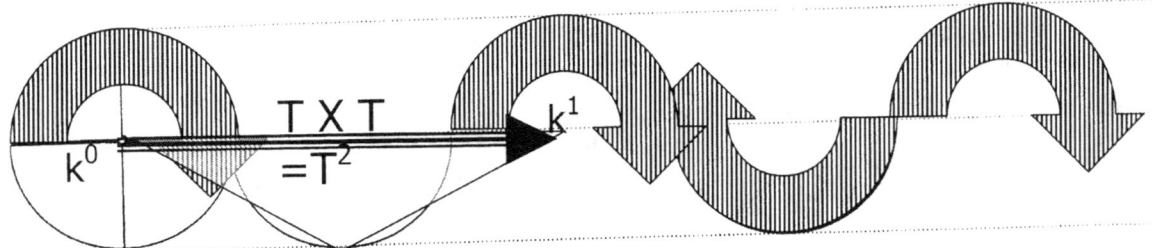

The movement of the bicycle concerns the unit as a structure that we recognise as the bicycle but the bicycle comprises of all the atoms forming the bicycle. The atom is in motion $k = a^3 / T^2$ but also within the atom forming the atom there is motion of the atom $T^2 = a^3 / k$ and the totality in reference form space-time $a^3 = (T^2) k$. The formula implicates that the space $a^3 =$ is equal to the position that the space in the next movement will have (T^2) **k**. The space becomes the motion of the space and in the space there has to be motion producing the liquid as well as rotary motion that gives the space a definition, which presents the space with motion in relation to the liquid, and then there is the space defining the purpose of the liquid.

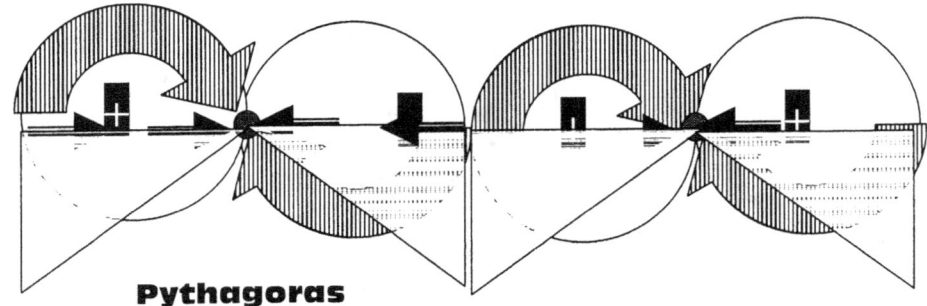

Pythagoras

$$a^3 = (T \text{ to } (X) \ T = T^2) \ k$$

Where there is a square present there Pythagoras is also present. I am sure Newton knew that...what is on the one side of the line forms a relation to what is on the other side of the square. The referring of motion in relation to a centre proves an establishing of a line in relation to a point. It serves a line being in relation to a specific centre and that centre projects the start of the line in a 90^0 angle with the point where the line must end. This is a mathematical reflex of Pythagoras using the line in forming a square that comes in relation to a centre.

That is the essence of Pythagoras.

The relevancy of k forms two correlating points where one relates to the centre and another relates to distance moved

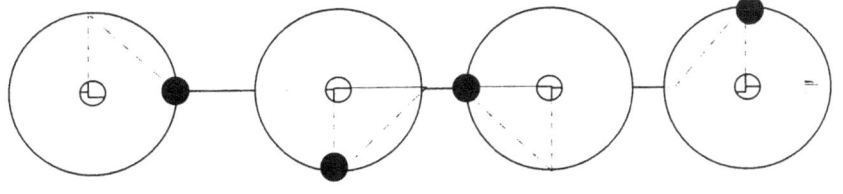

The motion of the atom reposition the outside to a point in the centre of the atom and since the electron is honouring such a centre any moving of the centre has to effect the atom in according to the dynamics Pythagoras

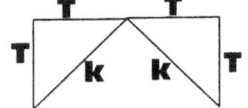

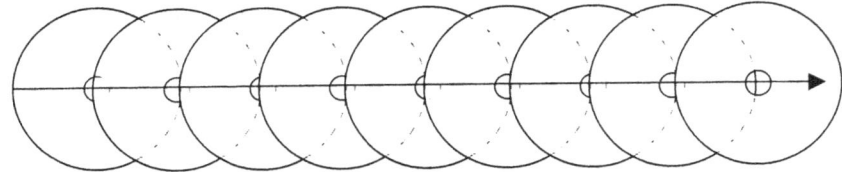

demands.

The moving of the atom is formed by a specific relocation as well as a specific reproduction of the atom in structure as well as in allocated positioning in accordance to every aspect of its surroundings. The atom must remain precisely associated with all aspects outside in order to find the precise duplicating of time in every aspect. That is gravity.

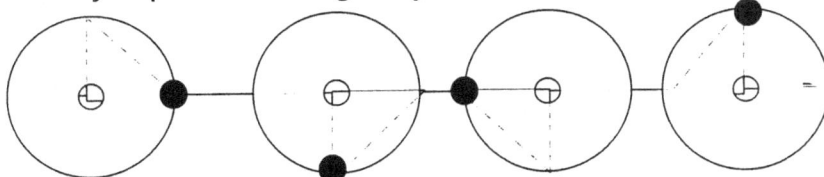

From this point the atom orbits as it honour the centre around which it focus the spin. There is a relation by four points where the distance or the time will align with the centre as rotation focuses on such a centre. Every time the realigning forms 90^0 with the centre we will find Pythagoras applying the law aspect of the aligning.

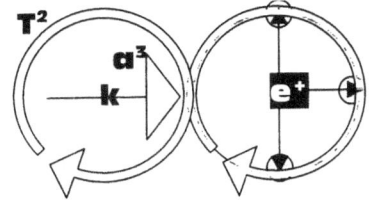

In relation to the outside there is more motion that has an effect on the atom where the center point that the electron focus on shifts in accordance with the atom motion as a unit. It takes the atom time in duration to remove all the material forming the atom to shift from one point to another point in order to allow motion to take place.

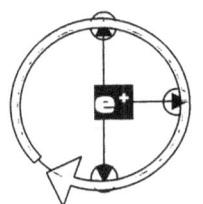

At the same time it takes the electron time to move around the centre and the distance the electron is from the proton serves as the measure of time that will apply when on cycle of rotation is completed. The rotation connects to time and since it takes light 3×10^5 km / sec to reposition the light photon the speed of light is not time but is in relation to time. Einstein proved that the speed of light is time but that cannot be since it then must take light zero time to cross the entire Universe. Since light can only travel at 3×10^5 km / sec and the Universe is slightly wider than that, Einstein's argument goes up in smoke. To light, as it is the photon, the light finds time standing still but that explaining is long and is off the point as far as this discussion goes. Since light is not time but is connected to time like the rest of all material it takes the electron time to circle the atom. One may argue it is insignificant but to the atom the cognisance is one year or one day or one period of time achievement. It is not only beauty that is in the eye of the beholder but everything about anything comes down to the measure of the beholder. To produce the space a^3 of the atom the atom must have the electron circle T^2 once around the atom border at the relevance of k. The electron is focussing the relevance of the atom a^3 where it secures the border T^2 at the relevant time position of k. However k as a factor is in relation with the start and the start of any line is not zero but it is infinite. Therefore k comes all the way from k^0 to where k ends to position T^2, which then secures the border of the atom in a^3.

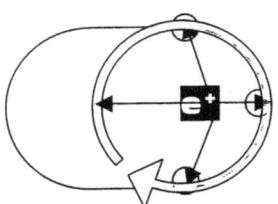

When the atom centre moves while the electron rotates such movement of the centre has a direction and this direction must influence the rotation because it will change the position of the rotation in relation to the centre of the atom. In the direction the centre moves the space between the electron and the centre will reduce in accordance with the direction the centre moves. At 180^0 from that point the distance between the centre and the electron has to increase by the margin the distance decreased 180^0 from there. The shift of the electron will be influenced by

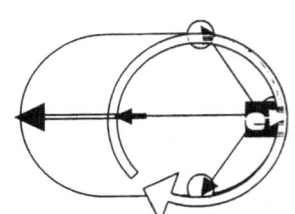

the intensity of the motion of the electron in the direction the electron moves. Since all the value of all the planets in the solar system mounts to $a^3/T^2 = \pm 300$ that means all the motion of all the planets related to the Sun is moving at an equal pace in time with the directional flow of time. The rotations around their individual axis might be different but the rotations they have in relation with the Sun are all equal. That proves that from the Sun we have all planets orbiting equal at $k= a^3/T^2 = \pm 300$.

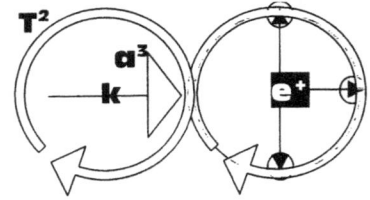

The fact of the relevancy is not merely argumentative but is a mathematical reality. Changes in the motion of the atoms will affect the atom in the rotation time it takes the electron to circle bout the centre. By shifting the centre the time will reduce the ration period in one direction as much as it will increase the rotation in the other direction. This will have a sever timing implication on the atom as much as the time in motion increase the space it holds per time unit in rotation will decrease.

As the motion of the atom increase the electron allocate position

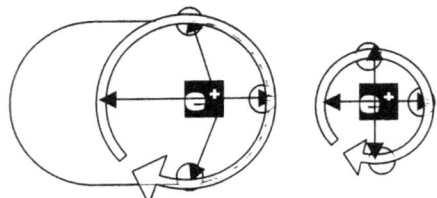

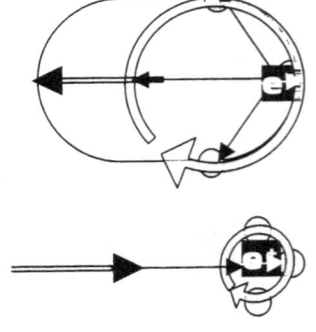

while rotating in alignment has to readjust with the shifting centre. If k shifts then k will insist on T^2 readjusting and with T^2 readjusting we will find the borders of a^3 reposition the circle of rotation. The more motion there is in relation to another fixed centre that represents the starting point that serves as k^0 the smaller will the allocated distance k be that we find between the rotating electron T^2 and the shifting proton in the centre. In this the size of the electron will reduce the size of the atom by the margin that the proton comes closer to the electron. The time it takes the electron to orbit is reduced by a ratio of four to one as the motion of the moving proton centre increases.

That is why gravity or motion reduces the atom as it accelerates the atom in relation to a more compact space. As the space in which the atom moves through time by duplicating increase in density that projects the ratio to reduce the rotating

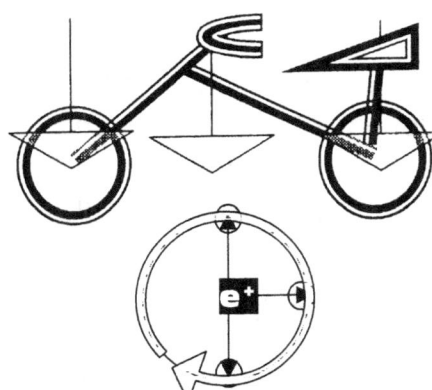

of the electron and with that the electron reduced the claimed space the atom holds. But it is not where the significance ends. In that way gravity increase time as gravity reduces space where gravity is motion accelerated or decelerated by change of the density of the heat through which the space moves.

By reducing the atom it relieves the atom of controlled heat within and such heat is ejected to the outside of the atom. The atom heat reduces by claiming less space because it holds less heat.

Since the movement aligns with another centre k^0 that represents the link between k^0 and k and therefore institutes the triangle the space held by the electron remains an allocated value in relation to the Earth. Bo body shrinks because it removes heat from the atom but the atom still hold the heat in relation to what the Earth measures. That is the essence of the sound barrier. That is the essence of the heat blanket that covers the aircraft.

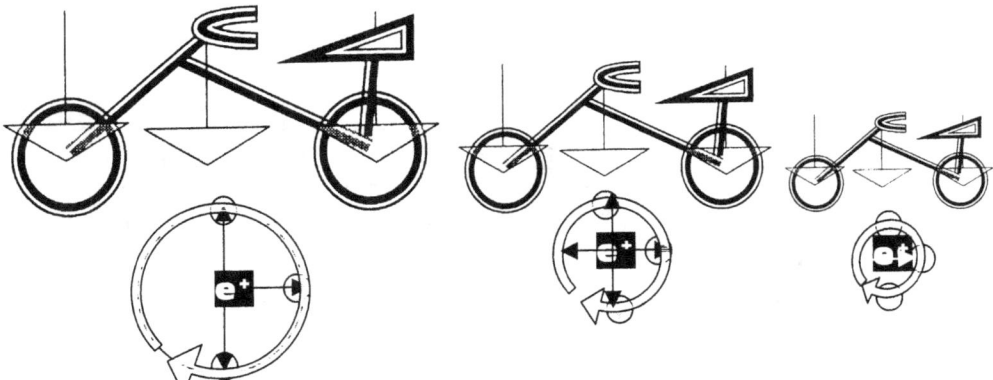

The atomic size of the aircraft reduces while the structure space of the aircraft remains in accordance with the size that the Earth permit and heat fill the vacancy that motion produce. The aircraft has no mass since the aircraft has motion. Newtonian misconception can never establish any reason for the sound barrier.

That is what time implicates but all that is lost to Newton trying to fixate every aspect in cosmology on mass and by making the value of k redundant he removed the concept of time from the cosmos altogether. The value of k is space-time and that is $k = a^3 / T^2 = \pm 300$. **What this suggestion implicates would put the cosmos in a bowl of liquid where all swim alike**. That will explain gravity. That proves outer space has a specific density relation with the Sun and a density

In as far as material within outer space goes. Every object floats just like the rest with no differences in as far as size or mass goes and outer space is liquid/gas.

In accordance with and depending on what Kepler's law indicate, we find space-time is $a^3 = k\,T^2$.

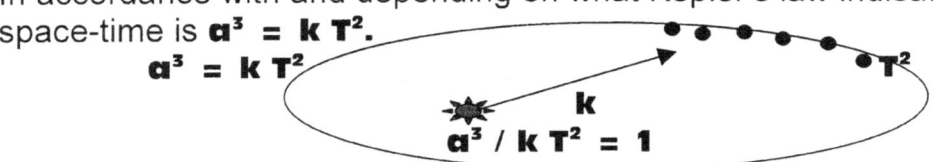

$$a^3 = k\,T^2$$
$$k$$
$$a^3 / k\,T^2 = 1$$

In the relevancy we have space a^3 that is equal = to the time T^2 that stand related to a specific centre k. The relation is time and space forming the interchangeable unit a^3/T^2 that stands in ratio to a centre k. The ratio between the space a^3 and the time T^2 totally depends on the measure of k.

The formula that the cosmos gave Kepler reads that the space a^3 is equal to the motion of the space in the time that it takes the space to move. That means the space is duplicating every second it moves and stopping will destroy all moving of the space. Since Newton became an institution forming the King bee of the academic cartel world wide The Brainy Bunch had Newton's vision written in the minds of the future generations almost at gunpoint...well definitely at an academic gunpoint.

$r = dJ /$

$$r = \frac{dJ}{dt}$$

In the case of planets in orbit zero because $dJ / dt = 0$.

around the sun r forms a value of

This goes directly against what the cosmos told Kepler when the cosmos told Kepler that all space moving through time is equal to the movement of the space in the time it is moving through. That means by suggesting time may stand still such a suggestion throws the entire Universe into singularity. You cannot stop everything that moves into a sudden halt because the momentum will destroy every aspect of such motion. Therefore one cannot remove the relevancy of motion by mathematically stopping time.

$r = dJ /$

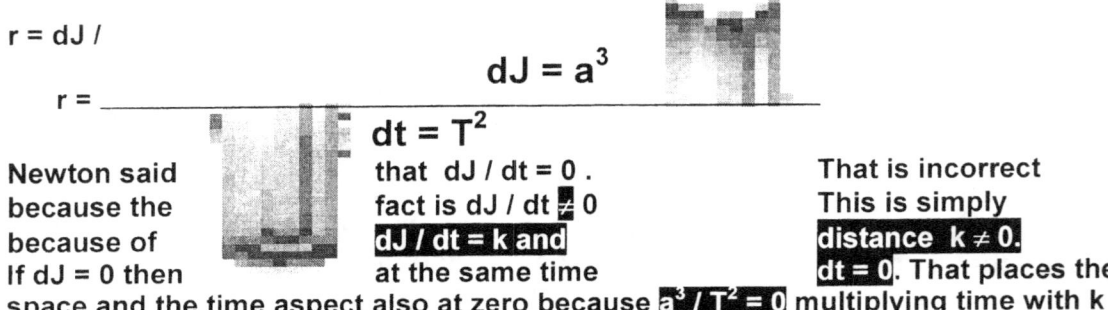

$$dJ = a^3$$
$$r = \frac{dJ = a^3}{dt = T^2}$$

Newton said because the because of If $dJ = 0$ then that $dJ / dt = 0$. fact is $dJ / dt \neq 0$ $dJ / dt = k$ and at the same time That is incorrect This is simply distance $k \neq 0$. $dt = 0$. That places the space and the time aspect also at zero because $a^3 / T^2 = 0$ multiplying time with k being zero puts all there are at zero.

By reading what Kepler said correctly so many centuries ago the effort brings all the answers to so many unrealised questions…but it does not involve looking at what Newton said about what Kepler said… it is all about looking at what Kepler said. Not only did he introduce space-time a^3 / T^2 but he also placed space a^3 and time T^2 in a relevancy long before Einstein did and placed gravity in space-time a^3 / T^2 even before Newton named gravity. Kepler was the person who placed gravity as the ingredient in the universe that determines space a^3 and time T^2 and much more.   Kepler was the first one that saw that gravity comprises of two factors being **k** or linear gravity and circular gravity or

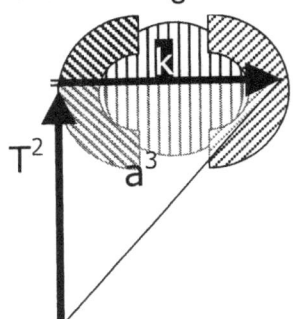

Kepler was the very first person to mathematically introduce space a^3 in a time relevancy by **k** and related directly at all time in the universe that determines space a^3 and time T^2 and much more. Kepler was the first one that saw that gravity comprises of two factors being k or linear gravity and circular gravity or T^2.

Kepler was the one that discovered space / time as
$$k=a^3/T^2$$
Kepler was the one that discovered gravity holding space-time relative
$$k = a^3/T^2$$
Kepler was the one that discovered there in the centre of a sphere is singularity as $k^0=a^3/T^2\,k$

Gravity then is $a^3 = T^2 k$ where space is equal to the time it moves in. Space has to move or space will collapse. That we see in the case of a Black Hole. Gravity then is also $T^2 = a^3 /k$ Time forming is valid since the object is unable to depart any further distance and is captured at that distance by the restricting of the space in motion. Gravity is keeping space relevant to determining the relevancy of motion in relation to the centre. Gravity then is also $k = a^3/T^2$ because gravity is space in division of time which comes down to space having a velocity that time holds space in motion too. Gravity is motion of space at a specific speed or velocity, which is moving across a distance in relation the time it takes. Again that is space-time. Gravity then is also $k^0=a^3/T^2\,k$ where the Universe is being in the centre of space from where the Universe can and is claiming and controlling space- time. Translating Kepler's mathematical expression $a^3 = T^2 k$ correctly to the verbal statement in English Kepler said that there is a space a^3 which is equal = to the motion in the time duration T^2 thereof where the motion of space takes up the time it uses to go between two specific points which holds a relation to a centre where from where there forms a straight line **k** and is located on the spot where space begins the circle therefore that centre spot has the least space. Forget for one minute what Newton said. Then take my challenge and prove what I said is not mathematically sane while you also know that $\dfrac{dJ}{dt} = 0$. I challenge you to prove how, where and why this can mathematically be sustained anywhere and under whatever mathematical law.

Place whatever you wish in a relevancy and have to outcome mathematically reach zero and see how mathematically appropriate that will be.

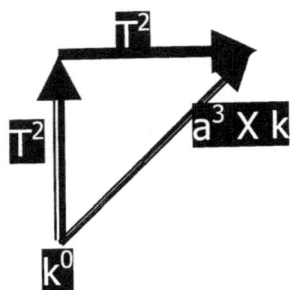

Kepler stated the very opposite of what Newton saw. Kepler had direct opposing ideas about the circle and what factor should be using the square because in Kepler's method of expression the circle indicator T^2 goes square ^{2} and the diameter indicator **k**, which replaces r remains single in the face of the volume being a cube a^3.

The two masters was using different dialects of the same mathematical language spoken by all

It is about translating mathematical equations and being correct in interpretations of mathematical expressions. Other factors are about certain mathematical deductions that were made in the past but were incorrectly presumed. A line cannot and therefore does not start with zero. Should you think such a statement is trivial then this book is even more especially for you because that changes where one presumes the Universe came from at the very beginning of the cosmic conception? Kepler answered all the questions we have…and much more!

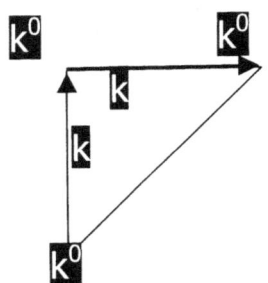

In order to have any line running there is a vertical response to the horizontal implication as much as there is a horizontal response to any vertical implication. The one response implicates the other. What ever there is there has to be another to implicate, which is. If there is **k** there has to be k^0 to begin **k** as much as there has to be k^0 to end **k**

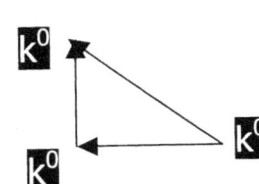

It is a mathematical fact that where there is a line of any description such a line will stand between two points holding infinity because a line can only start at infinity. That fact of a line puts a triangle in place because a triangle has the equal value of a straight line. If there is a straight line there has to be the presence of a triangle because both are equal to 180^0 and therefore are equal in all respects. Also in response there will have to be a responding line of 180^0 crossing by the value of 90^0 in relation to 180^0

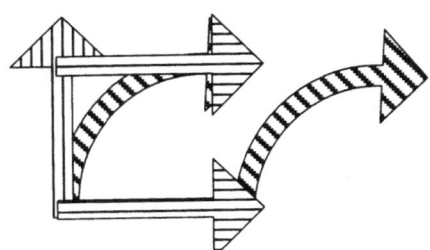

There can be no line between two zeros because only zero is between zero and another zero. All lines have to start at infinity and end at infinity and with infinity connecting lines all lines stand related to each other at the least of the value infinity. If that was not the case then the Universe had no validity. The first line 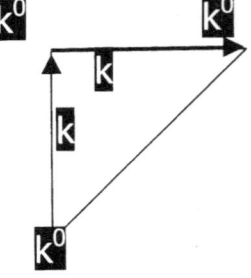 assembled a Universe of possibilities. The universe is possibilities waiting for a chance to be whatever it can hold. Where you now are could and will become the centre of a raging star at a time when life is not even history. For all we know the Sun could have been a planet teeming with life and as the Sun developed into a star life was desecrated by development. The Universe is not nothing but is every possibility waiting to happen.

The moving of one line establishes another line because by departing from singularity such motion establishes two dimensions. Motion goes by the square, which means that no matter how long the line is the line also has to be wide to be a line. Being wide in the face of being long puts the line in a square and with the line in the square the line immediately represents the other factor of time where time is in the square. Therefore drawing a line has the same implication than enticing Pythagoras because every aspect of the cosmos is the cosmos because of the law of Pythagoras. The cosmos is Pythagoras but that is not what the cosmos introduced when the cosmos introduced it to Kepler.

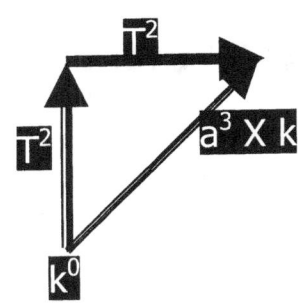

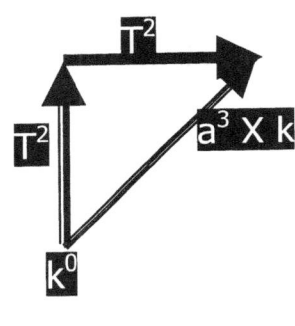

The cosmos presented gravity, the glue that keeps the cosmos together.

Pythagoras and the cosmos is a^3 X k = T^2 X T^2 and that is putting the three dimensions of space in motion that is in relation to time moving from where it was (T^2) and towards where it will be (T^2) while it is at where it is at that moment (a^3X k). However in this book we do not enter that arena because this book as a letter is dedicated to collectively indicate Newton misgivings. Yet that is the Coanda where the double square of time instates the space by three connected by line to the centre holding singularity. All of this is meaningless if it is not connected with a line to the space that is defined by time

That is the atom and that is time and that is gravity that is forming material in time.

   Planets orbiting a given centre move by a line that measures by a triangle, which forms a circle. That is the conclusion of Pythagoras but that is far better explained in another book.

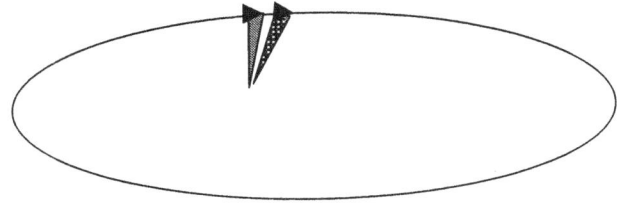

The gravity controlling stars work different from gravity working planets but gravity is the motion of space in time in relation to time. That puts all motion in context to the triangle and that puts all motion in relation to Pythagoras. The gravity, which was referred to when Kepler's studies revealed gravity was the motion of space a^3 through time T^2 in relation the a centre. The line running from the centre confirms the square of time by the distance. This is what Newton did not see.

Taking the argument back to Kepler's law,

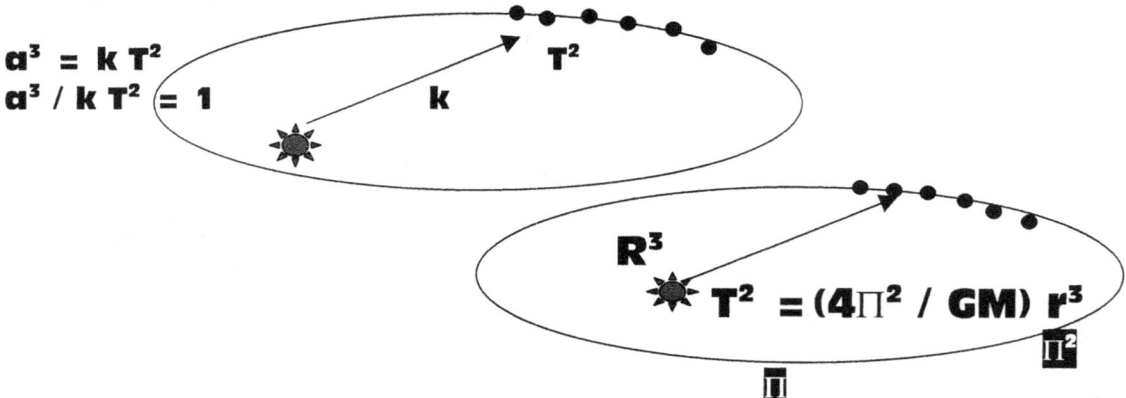

$$a^3 = k T^2$$
$$a^3 / k T^2 = 1$$

The formula that the cosmos gave Kepler reads that the space a3 is equal to the motion of the space in the time that it takes the space to move. That means the space is duplicating every second it moves and stopping will destroy all moving of the space. <u>Since Newton became an institution forming the King bee of the academic cartel world wide The Brainy Bunch had Newton's vision written in the minds of the future generations almost at gunpoint...well definitely at an academic gunpoint.</u>

$$r = dJ /$$

$$r = \frac{dJ}{dt}$$

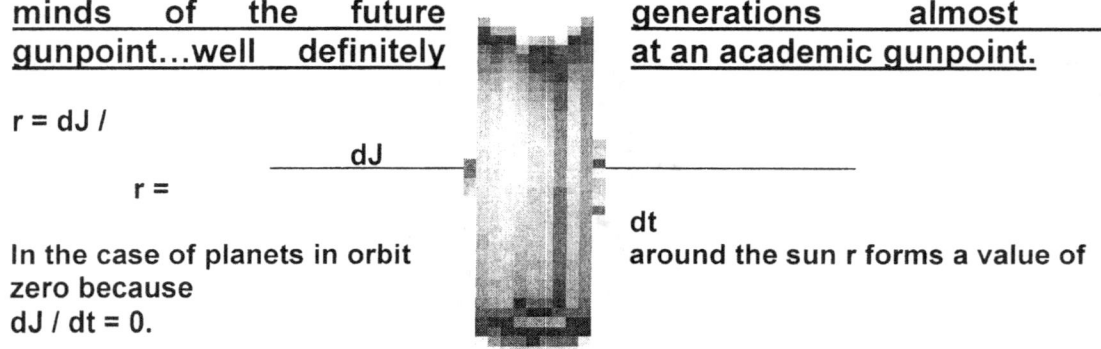

In the case of planets in orbit around the sun r forms a value of zero because dJ / dt = 0.

<u>This goes directly against what the cosmos told Kepler when the cosmos told Kepler that all space moving through time is equal to the movement of the space in the time it is moving through. That means by suggesting time may stand still such a suggestion throws the entire Universe into singularity.</u> You cannot stop everything that moves into a sudden halt because the momentum will destroy every aspect of such motion. Therefore one cannot remove the relevancy of motion by mathematically stopping time.

$$r = dJ /$$

$$r = \underline{\qquad\qquad\qquad}$$

$$dJ = 0$$
$$dt = 0$$

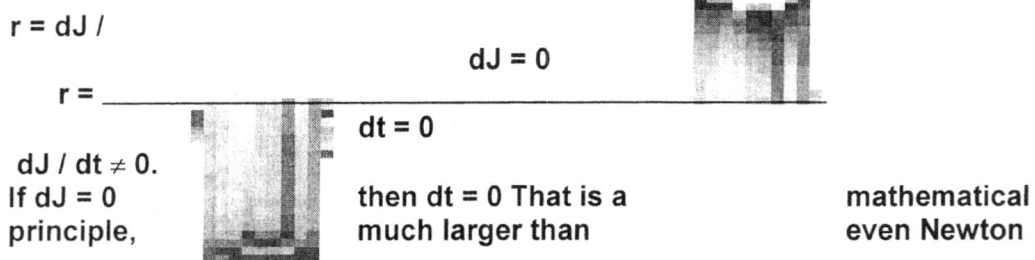

dJ / dt ≠ 0.
If dJ = 0
principle,   then dt = 0 That is a   mathematical
                     much larger than   even Newton

One cannot claim there is a wheel and then remove the spokes because according to you taste, too do not like the spokes

If we investigate the Coanda effect it is clear from what effect the Coanda principle institute that the space of the circle depends in size on the distance that **k** forms but that distance reverse with the action of the motion, which the liquid produces.

I challenge you Academics in physics too show how much the Moon , is coming closer as the product of the mass of the two objects namely the Earth and the Moon having a mass that is pulling the Moon towards the Earth, and the Earth towards the Moon bilaterally. I challenge you to show how mass influences orbiting planets by contributing or increasing / decreasing or to what degree the mass is influencing the orbit velocity of any planet in any way or manner. The orbits are in ratio. Notwithstanding whatever mass, the lot are still very alike and in fact they are the very same. Notwithstanding the enormous mass discrepancies between the planets where Jupiter is 318 times the mass of the Earth and Pluto has only 0.0025 the mass of the Earth they all orbit at a shared equality in ratio of about 300. I challenge you to mathematically prove that there is a gravitational constant. I challenge you to prove that the gravitational constant is responsible for evening out the mass differences between orbiting planets and that this is the cause why planets all orbit in a compatible fashion using the same velocity notwithstanding mass differentiations. Remember this evidence I mention opposes Newton in no uncertain way by contradicting Newton at the very heart of physics and yet with all the contradicting evidence coming from the cosmos as such Mainstream science remain of the opinion that Newton was never yet contradicted!

The force is the measure whereby the radius in the square will vanish and this argument becomes the cornerstone on which all physics hold its foundation. The gravitational constant is used as the complement of the mass in conjunction with the gravitational constant to produce such a force value, which then supposedly produces a force able to destroy the radius between the objects. Then Newton went even further at the time and gave the concept a name: he called it gravity. In the way the formula is presented the gravitational constant increases the effect of the mass of all the planets, which is forming the force of contraction between the various objects and the Sun.

The only way this can be present is when outer space is equal in relation to the Sun while to the Sun planets are just floating in outer space and is the same as outer space is. The Sun holds no planet in any different way that it holds outer space in regard to the Sun. All planets are at an equal distance where the planets are point in an area where no point has special value. There is no mass in any consideration that the Sun gives credence to the planets.

Considering that neither distance nor size proves to be any different in relation to the Sun the only conclusion that can come from this is that the part outer space must perform as a liquid or a gas of sorts giving the planets a buoyancy regard put them all in equalisation. That can only be true if there is buoyancy in outer space, which makes the one part perform as a liquid T^2k while the other part serves as space a^3. This brings the Coanda effect to mind. It clearly shows a density factors as all the evidence of mass or size disappear. The only way that can happen is when the Sun becomes the space a^3 part and outer space becomes the liquid T^2k.

By expressing the statement in a mathematical equation it reads as follows $F = G\dfrac{M \times m}{r^2}$ F=G (M X m) / r^2 The evidence that this is a fable and that mass is not pulling the objects closer was already established with Kepler introducing his work. Newton knew that the planets orbit in equilibrium by the measure if T^2/a^3 = **300** and all of that is in a ratio of about 300. That disputes Newton claim on mass forming influences because it clearly shows equanimity.

Mercury

Venus

Earth

Mars

°From what we can see the Sun holds the space a^3 equal to the time T^2 at any given distance from the Sun . The ratio of space in relation to the motion in time is at an equal footing. That means to the Sun the only thing that is out there is outer space, which is

Jupiter

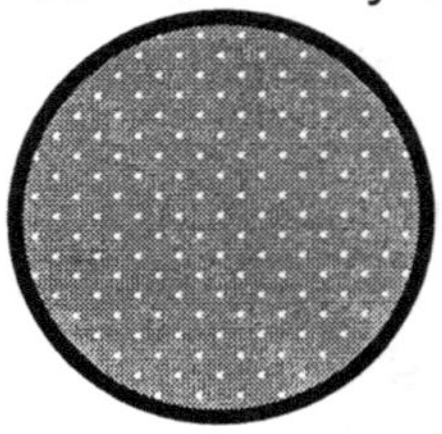

representing a unit in relation to the Sun forming

Saturn

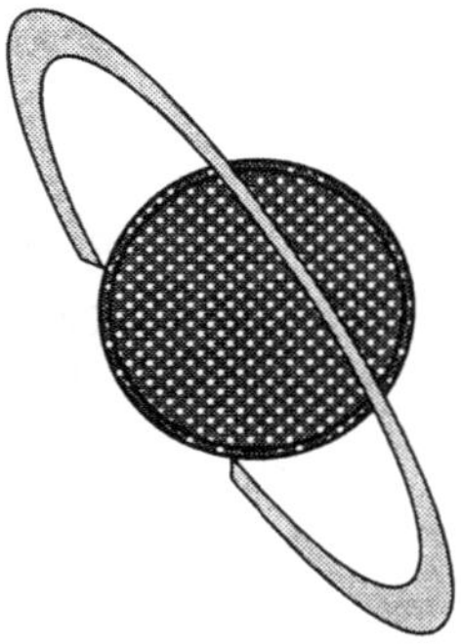

another unit. That is exactly what Kepler inferred with

Neptune

his formula $a^3 = T^2k$. From the space the Sun holds the Sun is

Uranus

equal to the motion of the space and the Sun forms the one

Pluto   ° factor of space a^3 where the other equal factor is then that, which is moving. Every planet is in ratio as far as any other planet, which puts all planets on equal footing. From the time of relation to the size in equality to the distance they are in relation to the Sun all proves a marked pattern of evenness.

Then when all the planets orbit at an even pace and all being the same ratio, where not one is found to come any closer to the Sun or to each other, your officially accepted reason for explaining that, is that the gravitational constant is evening out the mass differences and yet, according to the official formula you use, the gravitational constant is not evening out the mass but is complimenting the mass by improving the so called force. If the gravitational constant is evening out the mass the constant firstly has to divide into the mass to wither the effect that the mass produce and by nullifying the mass to the extent that Kepler's work proves, there is no mass factor present left to use. The formula then should read $F=\dfrac{M \times m}{G \times r^2}=1$ F=(M X m) / (GX r²)=1 which in normal verbal English states that the product of both masses is in equilibrium when divided by the conjunction of the gravitational constant and of the square of the distance parting the objects where the product of the gravitational constant and the square of the radius between the planets is in division with the product of the mass which then will produce the force that cancels any influence of the mass factor that both of the planet's will have on the orbit of every planet orbiting evenly around the centre structure.

That is very much not what your formula you use $F=G\dfrac{M \times m}{r^2}$ suggests. Your formula shows that the gravitational constant is helping the mass product to destroy the square of the radius. In the cosmos however and in reality that destroying of the radius is the last thing that is happening. To explain this ridiculous deception Mainstream physics go on to explain that the gravitational constant which is supposedly the gravity that is keeping outer space in a unit, where that gravitational constant nullifies the mass influence on the force. Newton said mass is all-important but this explanation now suggests that the mass is nullified by the intervention of the gravitational constant.

This is a statement that is not in line with the mathematical equations Mainstream science present to serve as truth. Making such claims that is so far away from the actual mathematical statements, which came into place serving as yet another attempt to cover up Newton fraud with more deceit and becomes compatible with mafia like behaviour. This criminality is the behaviour of all physics paternity in control of all physics worldwide! This is the group of men that is in control of the thoughts in the minds of the brightest developing brains in the world. They harvest the best intellectual group of all students this Earth can offer! To the students questioning or rejecting the Newton claims is not an option.

The students accept what is taught or face examination failure and total rejection of the institute where they are educated. If that which is presented that the statement would suggest about their explaining how there is such planet mass apathy that will not influence orbit details of planets making them go faster or slower it is based on yet another Newtonian lie. If the gravitational constant were evening out the mass then it would have the gravitational constant in multiplication with the square of the radius in dividing the product of the mass. This will then be evening out the product that the multiplying of the mass put in place. If that was said and translated in mathematical terms, then it would mathematically equate to showing where the force will be in equilibrium because the radius is supported by the addition of the gravitational constant which increases the radius in the square and the total of that is in equilibrium with the product of both object's mass. If that was true then the gravitational constant must be placed where it increases the radius in every event and it would be stating that $(G \times r^2) = (M \times m)$ (G X r²) = (M X m) is applying in all the cases so that the mass has no longer an influence on the outcome of the equation because out there in the real cosmos Kepler proved that the mass has no implication on the orbit of any planet $T^2 / a^3 = 300$.

Kepler never even thought of a mass of any description and I can assure you that his not mentioning any mass was not the result of his retarded mind or not because he was mentally inferior to one Isaac Newton. With the corrected implementation of Kepler I can and I do support such statement of equality but then in that event the equation then should read that the gravitational constant is complimenting and supporting the square of the radius and mathematically such a statement is $(G \times r^2) = (M \times m)$ (G X r²) = (M X m). That removes mass as a cosmic factor. There is nothing wrong with this supposition but the way it is presented. The way it is presented dies Newton's claim on mass, which then should run through to the abolishing of his formula and the discarding of his theory. It shows that firstly k is not insignificant as zero but presents much significant proof that the Sun places all planets on an equal footing of 10 / 7. I shall explain that statement in a short while. Every distance of every planet in relation to the Sun is equal and every orbit of every planet is in harmony. Mass does not implicate or profess to have any influence which is strongly suggesting the characteristics of a density factor as particles would have when they display buoyancy characteristics in water.

You know very well that in the planet orbit ratio the space factor (space a^3) divided by time T^2 leaves a space (a^3) – time (T^2) displacement factor ratio between space and time of round about three hundred in all cases, $\left(\dfrac{T^2}{a^3} = \pm 300 \right)$ (T²/a³ = ±

300 , which disqualifies mass discrepancies altogether. With that the cosmos put Newton in dispute! So why then bring in the use of mass in any event if it does not change the result but has only one purpose and that is to cover Newtonian's misconception and to hide Newton's fraud? However, the formula used by mainstream science clearly shows that the gravitational constant is in effect increasing the mass by multiplication of the mass. The formula undeniably put the gravitational constant in a multiplying position with that which the mass of both structures hold.

$F = G\dfrac{M \times m}{r^2}$ F= G (M X m) / r² which then in ratio is decreasing the square of the radius. That is pure rubbish and I challenge you to prove me wrong by proving your mathematical interpretation. Then all the while, while promoting such rubbish you still maintain that Newton has never been proven incorrect. When Newton is taken into the cosmos Newton physics goes bizarre. There is no evidence of mass influencing the motion of planets in any manner. On the contrary all evidence show a clear equilibrium present in the relations of the Sun and planets.

Newton is not incorrect as it is used in normal applied physics but in cosmology there are no grounds in supporting of mass or gravity in the manner Newton saw or in the way Newton suggested mass to apply. On Earth and on the ground Newton's work is impeccably correct and that I admit without any reservations on any aspect of physics in whatever manner. However taking Newton's ideas into space becomes fraud. Mass

only serves as a human concept to allocate differentiation when required when gravity is obstructed and in the cosmos gravity is not obstructed.

In the Coanda effect we find that when a liquid such as water runs down the side of a round shaped surface, which is a glass amongst others the water follow the contour of the glass. Where water normally is used, the water would fall directly to the ground as gravity effect the water and only allow the water to run in a straight line towards the Earth. When the water is poured over the round surface that the glass provides, the water follows the contour of the roundness of the glass.

This part Newtonian physics admit and they put two forces to explain the action but fail to give any further explaining as to why the forces come about or what would initiate the forces to counteract and disturb the gravity pull that should have the water running around the surface of the glass at the top end and then as soon as the water would find the opportunity, it will get a gravity pull and run straight down ignoring the interrupting of the glass contour since it then no longer blocks the path of the running water.

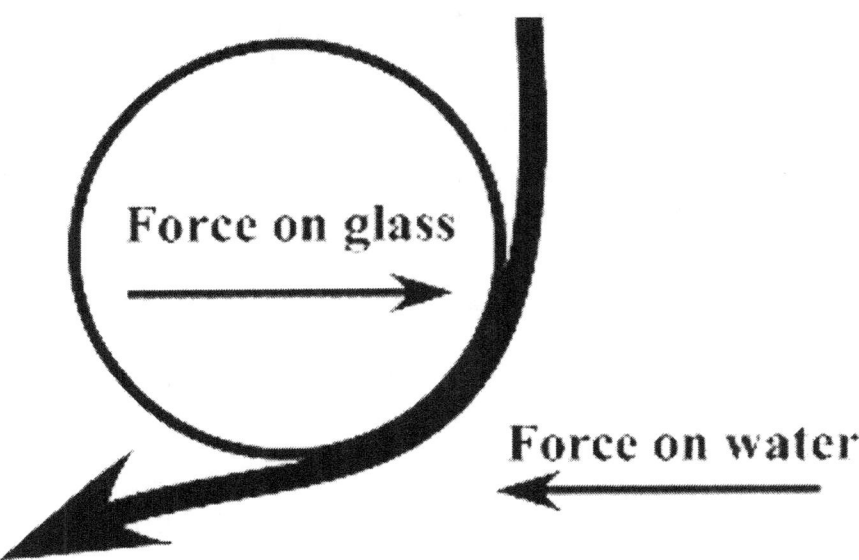

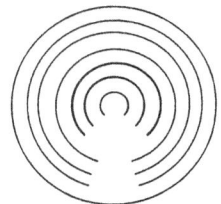 If we return to Kepler's formula we find the answer, but the answer again totally contradicts what Newton claimed about a circle having no effect on motion. According to Newton the rotation of an object is not time related and has no influence on the surroundings of the object.

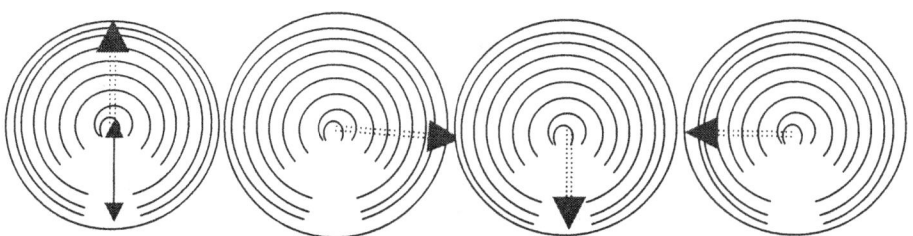 When a circle of an object rotating produces the motion there are four points on

each side that plays a part. Each point produces a direct relation as to the coming towards / the going away, the passing the front and the crossing the back marker.

The Coanda effect proves otherwise when studied closer.

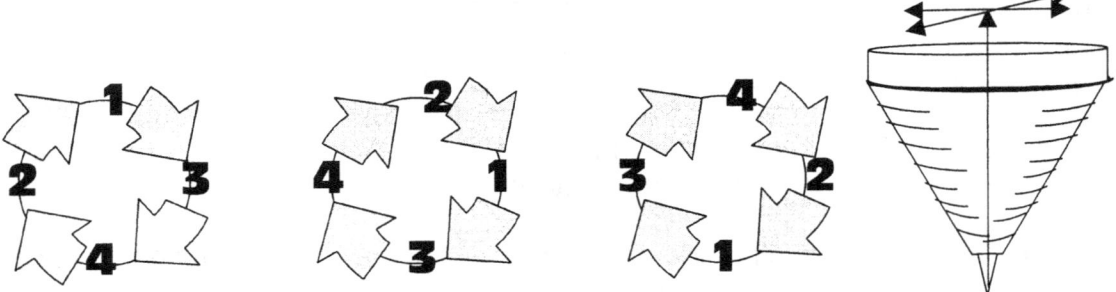

We can see that there are some coherency between the time and the rotation of objects.

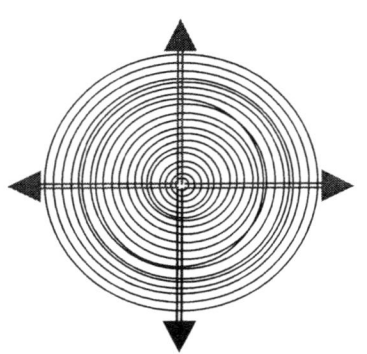

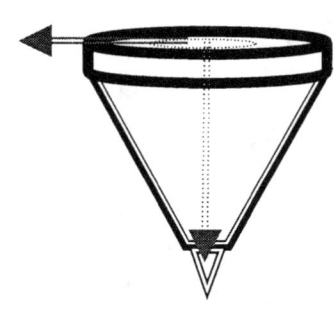

That coherency we think of as time. The rotation of the object finds new alliances with the surrounding and if that was not the case then no fluid drive of objects was possible. You may discard this remark as not applying to the cosmos, but later in the book I am about to prove that outer space is lesser-concentrated form of heat and the atmosphere of the Earth is just a higher concentration of heat.

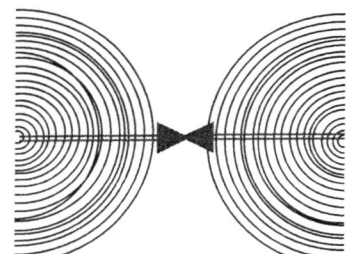

Please allow me to explain the relation between time and rotation by a simple experiment. Let's take a top and think of the same top as having two ends in opposition to each other.

When we take an object in rotation as having one side running left which it has, and another side running right, which it also naturally has. The side heading left is totally opposing the side running right because of conflicting directions in the spin.

If we put the same objects two sides in position, as they are when they rotate we find the tow sides is contradicting each other by not being the same as one another. The one is a clone of the other but the clone is an opposing duplication of the other. This is part of the characteristics of all rotation and in this rotation it entirely depends on a liquid state that limits the space where the space includes the liquid $a^3 = T^2k$. **Gravity is merely the motion of particles in space.**

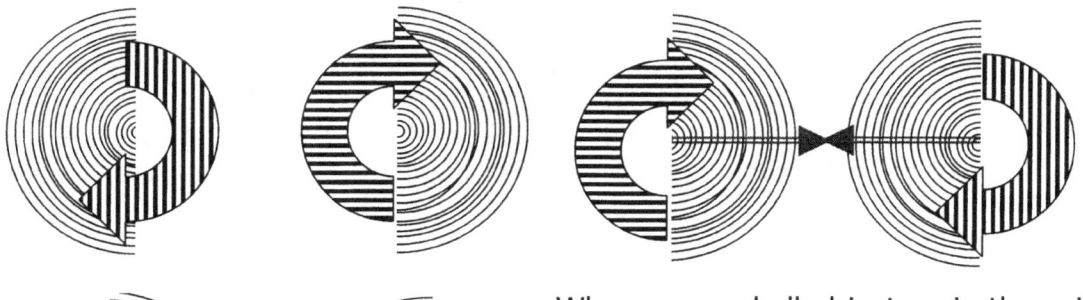

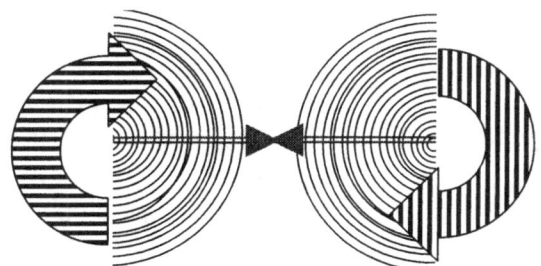

When any and all objects spin the spin is in opposing relevancy notwithstanding being part of the same object or being the in the location where the next allocated position is precisely where the opposing direction at that point forms the opposite of what it will be as the

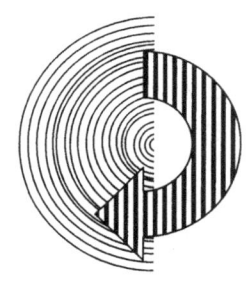

next part of the same but ongoing rotation produces the rotation graph. That is because no object can be twice in the same Universe and every object is in two parts of the same Universe in different locations. No object share Universes but forms duplication by motion of the same Universe on two sides of the same

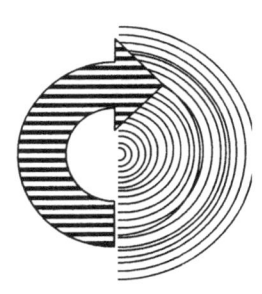

and one unifying Universe.

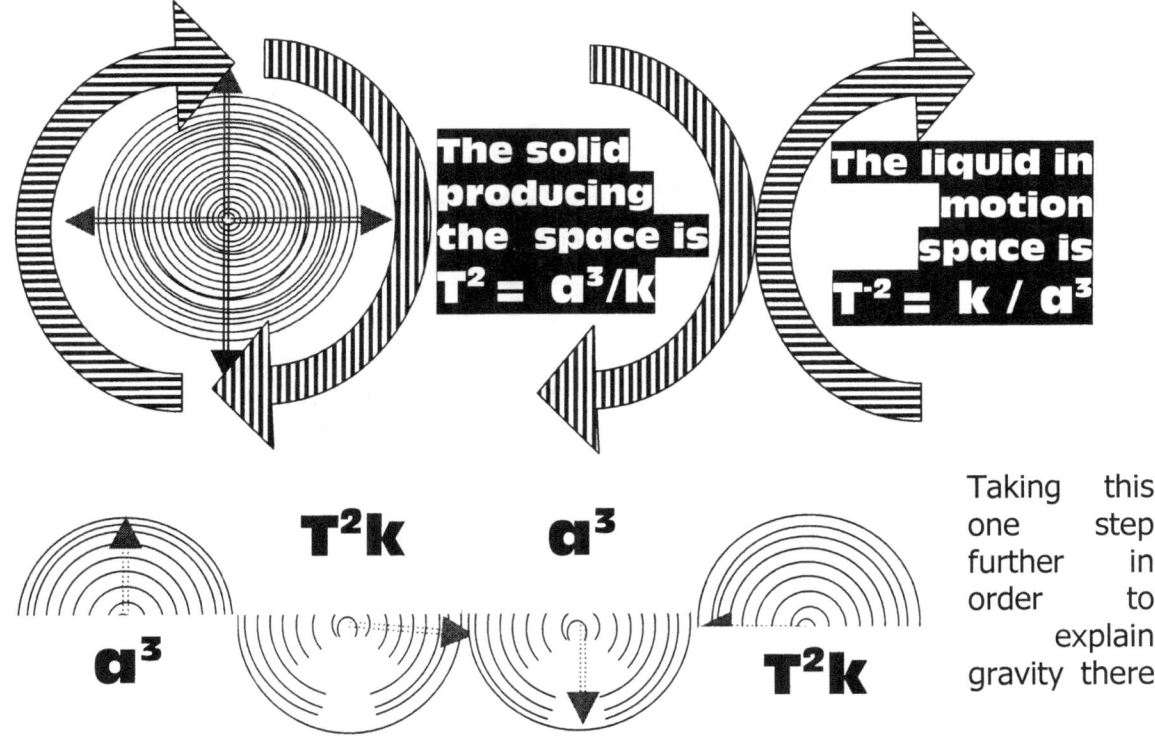

The solid producing the space is $T^2 = a^3/k$

The liquid in motion space is $T^{-2} = k / a^3$

T^2k a^3

a^3 T^2k

Taking this one step further in order to explain gravity there are two factors found in sharing the same unit. The one factor is the liquid to motion in relation to the other substance forming in the relation the part of the solid or the stable not moving part.

Taking a considered look at the spin procedure we find not forces but a rotation ratio at work which produces the opposing of two sides in exact opposition of each other where the sides in opposition will cyclic interchange and become the other party as the direction of the spinning objects change. In the spin we find a divide running as a line in the centre of the spin and is the point on which the ration is grounded. From that line of division two sides oppose each other all the time. On the one side we find **k** being the distance having a relation with time in division of space and on the other side we find space being in division with time.

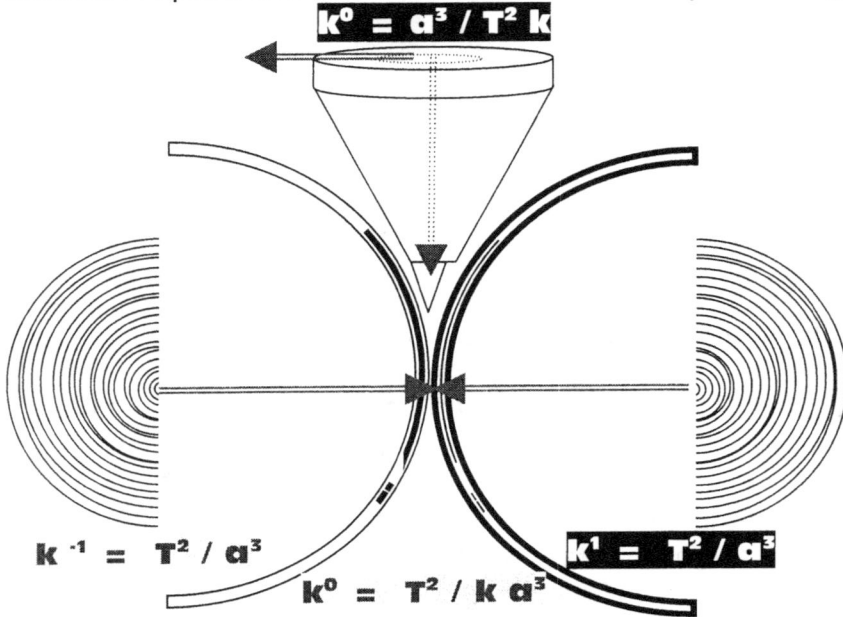

$$k^0 = a^3 / T^2 k$$

$$k^{-1} = T^2 / a^3$$

$$k^1 = T^2 / a^3$$

$$k^0 = T^2 / k\, a^3$$

If we take Kepler's formula on individual merit the cosmos told Kepler that space is equal to time by motion form a specific centre. It is well known as   , but at the same time very badly understood by the science paternity.

Reading the mathematical statement correctly we find that there are two opposing parts of the same unit contradicting each other by being on two sides in a total equal balance. On the one side is space a^3, **which is equal** to the motion thereof   .

$$T^{-2} = k / a^3$$

> **The sun is in a position of contracting by preserving the limits of the borders of the duplication limits.**

$$T^2 = a^3/k$$

> **The comet is in a position of expanding by setting the duplication limits.**

It is documented at this point that any object spinning has two sides directionally opposing one another and we have the cosmos telling Kepler that space divides in opposing sides as space a^3 and T^2k motion which forms the time aspect. We even use that side as time by indicating a cyclic year from the completion of one rotation and monthly and seasonal positional points indicating the different periods.

In this we have the one side holding   related to space   in division of time being the relevancy of $k = a^3/T^2$ There then is an opposing relevancy to this because of the opposing we find in the direction of the same object and being in opposition it then must be $k^{-1} = T^2/a^3$. However with that being the case we then find objects are in opposing relations to time as well with time going positive and negative or going bigger in relevancy or smaller in relevancy. $T^{-2} = k/a^3$ and $T^2 = a^3/k.$ This proves that the object in rotation around a specific centre will reduce the relevancy or distance by reducing or increasing the time and space relation. That is the reason why comets do not slam into the Sun by force instigated on grounds of mass as Newton would suggest but rather cyclically circling the Sun in rotation by going away further and afterwards coming back to come much closer. Guarding this very jealously is the balance of the line being $k^0 = a^3/T^2 k$ and every time the rotating object crosses this dividing line the relevancy changes to the opposite because the direction changes to the opposite.

This we find as a principle in the orbit patter of the Sun and all planets where the relevancy shifts from the one onto the other. But mass has no role to play in this because the structures do not have mass, they have gravity.

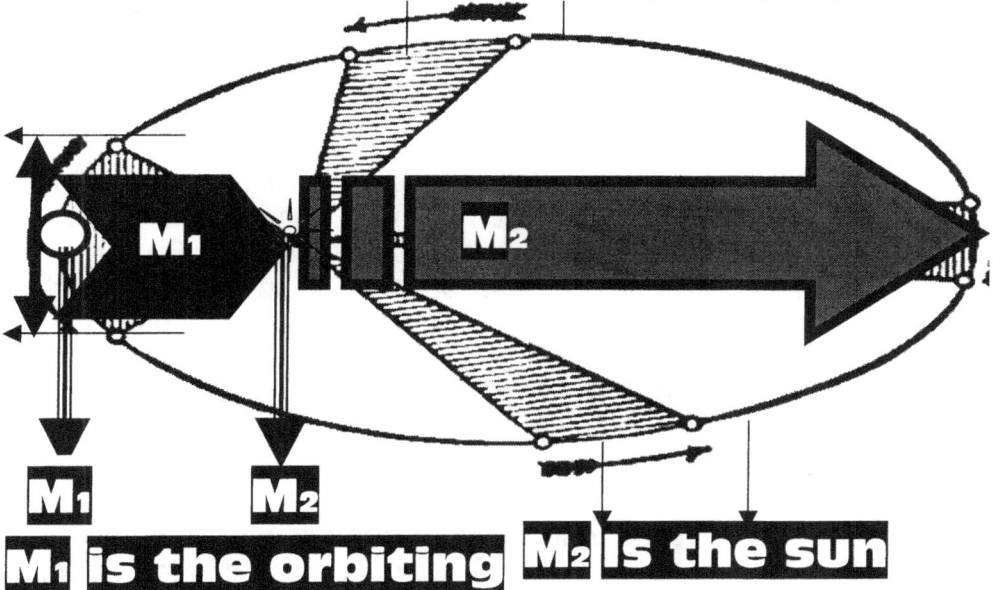

M_1 is the orbiting M_2 Is the sun

What Kepler said in hid formula $\underline{a^3 = T^2 k}$ is directly opposite to what Newton suggested with $\dfrac{dJ}{dt} = 0$. This means that the spin is uninfluenced by that which is forming the surrounding of that which is spinning.

Newton stated that $\dfrac{M_1 \times M_2}{r^2} G$ = Force and the mass in each case remains the same as well as the gravitational constant, then why would the radius change every time of each year and not draw closer (or further).

F = G(M . m) /r^2 where:

G = the gravitational constant,

M = the mass of the body,

M = the mass of the lesser body

r^2 = the radius between the two bodies.

Newton stated that $\dfrac{M_1 \times M_2}{r^2} G$ = Force, because of the fact that:

1. The value of M_1 in both cases is equal because that is the mass of the earth.

2. The value of r^2 is equal because the two objects have been dropped at an equal distance.

3. The value of M_2 is still a mass filled with gravitons and therefore has to effect the relation in the same way as both objects consist of different compositions of materials that are used to manufacture the different objects.

4. That means Force one has to have a different value to that of force two (F_1 ≠ F_2). In contrast time duration P_1 = P_2 and this cannot apply in the case of gravity because the greater the force are through more mass, the bigger the impact has to be on the time duration.

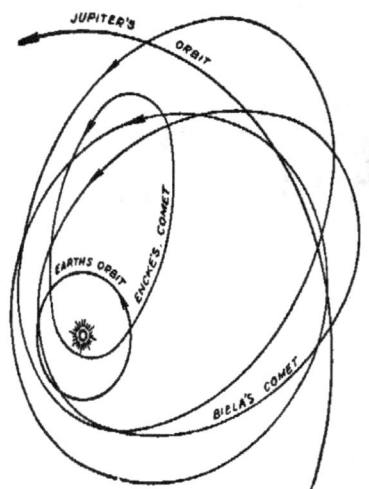

I think in all fairness it now is rather obvious even to children that orbiting comets do not destruct as they collide with the Sun by the pulling force of gravity. Except in extra ordinary cases where some interference changed the orbit cycle of comets we find that comets passes by the Sun unscathed to go into another cycle.

In such manner Halley was able to predict the cyclic return of the comet named after him.
As far as cosmology physics is concerned, little can be further from correct than the idea that just it is the mass bringing destruction by the pulling power which it unleashes instigated as a force by the mass in relation to the product that the mass on both ends exchange for becoming the force of gravity. The comet misses the Sun every time without failing. That means if it is occurring in repeat there is a balance in place.

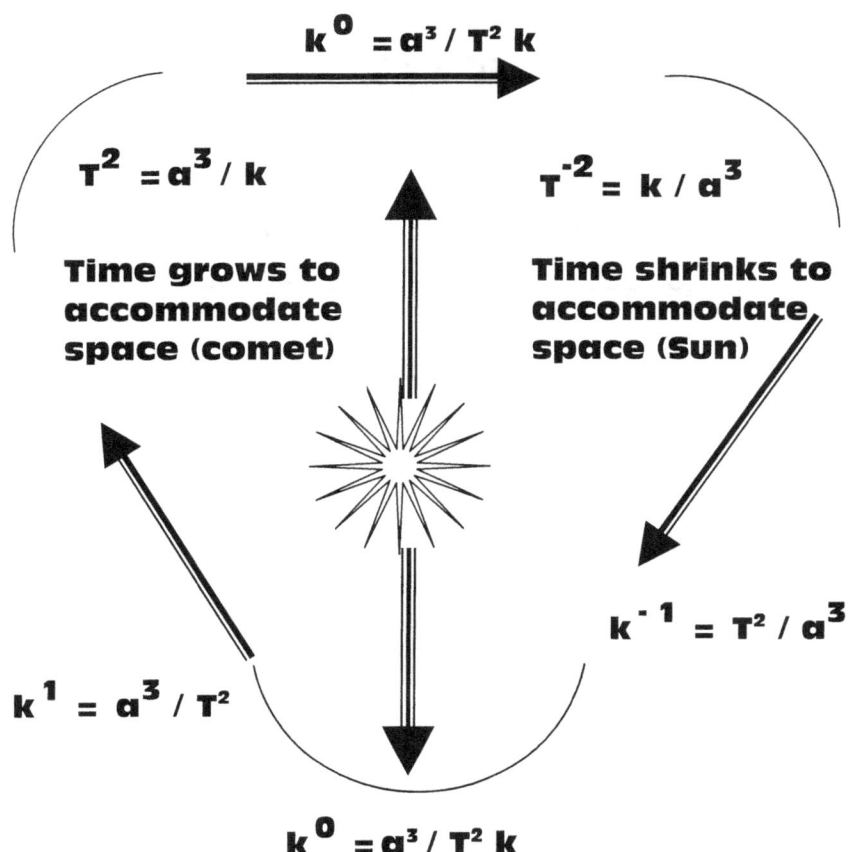

$$k^0 = a^3 / T^2 k$$

$$T^2 = a^3 / k$$

$$T^{-2} = k / a^3$$

Time grows to accommodate space (comet)

Time shrinks to accommodate space (Sun)

$$k^{-1} = T^2 / a^3$$

$$k^1 = a^3 / T^2$$

$$k^0 = a^3 / T^2 k$$

On the one side we find that the structure controlling the contracting aspect holds dominance in providing the pivotal circle relevance $k^0 = a^3/T^2 k$, then this switches around where the other plays the key factor providing the gravity as $k^0 = a^3/T^2 k$. Seen from the Sun the one side of the rotation would be time in progress of becoming more by applying $T^{-2} = k/a^3$ and in this the other orbiting structure would provide the relevance. The factor then would be in relation $k^{-1} = T^2/a^3$. As the orbiting object crosses the centre divide the complete opposite comes into action where the time factor once more grows and $T^2 = a^3/k$ applies. However at that point from the other side we find that the factor also grows in $k = a^3/T^2$ Since the factors inter act by taking on different roles through the orbit we do not have collisions as Newton suggested but controlled orbit interacting as Kepler suggested.

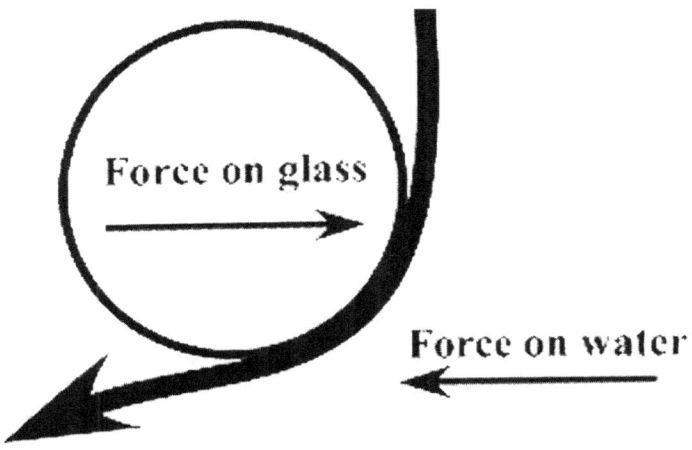

Force on glass

Force on water

By applying the Coanda principle there are two relevancies at work and the motion as well as the direction time takes the motion will place either one of the two relevancies in prominence at any particular given time. There is absolutely no chance that both the factors may apply simultaneously and either of the relevancies has an opportunity to dominate but with motion applying stronger, there is a situation that presents the contracting to overshadow the expanding but that I explain in a later stage of this letter.

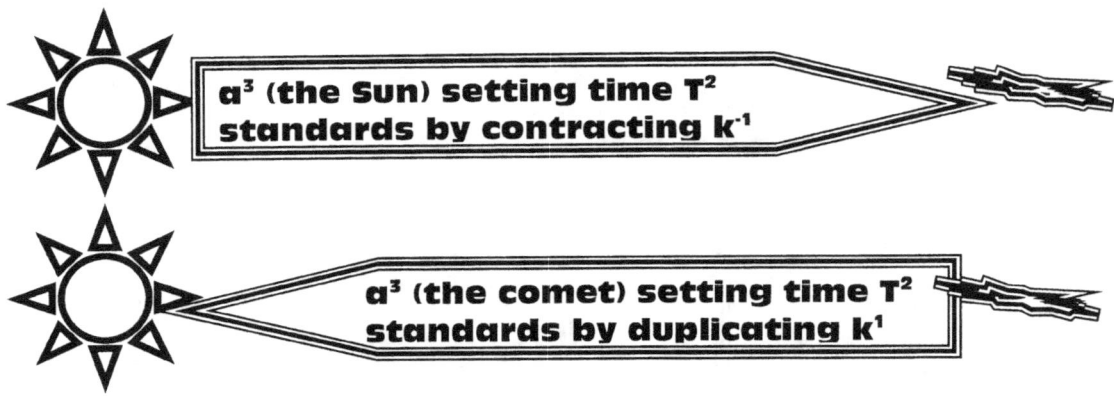

From this comes the working principle we at present gave the name to as the Coanda effect. In the Coanda effect we have full gravity working as a compliment when the liquid provide motion to implement the gravity action.

It would be far more prudent if we left Newtonian forces (and this I say in spite of notwithstanding all the rejection I thus far encountered from the celebrated mainstream physics paternity by my criticizing Newton on cosmology) and witchcraft to the Middle Ages from where it comes and see what gravity is by standards applying in the modern age.

$$a^3 \qquad T^2k$$

$$T^2 = a^3/k \qquad T^{-2} = k/a^3$$

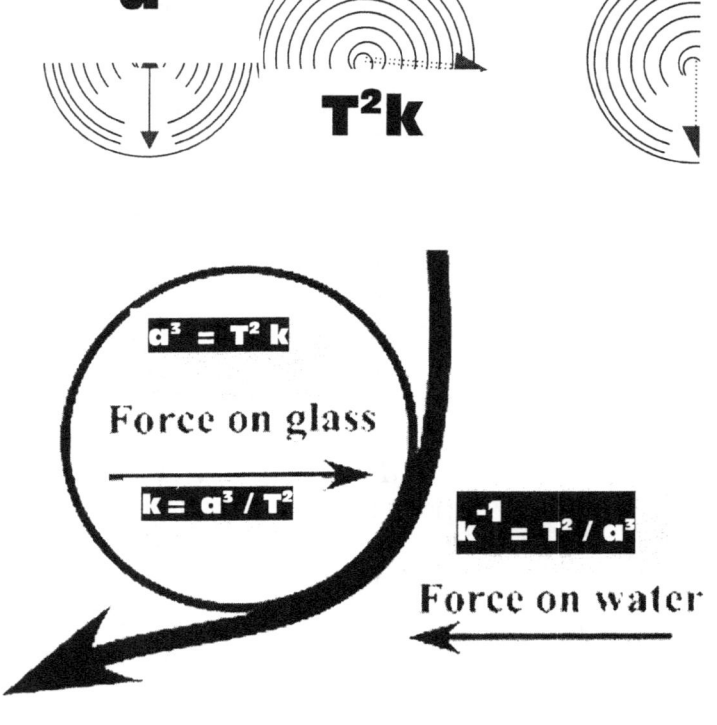

$$a^3 = T^2 k$$

Force on glass

$$k = a^3 / T^2$$

$$k^{-1} = T^2 / a^3$$

Force on water

In the Coanda effect we have the same opposition forming the same unit but two different substances taking on roles in the opposing sides and from this springs gravity.

Going back to the spin action we saw that one side is opposing the other side while in rotation and while forming a single unit. This we take back to gravity and in that we find what forms the motion we find in gravity.

As I explained previously there ate two action forming part of four factors bringing about a rotation action.

While on the one side of the divide $k^0 = a^3/T^2 k$ we have $k = a^3/T$ where the time factor provide space with a limit there are the other side where the time factor being on the other side of the divide $k^0 = a^3/T^2 k$ extends the limit of the space by the liquid to provide the limit to space at $k^{-1} = T^2/a^3$.

In all of this the lot that I mentioned prove Kepler different to what Newton portrayed Kepler to be and also it proves Kepler correct in gravity whereas Newton is absolutely incorrect in his surmise on cosmic gravity. The liquid is part of the spinning circle but is also part of the other side of the divide $k^0 = a^3/T^2 k$ and on that side the flow or motion $T^2 k$ establishes the limit to space while the solid forms the space.

On the other hand the space a^3 finds an extending of k by the liquid that forms a part of the unit as the motion part in Kepler's formula $a^3 = T^2 k$

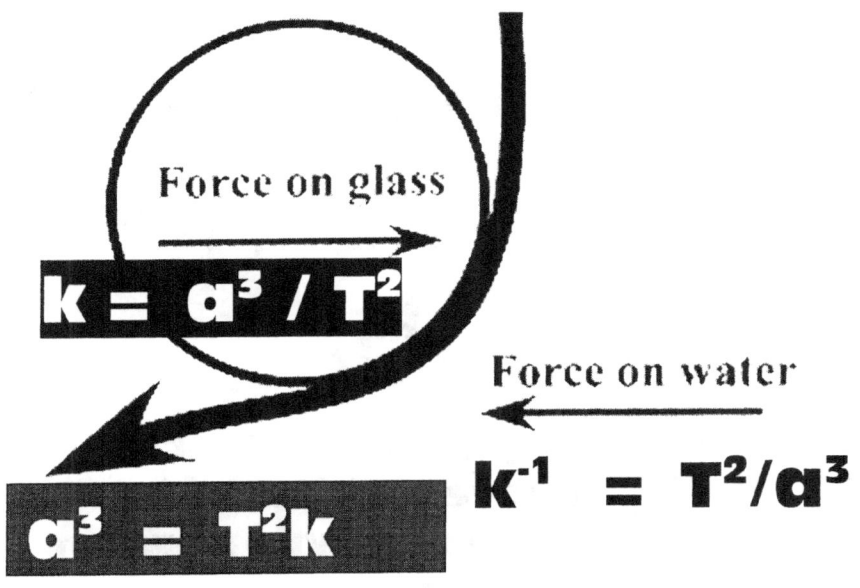

Force on glass

$$k = a^3 / T^2$$

$$a^3 = T^2 k$$

Force on water

$$k^{-1} = T^2/a^3$$

The **space a^3** of the unit is **defined k** by the **flow T^2** of the liquid

Because the liquid seems to run in the opposing direction while it is running in the same unit and therefore in the same direction it is in influence by the same relevancy factor k but in the other direction k^{-1}.

Although both the solid and the liquid forms part of the same circle and is in one unit the liquid as one part serves one side of the Universe or the divide while the solid serves another part of the same unit but on the other side of the Universe or the divide.

That is gravity. The rotation forms the gravity by forming a divide that initiates the start of a Universe. There is no possibility that time can ever stand still or that motion of rotation has no influence on the gravity bonding that forms because the gravity is the result of the divide that the rotation brings on.

This goes directly against what the cosmos told Kepler when the cosmos told Kepler that all space moving through time is equal to the movement of the space in the time it is moving through. That means by suggesting time may stand still such a suggestion throws the entire Universe into singularity. You cannot stop everything that moves into a sudden halt because the momentum will destroy every aspect of such motion. Therefore one cannot remove the relevancy of motion by mathematically stopping time.

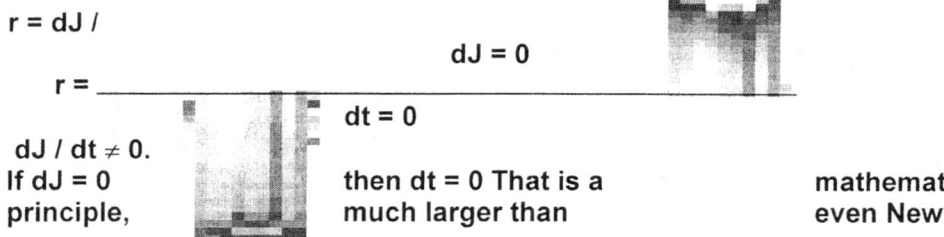

$r = dJ /$

$dJ = 0$

$r = $ _____

$dt = 0$

$dJ / dt \neq 0.$
If $dJ = 0$ then dt = 0 That is a   mathematical
principle,   much larger than   even Newton

The space a^3 is equal = to the motion T^2k

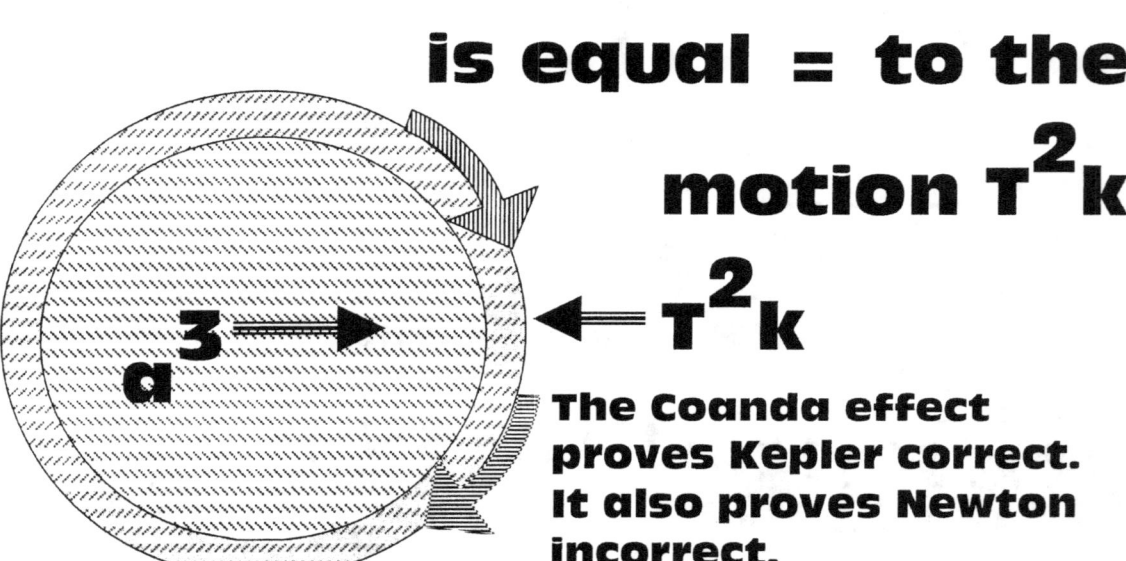

$a^3 \Longrightarrow$ $\Longleftarrow T^2k$

The Coanda effect proves Kepler correct. It also proves Newton incorrect.

Let us argue what Kepler said mathematically and without Newton trying to convince every one about his discovery of mass.

___

Sir / Madam, the fact that Newton present k as zero is, is quite impossible because of what Kepler said. Kepler said $a^3 = T^2 k$.

The Coanda effect is gravity and moreover it is the movement of a liquid in relation to a solid that forms gravity.

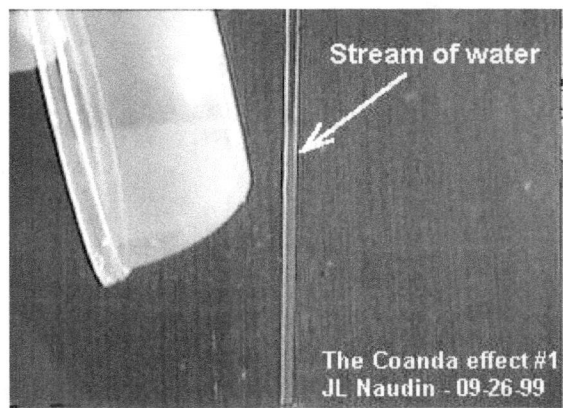
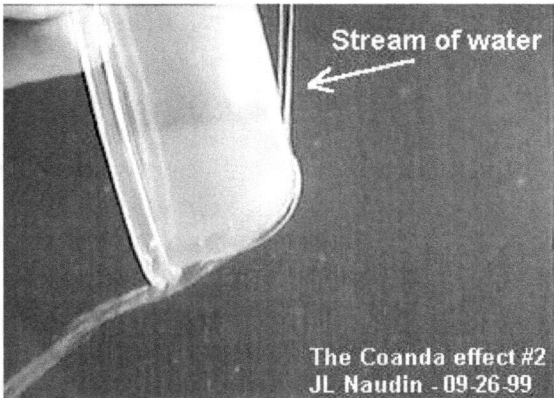

In the phenomena we named the Coanda effect we find that water runs down in a straight line seemingly following a direct route to the centre of the Earth or to where it will contact the soil.

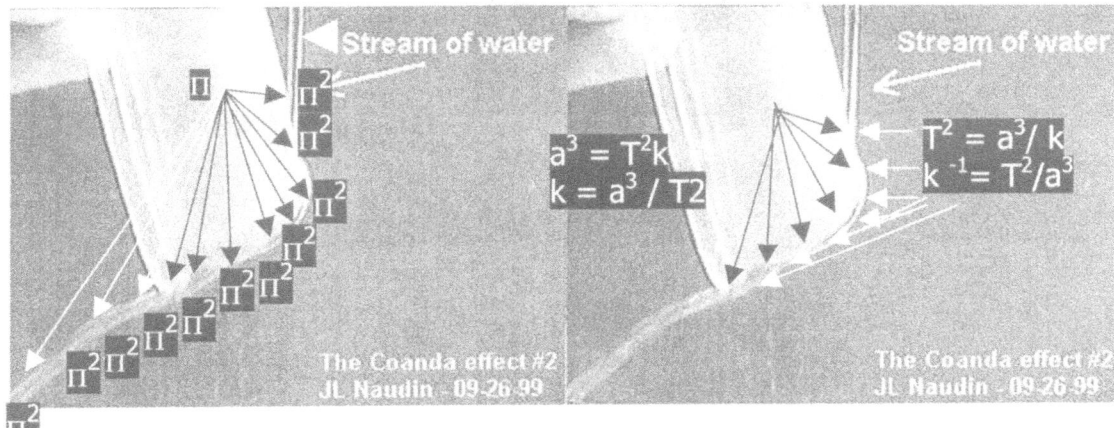

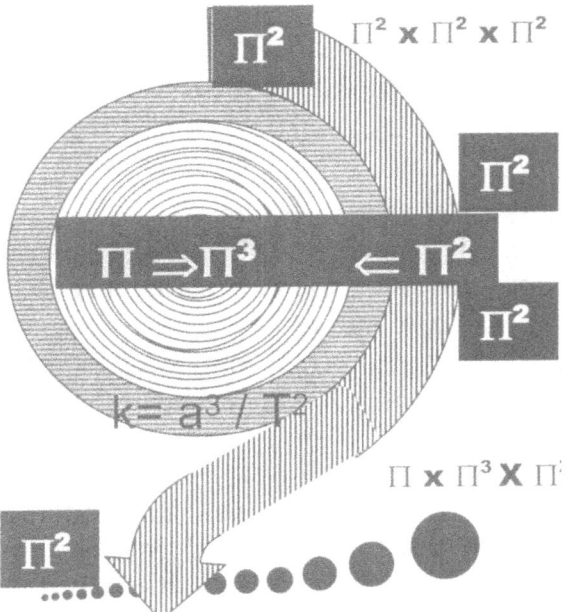

When motion produced, space and motion came because of singularity that is unable to produce any sort of motion of any nature whatsoever motion appointed three positions holding singularity that formed a relevancy with other positions that produced the space in relevance to singularity as a factor.

The reduction of space will bring about heat. Injecting fuel in a place where such reducing of space already exist it increases the heat level further and the fuel will "spontaneously" ignite. The igniting creates a rise in the heat level to a point where such a level that one can only find that level in the stars. Fuel will establish conditions, that is in accordance with the laws of cosmology such heat will only apply to stars where such heat levels will indicate the enormity of the gravity present that is generating the massive gravity accumulated in the absence of space. Gravity is the concentration of heat in the utmost reducing and concentration of space thus a star is born in the gravity on Earth. In this example, **k** increases as thrust pushing space **a**^{3}, which the **k** creates to a new **T**^{2}. That started the Coanda effect but that also started the Universe before the Universe and space – time. The Universe then was spinning at speeds faster that the speed of light and everything in the Universe only had form. Later on with

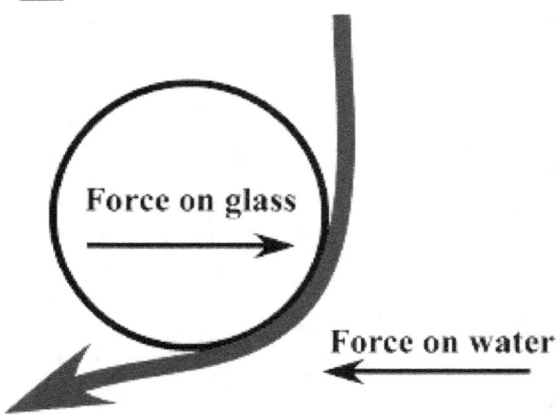

The Coanda effect is creating gravity. It is not replacing gravity. It is not recreating gravity. It is not substituting gravity. What the Coanda effect is is what gravity is. The Coanda effect is gravity.

the event of the Big Bang the Universe, received dimensions but form remains the template of dimensions and even today that is how we interpret timesaving three positions to a centre. We still have the evidence, which is even more obvious than most other

certainties we uncover in the Universe.

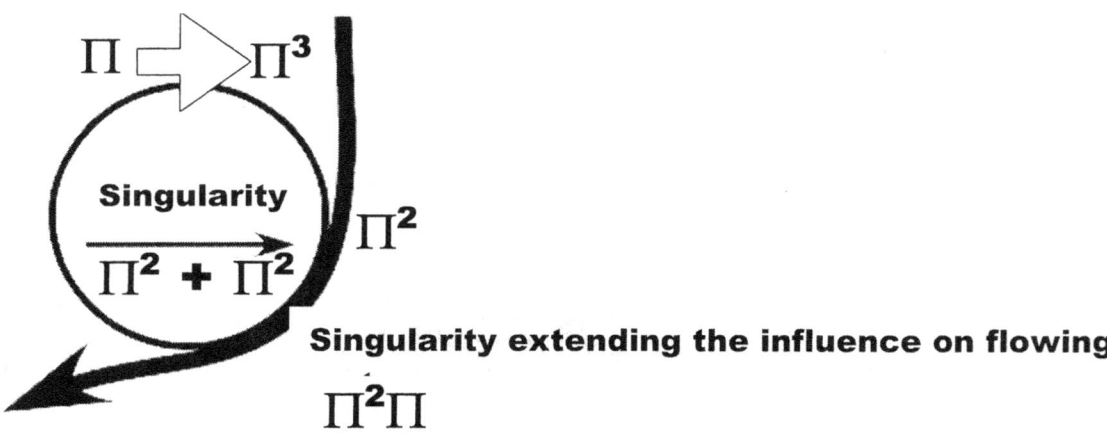

Singularity extending the influence on flowing

The Coanda effect is also the perfect example of the curvature of space-time brought about by the extending of singularity influencing due to the shape that

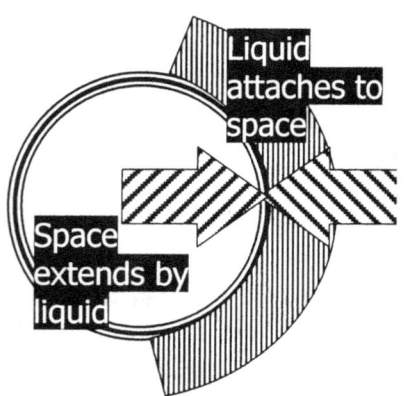

imitates or duplicates the value of singularity and again conforms Π. By establishing a new value of singularity as Π, singularity can once again take control and establish a new Π^2 as gravity in the new Π^3 forming space.

Looking at the Coanda principle we find two distinctly separate parts forming one Unit. Looking at Kepler's formula there too we find two distinct sides forming the same unit. In the Coanda principal there is a space   forming a basis to which a moving   attach. The   extends the space limit while the liquid provides the motion, which the space uses to become larger. The space   is one side with the motion of the liquid being the other side. As I shall indicate later on is that Kepler did not introduce a mathematical calculating measuring formula as Newton would have suggested, but it is a cosmic principle on which the entirety of the Universe is built. It is gravity by all measures there can be. It is does not mean that the one side is in precise measure the other side but it states a principle on which all aspects of the cosmos rests.

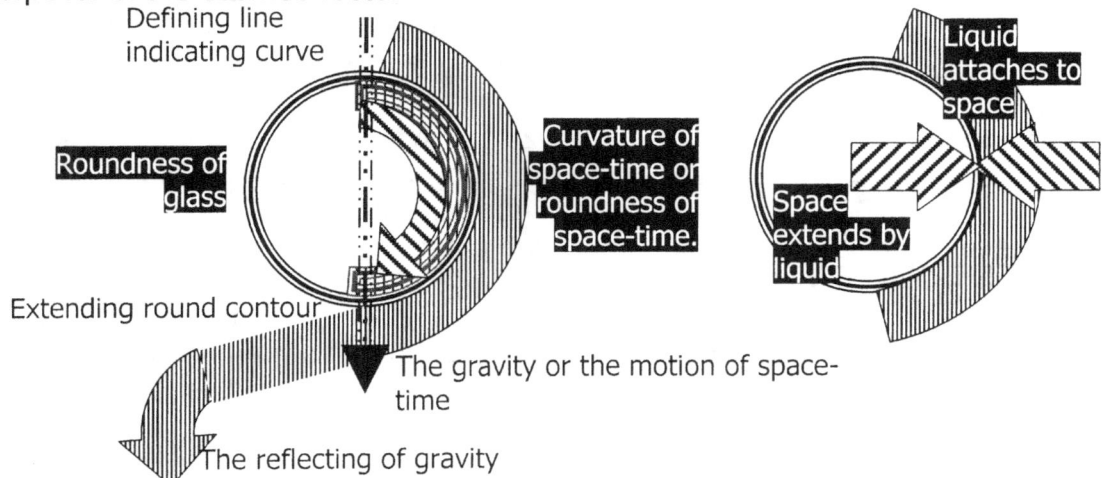

Defining line indicating curve

Roundness of glass

Curvature of space-time or roundness of space-time.

Liquid attaches to space

Space extends by liquid

Extending round contour

The gravity or the motion of space-time

The reflecting of gravity

$$k= a^3 / T^2$$

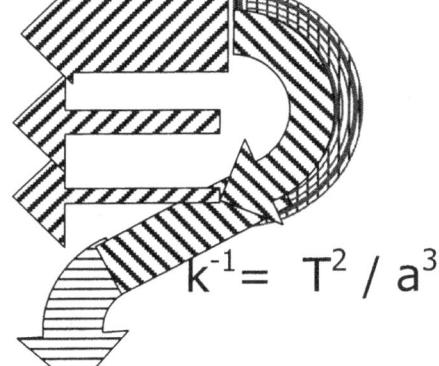

$$k^{-1}= T^2 / a^3$$

We find a line running through the middle of the round glass, which is dissecting the two parts in two directions. The line is where there is space on one side of the divide and a substance providing motion on the other side of the divide. On the one side there is an attraction where the liquid is controlling the space border and on the other side there is where the space is identifying the position the liquid has as the liquid then becomes a part of the solid space.

Looking closely at the principle there are two principles where the one involve a contracting to the centre and the other form an

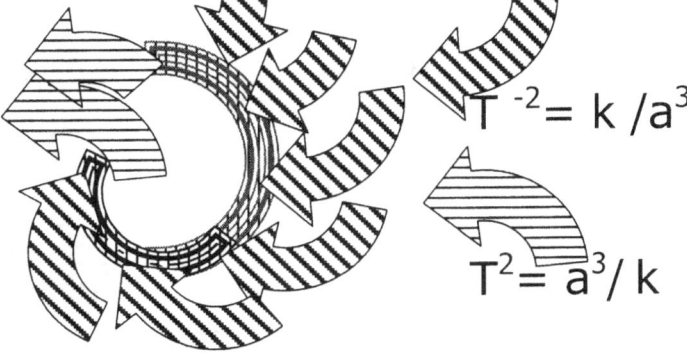

$$T^{-2}= k /a^3$$

$$T^2= a^3/ k$$

expanding or a rejecting of the centre. There is one side that openly and exclusively favours contracting while there is the other side that provides expansion the possibilities to be.

In the two contradictions there normally is a fair balance but as life interfere with physics we find a shift in favour of contraction as the motion life provides throws the balance towards the motion part.

The contracting and expanding balance is totally in line with the motion of the liquid that provides the balance. In normal gravity motion the balance between contracting and expanding is divided in the line that puts division in place. As motion excels it favours the contracting to the decline in the expanding.

However when saying this we have to realize the symptoms we find is in relation to the Earth providing the gravity conditions and by excelling the normal motion the conditions then would amplify the contraction.

The one part is forming a motion and the motion is attributed to the fact that the part is a liquid. By improving the motion the contraction benefits but that is because the motion that favours contraction turns the balance in the favour of motion. The higher the motion the more the motion will overburden the expanding and reduce the solid factor. With us completely engulfed and in the control of the atmosphere we experience a complete contracting with no possible expanding.

The slightest inclination of the solid object also spinning that will favour the contraction and reduce the expanding part.

This is a fundamental part of the nature of motion as Kepler's formula introduces such a principle.

It is the motion of the spinning that produces and substantiates the attracting or the expanding.

With only gravity working in the motion sector will reduce the contraction to only one sector while moving the solid as well as the liquid will bring altogether favour to the contracting part.

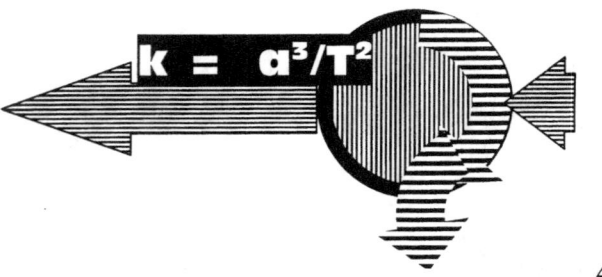

$$k = a^3/T^2$$

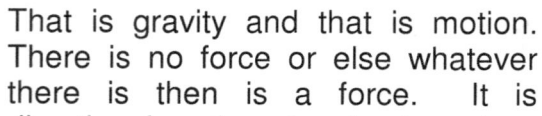

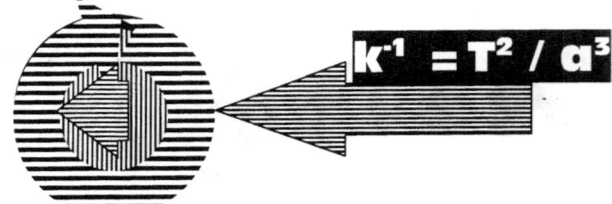

$$k^{-1} = T^2 / a^3$$

That is gravity and that is motion. There is no force or else whatever there is then is a force. It is directional motion of a circular nature that govern the relevancy in strength as well as direction of inclining. There is an expanding to the contracting and the expanding is in direct relation to the motion of the liquid, which brings about the contraction

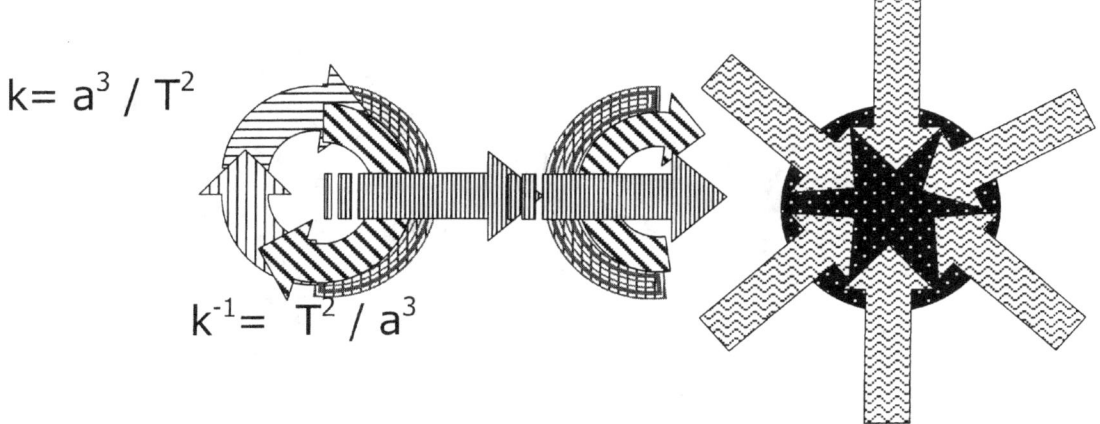

$$k = a^3 / T^2$$

$$k^{-1} = T^2 / a^3$$

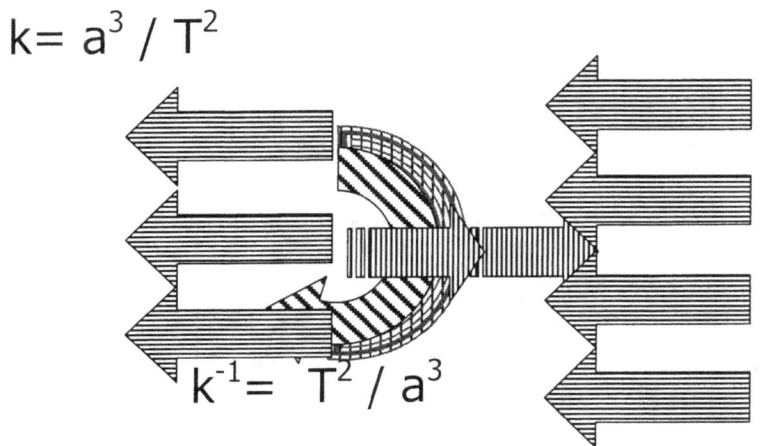

$$k = a^3 / T^2$$

$$k^{-1} = T^2 / a^3$$

Even the direction of motion needs not to change because it is the relevancy when the motion crosses the divide that sets the motion apart from what it previously favoured. It is always a liquid that brings on the motion when the liquid in motion is connected to a solid that gives the liquid stability and the liquid provides the space in question with a limited definition of securing a precise border to end the space. It is a liquid in defining a solid that produces the influence of the contraction in relation to the expanding and in that we find gravity. It has no bearing on mass what so ever. It is motion of a liquid relating to the solidness securing the position of a solid.

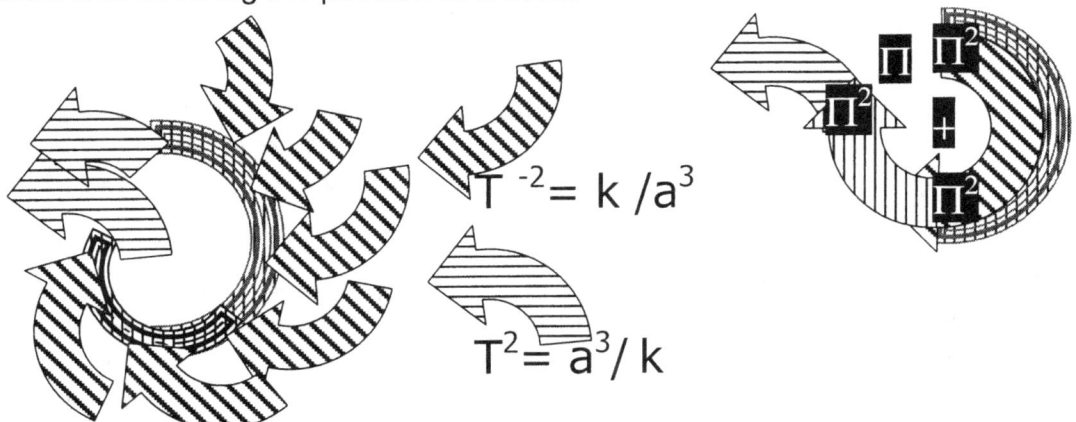

$$T^{-2} = k / a^3$$

$$T^2 = a^3 / k$$

At this stage it is not important to study the direction of motion but it is vital to acknowledge that contraction comes about from one side where as where contraction does not totally dominate we find expansion also part of the equation.

From Newton's time we have observed the gravity from the position we hold as trapped belongings within the Earth's liquid motion.

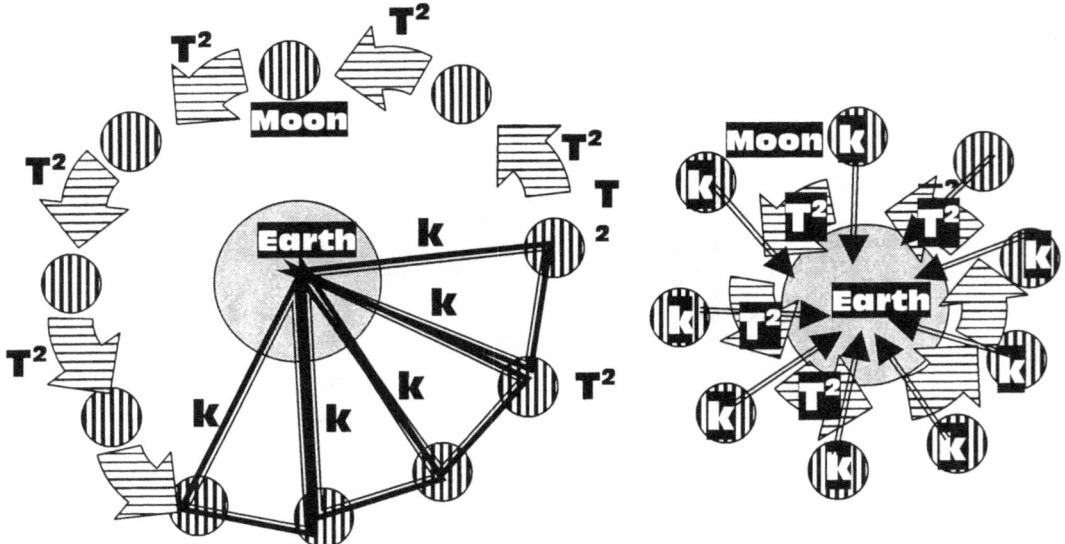

The Moon and the Earth is in a Roche connection and are sharing some minor sort of a Roche lobe with the two being slightly outside the limit. What connects the Moon and the Earth is inverse relevancies from either side.

Seen form the Earth there is a line running from the Earth to the edge of the Earth. Seen from the immovable centre the Earth cannot spin just like the centre cannot spin and the centre connects to the Earth edge without variation of any sort that would implicate changes in position. The centre point at one point on the Earth horizon and that line secures a connection that never fluctuates or change on the same point on the edge every time. The connection proves to the centre as a solid according to the Earth centre. The fact that the Earth spins around the axis is to the axis neither here nor there because the line **k** connects to one secured point on the Earth edge at all times. While the Earth rotates the Earth seems from the centre to be the solid stable partner, which puts the part of being in motion or liquid to the other partner that is in outer space. In relation to what applies to the earth, the moon is outer space and the Moon holds the earth in the same regard or disregard.

The Earth face the Moon in a new allocated position every time and seen from the centre of the Earth the Moon is a liquid that is moving in ratio with the Earth time. It is holding a position of time as $4\Pi^2$, which represents time in a complete circle in relation to seven degrees of change per rotating interval. However we are on Earth and we look at matters from such a vantage point. From the centre we find that the Moon rotates in the boundaries of time (7) $4(\Pi^0)^2$ because it is the singularity Π^0 at the centre that holds the prominence from which the cyclic days arrive. From the Moon the cycle lasts one day and from the Earth the cycle lasts 28 days. On the other hand since the moon is still part of the atmosphere of the Earth, therefore, according to the Earth we find the Earth apply another type of Roche limit being $\Pi^2/2$ in relation to the Moon and that means the Earth day gets one cycle $(\Pi^0)^2/2$ from the centre singularity divided into two entities of one cycle considering the Moon.

From the centre of the Earth the Moon is as much liquid as the space it is in and by rotating all the motion belongs in the factor, which the Moon represents. The Earth holds the Moon in a Roche lobe and the Moon is to the Earth, just more liquid rotating. It is important to note that the Earth regards the Moon as liquid and with the Moon already being liquid the Roche limit does not come into effect. It is also very important to look at the Moon and find that the Moon is where it represents the Earth limit at the end of its atmosphere because the Moon is still in a Roche limit with the Earth while the Moon is very much acting as liquid would that carries all the moving abilities.

   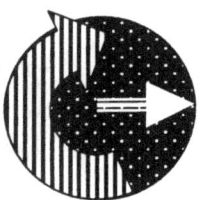

Seen in perspective one arrive when taking a vantage point from the centre of the Moon, the Moon is as solid as the motionless of a solid needs to be. The Earth represents all the liquid and all the motion since the Earth is representing another point towards the Moon every time the Earth rotates.

The Moon is standing dead still as any solid requires to do because the Moon does not rotate around its axis. All the rotating belongs to the Earth therefore the Earth represents all the motion and therefore the Earth is the liquid while the Moon represents the centre of the Universe being as steady and as sturdy as it seems to be from the Moon's centre of the Universe. The one to the other is in liquid and the one to the other is liquid. The liquid it represents is outer space it is within.

The centre of the Earth is the immobile unmoving sturdy dead solid part whereas we know the Earth is spinning like the timing gear of a Swiss clock. Again what the Earth uses to focus when it takes regard of the Moon it applies the very same manner when regarding outer space.

The line coming from the centre of the Earth and which is going to the edge of the Earth is always connecting to one spot, which also is always the same spot. From the centre the line never shifts and therefore the line is always consequent.

Therefore even though the outer space does move in relation to the Sun but maintains equilibrium in relation to the planets, outer space does in affect move according to the motion of the Sun but is completely motionless in regard to the position the planets have within outer space and outer space and the planets are one thing. All the planets are going as much straight ahead in relation to the Sun as they are rotating around the Sun and therefore in relation to the Sun outer space is coming towards the Sun at a pace of approximately 299 in ratio, but in relation to the planets, outer space is standing still.

From the centre of the Earth there is a line **k** that holds a position of 1^0 to 1^1 in all possible directions. However as said the points never reaffirm new allocated positions in relation to previous positions and no change comes into place as result of change. However that part represents eternity and we know the Universe is about patricians forming in between time and moreover between eternity and infinity.

From the perspective of the planet such as the Earth the rotating of the Earth does not implicate change in relation to the Earth centre. However change does come about and the Earth is rotating, which places a movement between the Earth and outer space. The new alliances with a changing relations between the earth and outer space find focus on the part of outer space where outer space are focussed on then being the liquid in relation to the moving solid being the planet. The planet is moving but represents the solid while outer space in accordance with the Earth is standing still but is focused as being a liquid representing the motion. The changes coming from the rotating of the Earth in

relation to outer space puts the liquid emphasis on outer space because it translates singularity from the edge of the planet to the connecting point between the planet and outer space.

Considering that outer space then represents 1^1 in relation to the centre that is holding 1^0 the change diverts to the changing position that outer space shows in the partnership. The connecting is always 1^0 extending space-time to 1^1 and therefore the Earth or all other planets and structures for that matter holds the solid position. The solid position that the Sun give to outer space for doing its job in housing the planets are transferred by the planets providing motion onto outer space where the motion is projected from that which move to the which in relation does not move.

 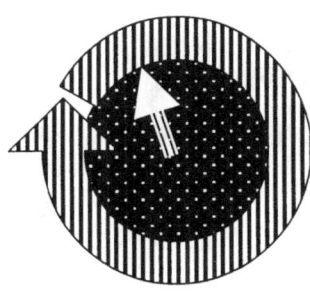

Therefore outer space is granted a steady position that it maintains in its relation with the planets when the Sun grants the security of a centre or perform the task of the solid, outer space should by any stretch serve as a solid in regard to the planets.

To outer space the planets are in equilibrium by motion therefore the partnership should not reveal a definite solid and a specific liquid. But by rotating the planet should be the factor that supports the liquid part. However in the sturdiness of the lien that comes from the centre of the planet to the edge such a solid line gives the impression of immobility and transmits such mobility to outer space, which is motionless in relation to the planet. In the relevancy there is between the Moon and the Earth the Earth moves in a circle and the moon moves in an orbit. This is very significant and is important to use as an example to understand how the cosmos functions. In relation to the centre of the

Earth the Moon is changing positions all the time while the Earth is locked in one position that does not change. The Earth stands steady while the moon is orbiting around the Earth. The Earth is the solid while the Moon is a part of the liquid. Then from the vantage that the Moon centre singularity holds the Moon is standing still because the Moon is not rotating around a personal axis.

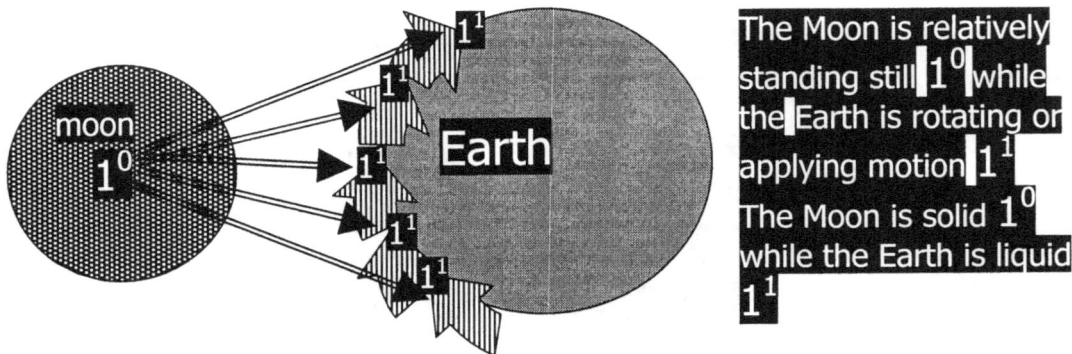

The Moon is relatively standing still 1^0 while the Earth is rotating or applying motion 1^1 The Moon is solid 1^0 while the Earth is liquid 1^1

The Earth however is rearranging its position constantly by providing motion, which is realigning with the Moon centre by every rotation motion in the minutes of moving. In this instance again the Moon is the solid while the Earth forms the liquid by securing a motion free centre in relation to the Moon and the Earth motion.

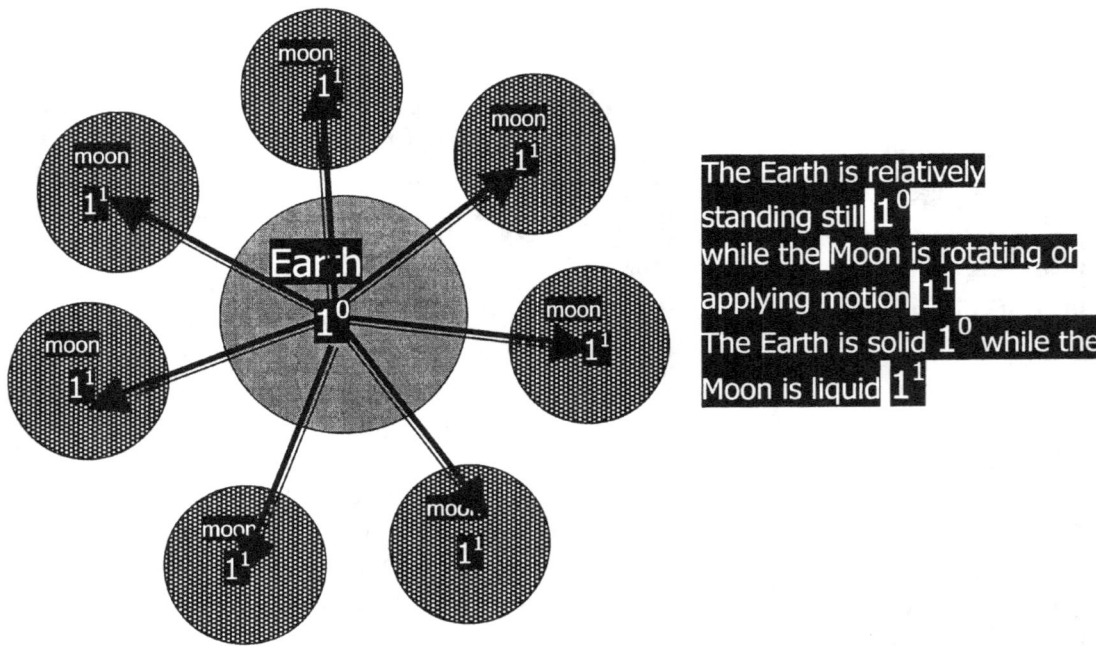

The Earth is relatively standing still 1^0 while the Moon is rotating or applying motion 1^1 The Earth is solid 1^0 while the Moon is liquid 1^1

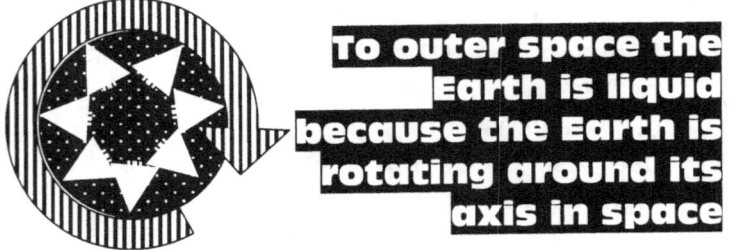

To outer space the Earth is liquid because the Earth is rotating around its axis in space

Being a liquid or a solid depends on what provide the anchor or pivotal role and what provide the motion within the relation. The planet moves in orbit as well as around its axis. This duel capacity is all motion while it is also all solidity. The cyclic rotation forms a liquid in relation to the point where the Earth meets outer space but since the Earth at that point is connecting to singularity by seven it is outer space that then carries the motion at the point. The contact point at the earth's end remains the same although it moves because it is solid, but the outer space are show a new point in relation every time although it is steady. To the Earth it is outer space that shifts while in fact it is the Earth that rotates but that rotation does not affect the centre since the centre remains directly aligned with the edge of the Earth solidity. However the Earth renew a contact point with outer space which is the motionless part in the relation at that point but serves as the changing factor in the relation.

With the Earth rotating while the Earth considered its position as stable and solid motion is reflected to outer space, which at that point is solid. Outer space, which is stable, is facing a new position in relation to the Earth but since it is not part of the solid structure of the Earth, outer space shows changes and diverts its position in ratio to the centre of the Earth. This ratio gives outer space the liquid partnership.

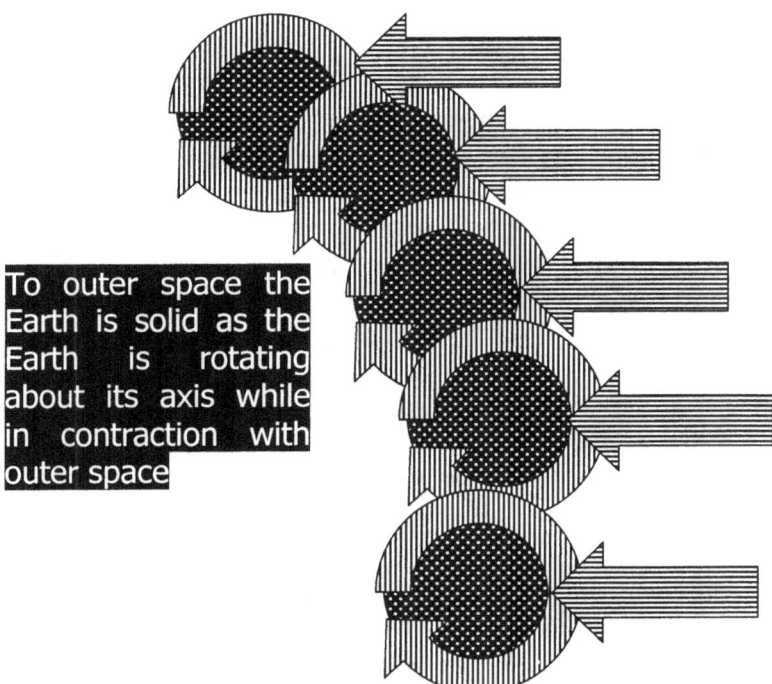

But then the Earth shows another side in the affair where the Earth in orbit tears through outer space being without motion. This brings about that outer space will allow the Earth to move and while the Earth is moving the Earth is aligning with the centre of the Sun. To outer space the Sun represents all that can be stable and in that regard it takes the Earth as another solid principle. In that way the Earth again proceed as the solid or stable factor while outer space holds the motion.

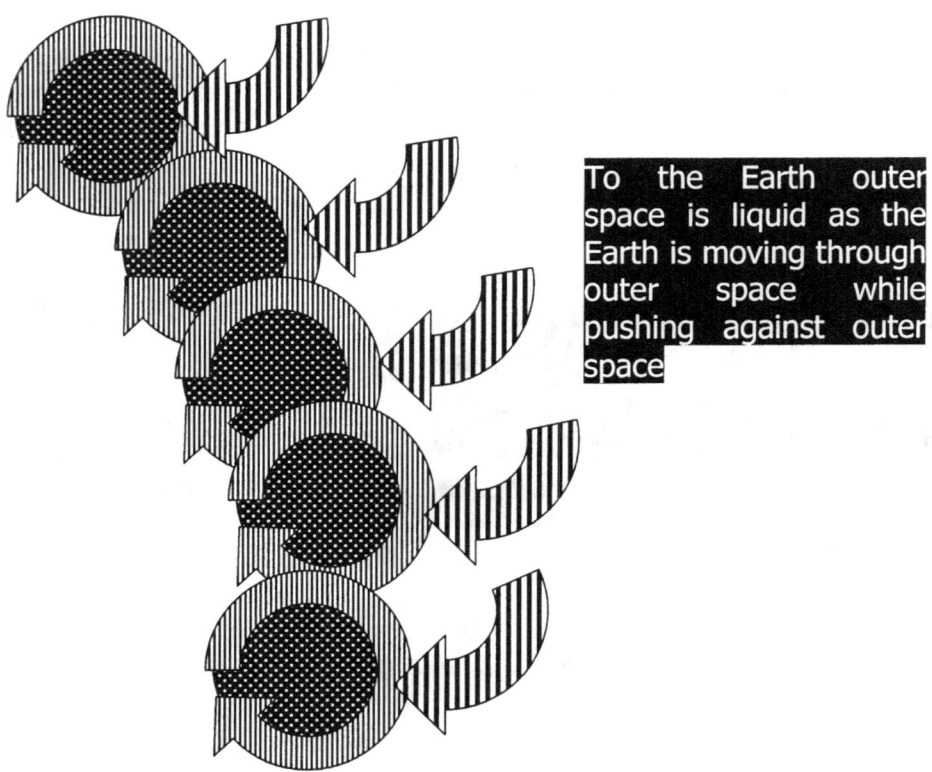

To the Earth outer space is liquid as the Earth is moving through outer space while pushing against outer space

That is not yet the end of the Coanda partnership. The Earth is pushing against the outer space and while outer space is inherently a liquid outer space give road to the moving Earth. Therefore while it is the Earth that pushes against outer space it is outer space that is moving away to allow the Earth the motion it insist to have. Therefore again outer space is moving in relation to the Earth which is pretending to be solid and the total result from all the activity is that outer space seems to move at a rate of 10 X 10 X gravity which is Π^2 and that gives space-time a value of space (a^3) / time (T^2) is (=) 298 but I shall come back to this when much more information is exchanged. The whole debate now in this part resonates around the fact that mass never comes into the argument and mass is no factor in the Universe. It is all about motion and being solid (not moving) while the other party in the equation is moving (being the liquid). It is as Kepler said $a^3 = T^2 k$ or then $k = a^3 / T^2$.

That is why in all cases the ratio of orbit in space-time is the same. The planet rotating the movement of outer space is in relation to the liquid position the Sun give outer space and from the vantage point that outer space holds the orbiting object is in space ratio performing as a solid partnership in relation to the motion it allows outer space to have as the liquid. The motion is a result of holding singularity steady as a solid while moving through outer space which then allows outer space to hold the object singularity as a solid reference point while taking on the liquid part of the relation.

My first nut I cracked in cosmology as an individual standing apart from what I was reading about cosmology through the avenue of mainstream science was concerning motion in relation to the speed of light. Today the fact that it took me a full six months to solve is a joke. That it took me so long to get to such a simple answer is in hindsight not very complimentary but please keep in mind that at the time I had a blank paper in front of me, which was blank in more than one way. However that was what set me on the way to be able to crack the first code. I must admit if I did not break the seal I would be totally lost in cosmology but still such a simple solution took me six months of head breaking arguments with myself. Einstein said that if light were travelling at the speed of light for one year it would be away from the source of origin by a distance of one year. That is acceptable even to a person with my mental capacity. Then came the jawbreaker.

The two photons travelling in opposing directions will also at that instant be at a distance of one year apart. It takes one light year to go in one direction while it takes the other photon one year to move in the other and opposite direction and yet the two is one year from the light source it left while also being one year apart from each other. That baffled me into almost madness. It was just way above what I could mentally cope with. The two photons opposed each other while travelling but at the same time moved apart only by one light year of total motion. The total that should add to a double was the same as the single, which was the same as three totally different points. Something told me in this was the key to understanding cosmology.

Then one day the simplicity about the whole argument hit me between the eyes like a ton of bricks. I was staring at outer space while viewing outer space as a distance. Outer space is time and not distance in as much as forming space. It takes space a^3 time T^2 to bridge time k^1 and reach space a^3. The time it takes a^3 is in relation to move at time T^2 to cross time k^1. We all confuse space and time. Time is what is between the Moon and the Earth and not the distance. The time lapse was one year and therefore the time that parted the objects and the source of origin was one year but also the time that parted the two photons travelling merrily was one year.

Light travelled in as straight line as well as a half circle and where light is connecting to time the half circles 180^0 is equal to the straight lines 180^0 which is equal to the triangles 180^0 that means outer space has nothing to do with space but is all about time while time is all about motion. That is the key to solving the riddle we call cosmology and by using that key I found a way to unlock so many answers. The light flowed in a straight line, which is 180^0 while they move apart by a half circle also to the value of 180^0 and while being in a triangle position in relation to the point wherefrom they came which was also a value of 180^0.

Time was moving and if time was moving then time has to be liquid. Since time and space is not the same space then space has to be a solid holding time as a liquid in relevance and knowing that it is time that is between the Earth and the Moon it made that which is between the Earth and the Moon the liquid and that made the Earth and the Moon solids. How simple can everything be if one takes the correct line of arguing? Even I being who I am could start to understand what everyone should understand. I feel obligated to explain my referring to myself in the position I have. When saying this about myself I must request that you should never forget while reading that I am only a motor mechanic and that is all I can ever be.

We are part of a thin top layer, the part that is contracted without really being part but only part as an extended part of the Earth. That is part of Newton's gravity where the liquid we are in and the liquid that we are becomes part of the Earth as it is confined to the Earth by the atmosphere of the Earth that forms the liquid restricting us to the Earth. That is why Newton and his apple were on the ground very much secured.

We are in the motion, which forms a liquid. We are part of the liquid because we are part of what moves. Every aspect surrounding us brings proof that we are contained in liquid and we are preserves as being part of the liquid. Where the wind blows, it indicate a wave pattern and a liquid leaves a wave pattern When looking at a mirage we distinctly see that where the atmosphere becomes more dense as the heat at that point becomes more intense, that what we see and which we call a mirage is water, is a liquid substance floating in waves where the concentration in density varies.

The liquid engulfs us and in being part of the liquid we are experiencing only the contracting aspect since we are totally secured by and in the motion the Earth provides. The earth atmosphere puts us on the ground as the atmosphere secure our positions on Earth. We are part of the $k^{-1} = T^2 / a^3$ while the earth being in motion and providing the motion forms $k = a^3 / T^2$. That is the time aspect, which the Earth provides and that time aspect enable us to read the time by the measure of Galileo's pendulum.

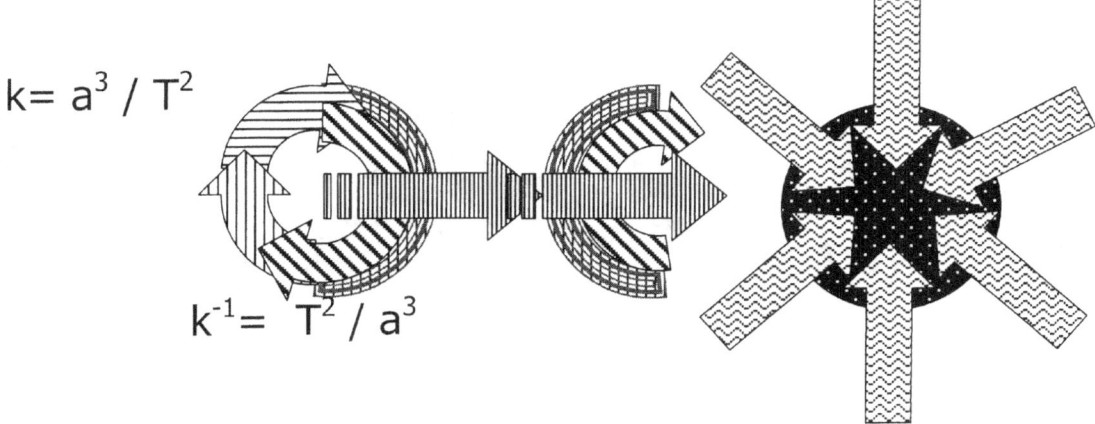

$$k = a^3 / T^2$$

$$k^{-1} = T^2 / a^3$$

By the motion in rotation as well as the motion in the lateral we find that the liquid being air confines us. It is not the particles in the air that is the liquid but the substance separating the particles in the sir that form the liquid. By moving around the axis while the axis is moving around the Sun forms motion, which confines us to a position that only Newtonian gravity, apply. There is only one direction flowing towards the centre of the Earth. That flow is the space a^3 that the liquid T^2 secures k to the space $k = a^3 / T^2$. In this however there are no grounds that support the suggestion that mass is producing the motion or that mass is indicating the flow.

By observing a fire we can see where and how the flames bring intensity to the heat in the air. The flames are the densest form that heat can have while being in a liquid form. The flames are so dense a liquid it provides light and light is pure heat in minute quantities of liquid space. In the picture we see three forms of liquid heat where in each case the element responsible for producing the liquid heat contains the heat in a different form. We see the flames souring and that is the responsibility of the nitrogen where the nitrogen being 7 expands heat into space. Then there is the oxygen that contains the heat in a dense material we think of as smoke. The next we can see is the wood or carbon 6 that keeps the heat contained in the particles. Every layer in a star has this duty that the substance hold in providing and managing the heat within the stars structure. Taking the idea of air (not particles in the air but atmospheric substance containing the particles in the air) to further proof of finding a liquid we again have to go to the Coanda principal.

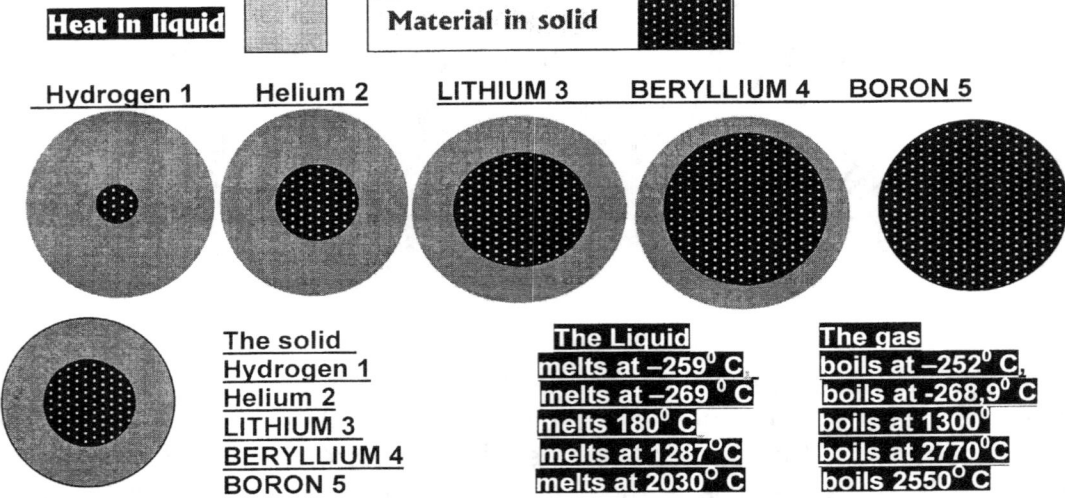

The solid	**The Liquid**	**The gas**
Hydrogen 1	melts at -259^0 C	boils at -252^0 C
Helium 2	melts at -269^0 C	boils at $-268,9^0$ C
LITHIUM 3	melts 180^0 C	boils at 1300^0
BERYLLIUM 4	melts at 1287^0C	boils at 2770^0C
BORON 5	melts at 2030^0 C	boils 2550^0 C

The particle (atom) secures the solid basis that provides the motion which enable us to catogoriser elemnts according to our percption. It is not the truth or cosmic reality but it is our perception through culture that tach us about gasses, liquids and solids, about noble gasses and heavy metals and non ferromagnetic or good

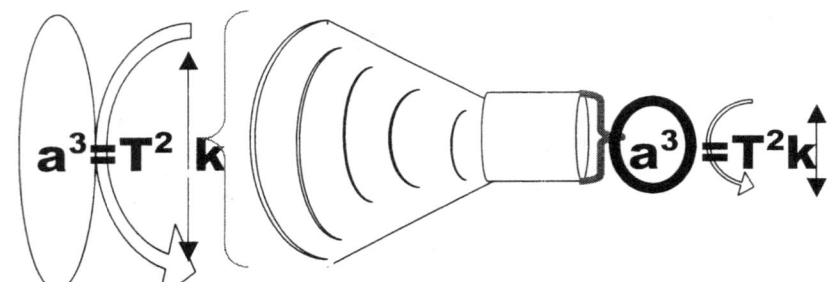

conducting and immesureble other characteristics we attach to elements except what trulyis importent as far as cosmology goes. All material are solids as much as they are gasses or liquids. It depend on the concentration of the heat surrounding the

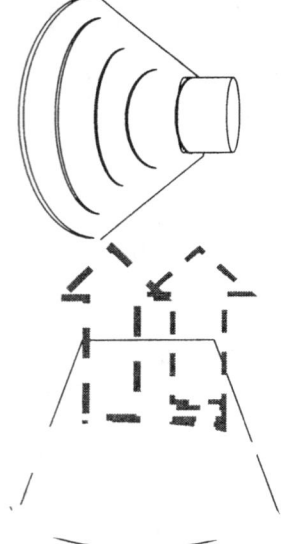

elment in that particular element that turns the element at that point and time (temperature) to be eother a solid, a liquid or a gas. The state that the elemnt is in is a responce to the conditions which the heat levels bring on and that is a responce to the Coanda gravity motion that serves the atom as a liquid at the moment of responce.

However when saying that it is a liquid in motion around a solid we also find that the principle drives turbine engines and the air compresses within the turbine to cause heat. It is the air that flows when driven by the turbine rotor, which provide the sold. Then from that we must deduct that the air producing the flames forms the liquid, which provide the contraction when the solid spins the contracting turbine and the spinning turbine serves as the solid part.

The Coanda principle works in two parts where there has to

be a solid securing space and there has to be a liquid performing the motion that result in contraction. The ingredient is about a liquid in motion T^2 capping a limit or an end k or k^{-1} where the space $k = a^3 / T^2$ holding the solid extends to appreciate space $k^{-1} = T^2 / a^3$ that holds the liquid. It reduces space by motion providing contraction to a point where the space goes liquid in finding flames. That is what Henri Coanda first saw when he tested his new enclosed propeller. It is all so exciting but what the Newtonians of the day and those today completely misses is that for the Coanda effect to be functional one needs a solid and a liquid. The solid part the rotary propeller provides and with such motion it contracts the atmosphere into a compressed flaming liquid. The atmosphere (not the particles in the atmosphere) compresses to become a liquid.

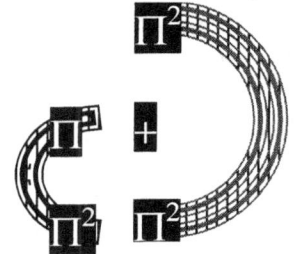

That is the conditions applying to the atom, which provides the conditions applying in the atom. We differentiate the particles by giving every part a name and try to find meaning in the particle combination. In a hundred years from now we are about to "discover" another million smaller particles and in a thousand years from now there waits another billion more, which are all smaller to be discovered. We will go on discovering until eternity once more meet infinity and still we would not trace and name them all. It is what the combination provides each other in supplying space-time that finds importance and not

naming the lot individually.

The major function that the combination of the electron provides is the expanding and contracting of solids in relation to time being a liquid that has a flow and a direction of flow.

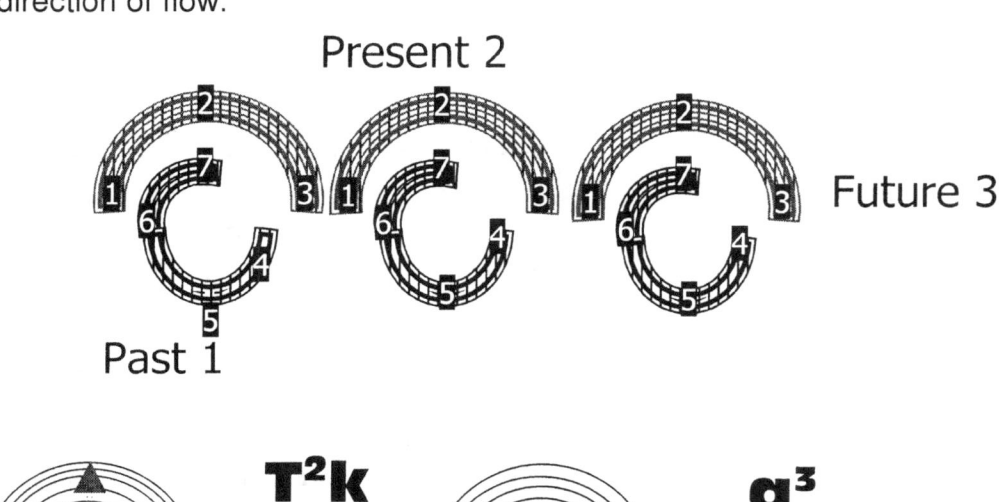

Present 2

Future 3

Past 1

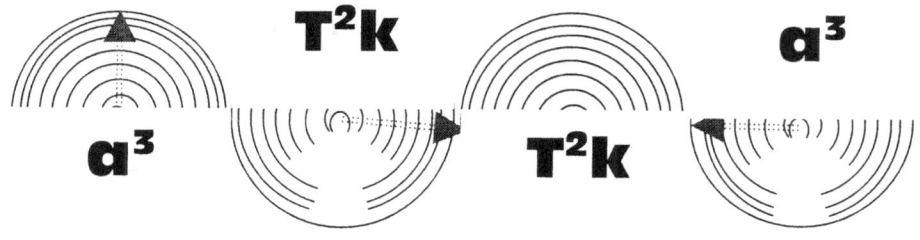

T^2k a^3

a^3 T^2k

It then becomes pertinent to find what is liquid and what is a solid in cosmology. The first aspect we have to abandon is our preconceived notion about what liquids are and what solids are. Everything that moves or forms part of that which moves or may move while another in relation is standing still becomes that which is a liquid while the part forming the motionless factor then becomes the solid. From that gravity becomes reality and gravity is motion that has no bearing on mass in any way possible. It is the restriction of mass that prevents gravity from becoming a reality and because the Earth restricts our motion as we find motion that the Earth prescribes we have mass and not gravity. An object may have some part vested in mass while having another part acting in gravity but mass does not bring about gravity. In gravity there has to be motion of space in time and when time ends space collapses. In this letter I am about to introduce you to gravity being the motion around a solid that provide the contraction.

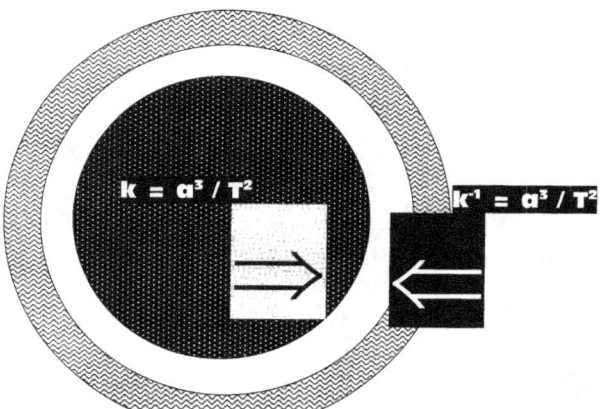

Gravity has two parts where the one part is expanding by motion and the other part is controlling the expanding by contracting the motion. It is $k^{-1} = T^2/a^3$ and it is $k = a^3/T^2$ but above all to find gravity we must find the centre of the Universe being singularity at $k^0 = a^3/T^2 k$

The liquid in motion provides the gravity and by initiating the motion the gravity contracts the solid as much as the solid is extended by the motion, which the liquid provides and the liquid is the gravity or the motion of the solid where the solid provide restriction or mass to the liquid.

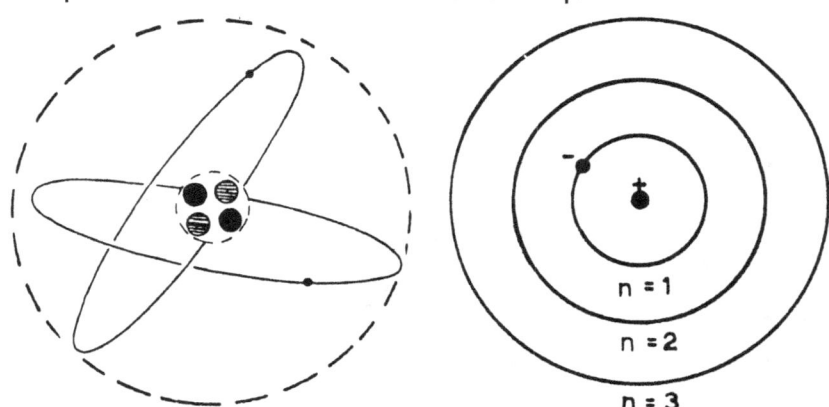

It is in this manner that the entire Universe operates because from the spinning atom to the most prolific galactica and to the other end of the spectrum where Back Holes destroy time the entirety out there works in this manner.

I first started my studies in the field of Cosmology as a spontaneous development of my natural curiosity spawned from childhood interests in the field of cosmology, which I developed even before I went to school. The studies were a reaction (I would imagine) that was part of my personal childhood development in how I was forming a personal concept of a lifelong interest that followed me into my future. At first I conducted all my earlier studying mostly on the basis that inspired me to find out more about what made the Universe tick, with no intention ever on my part to reach a point where I would be writing books on the subject. At first I was investigating cosmology on a part time basis. This went on, on and off, or the best part of twenty odd years (*as* time and *when* time would permit).

Then in later life with my health deteriorating I committed myself to more intense investigation and my effort developed onto involving a study using time that is only permitted by a person when that person is involved in such a quest on a full time basis. That quest has now been going on for the last seven years in full devotion and if one includes all the years invested on my part including the twenty odd years before, part time, then the time I have spent in completing my theory when adding all in comes down to almost twenty eight years. This is to say that I did not come to realise what I am about to introduce on a light-hearted conclusion. I mention this because I wish to ensure the reader that he should have no doubt about my most sincere commitment in producing a cosmic theory on matters concerning the start and the working of the Universe during and before the Planck era. At first I began by arguing that there is something that is blocking our progress.

There is some barrier preventing humans passing a threshold whereby our understanding will pass such an obstacle. If there were any way that any one may break through that barrier there is that is preventing normal research to go pre-Big Band, it would be accomplished by finding the barrier whereby then our vision we use to focus would pass such a limit. If we wished on progress in our pursuit of the very first cosmic moment then we have to find and cross the barrier that blocks our view. We have to look deeper and in another direction should the desire driving us be strong enough to commit us to reach into the very birth of the cosmos. We have to rethink the strategy that we use. Max Planck was one of the most brilliant men of all times and even he, notwithstanding all his personal brilliance, accomplished little.

There are parts missing in what we have and that which we have at our disposal to use, because if there was no such an obvious barrier then the Wise-Men involved in science would by now have found the way to break through the seal that is locking us out of the critical past which will uncover the origin of the Universe's infancy stage. I went about trying to find what everyone since Adam, (meaning all of the rest of mankind and myself) were missing throughout the ages of speculating and interpreting while philosophising about whatever we find inspirational. The obvious we saw; that was clear. Therefore I had to find a route that would lead into the not so obvious that all of us were missing, notwithstanding the best efforts of the best qualified to accomplish such a breakthrough. My effort involved trying to accommodate that which was in the cosmos available to use by the cosmos in all phases of developing. If I had any hope of finding the answer, such an answer had to be simple because I am not very inclined to unravel what is deemed as complicated. The simplicity had to be locked in what was not yet understood about that which was in the cosmos as it formed part of the process used in forming the cosmos. My realising this brought me to focus not on that which we understand.

There is not a lot we actually understand because even gravity is very poorly understood. In fact gravity is so poorly understood that there is not one person alive that can claim the prestige of understanding gravity and among the dead there is even less that can make such a claim. There are several phenomena that are presented in nature and acknowledged by science but also discounted by science and therefore not presented as accepted science. By admitting that that what we have available to us to use concerning our research of cosmology in an attempt to better our understanding of cosmology, is useless to use, then one realises that not having what there might be makes what we already have useless. It then is useless to use what there is as part of the big picture we are trying to paint because what we use is not really part of the picture. This leads one to believe that the picture of the cosmos Mainstream science is painting, is being painted without painting a full picture.

In my first attempt to understand the full picture of what science was painting I found so many colours missing there was no picture painted that anyone could appreciate. This is what made me decide to go on researching the 'unknown' in the hope it might clarify the 'known' and as the book unfolds. You as the reader may agree that I was correct in pursuing the misunderstood and rejected phenomena. Finding the missing phenomena helped me to place the phenomena mentioned above in a theory where the principles also mentioned above form a part of the overall gravity used in binding the Universe. I believe what is in the Universe is not able to be coincidental because of too many influences contributing to what there is - notwithstanding the fact that that is the manner which science uses when they refer to the Bode law. What is in the Universe has a role as it had a role, which is the same role that phenomena has had and in future will have. This is establishing a very new idea about the working relationship between particles and in explaining it by using Kepler's studies. Redefining the work of Kepler's views brings a new Universe to light involving new concepts that are based on old principles but principles in updating man's view about cosmology are very new in that capacity. Through that new vision I was able to come to realise what the reasons might be why Kepler never saw it fitting to include the measure of Π in his formula. I do not suggest his neglect thereof was intentional, nevertheless the formula he devised without using Π proved that there was no need for the inclusion of Π since his figures brought about a correct answer in the final end result leaving a well concluded fitting answer. The numbers he produced brought about a specific space $\mathbf{a}^3$ contained in a circle $\mathbf{T}^2$ at the distance of $\mathbf{k}$ from a defining centre thus the calculations did not require the use of Π to find a meaning. In that Kepler did not see a need to include Π. I would not go as far as declaring with absolute certainty on his behalf that he did it deliberately, however there never arrived such a necessity. It is prudent to agree on whether or not such a need is necessary, because if one is agreeing about such changing not being required a new Universe emerges. The circle that Kepler discovered came about without ever forcing Π into the frame because it is clear that the circle formation came about as a natural consequence and came spontaneously delivering an equation while he was working. In this book I prove that the reason for adding Π to the rest of Kepler's formula is unnecessary. This unnecessary addition is because when going one step further in the investigation one will find that $\mathbf{k}$ and $\mathbf{a}$ and $\mathbf{T}$ are symbolising the same value with the only difference being that each one represents a different dimension to our six dimensional or six sided Universe we enjoy.

In fact I shall show that Π replaces "**a**" and "**k**" and "**T**" and that Π is the true value that should be replacing each factor as to indicate the correct value to the sides nominating Π. We humans work on a numerical base using ten as a basis where we count to nine and re-establish a new decimal numbering line by adding a nought behind the number in value. This is using the numerical basis of ten, which I suspect we took from ancient knowledge about cosmology and not from using our fingers and toes as the earliest calculating processors. In this letter there is unfortunately no room to explain my suspicion but another fact I do prove is that the cosmos uses Π in the cosmic numerical basis as a means to measure and quantify. Therefore in fact the Kepler formula should read instead of $a^3 = T^2 k$ as it does it must be $\Pi^3 = \Pi^2 \Pi$ where I shall show that Π represents singularity wherefrom the entire Universe sprang from Π and by forming as $\Pi^3 = \Pi^2 \Pi$ it is confirming that space is equal to the motion thereof. Kepler's greatest achievement was showing that the cosmos is space –time $a^3 = T^2 k$ while time is the motion of space in space. The value of Π is the primeval and most basic of measures applying as an accepted cosmic legal value that the cosmos used exclusively in the very beginning and as it does today. The measure of Π in the Universe, values particle development that brought about all development ever conducted in the Universe. Only after this stage did the rest come including mathematics and went on to freeze spilled singularity into frozen material. Reading this statement may sound suspiciously senseless but as the book unfolds the sensibility will become apparent.

The full implication of such a statement will become clear when one dissects different facts coming from studying Kepler. My discovery of this fundamental basis of legal valuing ensured me again that there was no need for someone the likes of Newton to add Π in any form to the work of Kepler because Kepler discovered the ultimate Π in the Universe, the Π giving the Universe form and gravity. The concept of Π that is the only single form of all other forms available that can by duplication of Πs assemble the value of gravity. When replacing the symbols with Π the facts of the Universe become self-explanatory because the most basic form that forms the cosmos has a definitive and uncompromising value.

 But getting this far took me down roads overgrown by ignorance and which I had to uncover myself as if hacking away miles of overgrowth with a machete chopper. All of the disbelief science showed to my work in the past and their refusal to see past Newton made any and all attempts on my part as bad as they could be, strangling and smothering my attempts to announce my uncovering of the newly found insight on my part.

For decades I tried to come to terms with the inability there is in science to explain the cosmos in real terms, when using the science of official reputation. That which there is makes a mockery of science because the undisputable clues left in the cosmos makes what little correct explaining there is available, seem like a comedy of errors, when it is mixed in with all the other near Dark Age errors we still use after so many centuries that provided countless opportunities to revise the old muck. By applying current accepted Astronomy as such the phenomenon found all over the cosmos is still beyond the explaining ability of Mainstream science. This is true and it is a shame because it also is an undeniable fact in spite of the vast knowledge and progress in other forms of science taken in the manner science uses when it approaches cosmology. Cosmology truly lagged behind while the understanding and advancing of physics, mathematics and chemistry as subjects were flourishing. By comparison I saw how little there was available in explaining cosmic phenomenon and how much improvement in understanding the other departments such as chemistry, electronics, medicine etc. could offer as results were coming about from research. Even where there is a little explaining available in cosmology it turns out that such explaining is confusing to say the least and at best it highlights the manner in which science is applying double standards. For decades photographs were the only progress forthcoming as an addition to improve the meagre field in cosmology and that improvement was artificially stimulating cosmology. By providing a false impression of advancement, everyone missed what and how much was missing…To the connoisseur desperately looking for more than the obvious stirred in with some out-dated misinformation dating back to the Middle Ages, it all seemed as if it was a picture portraying the ridiculous to make the sublime look good. The pictures only proved the opposite of what progress in cosmology will represent. In truth and as such in cosmology the cover up that was hiding the lack of progress about the science of true cosmology was only forthcoming in the improving of electronic optical telescopic advances and spectroscopic progress. There were only photographs carrying beautiful pictures which pleased the less informed except the photographs did not bring progress to cosmology at any intellectual level by promoting insight. The explaining that the photos demanded about the subject had the opposite effect of installing hope because what it did do was underline what lack in any notable progress there truly is in our understanding of cosmology and laws in the cosmos.

While such Hubble telescopic images might seem to be as clear as daylight it was more than clear there was little academic value to them. To the person in need of more stimulation than being impressed with pictures of God's marvellous Creation and the sightseeing that always accompanies such pictures, such persons always felt very disappointed. The pictures did give satisfaction to those more easily impressed, but the rest of us seeking knowledge accompanied by understanding the images left us despondent. Although they leave the vast majority in total amazement there are those less impressed about not knowing the 'why' and the 'how' in such amazing pictures. I know the group I fall into may be the greater minority and the majority may only demand the portraying of the images, which is what that easily satisfied group demand. The rest of us rouse with anguish at the lack of information about what is known and what lies behind what those pretty pictures are conveying. Nevertheless there can be no real progress in scientific understanding about the images portrayed by the Hubble telescope, and others, if no one is able to show the slightest clue of a deeper understanding of what is going on in the Universe. Everyone is almost breathless waiting the commentating by the most informed which accompanies the magnificent cosmic portraying of God's Creation. When we are portraying the new images, we should also be investigating that what we see that the cosmos is at the moment portraying. The lack of actual believable explanation coming from investigating by means of telescopic imaging should impress one and all, but the impressing must not be based on the colours in the images but the sensible information attached to the image investigated. It is *that* that we wish to see. What we wish to see must at least be accompanied by scientifically backed information, which provides the proven understanding coming from science. When science is employing new explanations with such photos it should also be discarding senseless baggage carried over from the past. Most images contradicted Newton and for saying that, every Academic I ever came across in the past ostracized me. That bothers me little! I know I cannot possibly be the only person absolutely discontented with what Mainstream science accepts as science. Here I refer to the out of date theorising Mainstream science still accepts amongst many others as how they suggest stars and planets are forming. One cannot promote cosmology in honesty and advocate scientific fact whilst dishing up such fairy-tale nonsense to students. Moreover I hold the opinion that amongst Academics in particular there must be many if not most that share my personal serious doubts or have an inclination to share some of them. This I say when considering the overall doubtful picture painted about what there is and what one believes there should be. I just cannot believe those forming the most intellectual group of mankind are unaware of the mismatching facts seen over the broader picture because the contradiction and lack of a plan, makes what there is so very doubt provoking. Newton dismissed the formula Kepler presented as all factors forming motion. That is where the apple cart derailed.

In honesty we have to realise that we cannot dismiss the whole formula that Kepler produced as being motion. It is so much more than just motion. It is $a^3 = T^2 k$:

That is what Kepler brought into civilization for all time to come. He saw space a^3 being in isolation due to the time it uses to move T^2 claiming such space forming independence according to the lines k indicate.
Let us look at the factors in more detail before we proceed with the rest of the book.

a^3 symbolises a mathematical interpretation of implicating the three-dimensional space.

T^2 is representing the period or time that Kepler suggested we should use to calculate time that holds the orbiting planet in direct contact with the space in relation to a very specific centre.

k is the space taken from the centre to the end of the line from which the planets must have grown if one accepts the Big Bang growth of particles and the affect of the Hubble constant on all cosmos material. The specific value about the centre is most important because from the specific centre gravity always applies the strongest influence.

One cannot justify Newton's dismissing of Kepler's formula as that all factors only contribute to the motion indicated because that is misleading. We all accept that the true cosmic form *would be* and most probably *is* a sphere. Everyone accepts the Universe as a whole as a sphere...but why would the sphere form? What would be the reason why the original form that we devote to the Universe would take on a sphere as a natural form? Apparently our imagination grabs the sphere as form. In all natural events the gravity in that space which stands apart and independent from all other space takes on by cosmic pre-casting the sphere as form of shape ... **it is because gravity chooses the smallest space to hold the strongest force**.

I am of the opinion that gravity is about dismissing space to the advance of heat increasing in such a specific and concentrated space using the concentration as measure for the heat as well as the space holding the heat in space. According to Kepler that is what he found to be true. Space a^3 will always be circling space around as T^2 in any position from the centre **k**. That is what Kepler said when he said $a^3 = T^2$ **k**. Kepler indicated space a^3 will forever fight for independence and show separate individuality in remaining apart as identifiable cosmic components by means of motion. Every space will cling to independence indicated by **k** through fighting off the integrating of another coverall unifying unit by applying the motion of T^2! The problem we have to solve in this letter is what will the cosmos use to secure such independence between all particles? What sets space apart from the rest of space? First we have to admit that Kepler was the one that introduced the following.

Kepler gave us the answer to the following but no one ever took notice!

Kepler was the one that discovered **space / time** as $k=a^3/T^2$

Kepler was the one that discovered **singularity** as $k^0 =a^3/T^2k$

Kepler was the one that discovered **gravity** is holding **space-time** relative by the measure of distancing **k** as $k = a^3/T^2$ and $k^{-1} = T^2 /a^3$

Everyone able to read mathematics has to realise that Newton suggested collisions between cosmic structures must eventually come about as gravity erodes the distance separating the cosmic structures multiplied by the product of the mass of both structures from both ends. Newton said the multiplying mass of both structures destroys the distance between the structures by using the eroding force of gravity in the square. The cosmos then must end in a Big Crunch with all material joining together but that joining is not forthcoming at all...and that only indicates how much insufficient understanding there is on offer in cosmology by the educated–to-be-wise-about-these-matters. There is precious little available to explain about their field of cosmology amongst the ranks of Astronomers. So...let's us return to the beginning of cosmology before every one became oh so wise and see what there is to see. Let us see how the humble bicycle can teach the so wise about what gravity is because it is easy to demote the prominence of riding a bicycle when it can be so easily explained as just being a balance.

$$a^3$$
$$\Downarrow \quad a^3 \Rightarrow (T^2k)$$
$$(T^2 k)$$

A person that acquired the skills of peddling while staying upright on the bicycle has achieved the method of rearranging gravity within singularity. Without motion the bicycle falls on the spot it holds. When the bicycle is put in motion the bicycle can maintain the upright stance as long as the motion applies. When the motion stops the bicycle drops. To introduce motion to the bicycle the motion brings about a stable unsupported upright stance where balance can result from the motion the Earth enforces to the balance coming about by the bicycle using independence gained from motion of the space holding the bicycle. The space that the stationary bicycle holds is the direct result of the Earth providing the motion.

That is why it then will adhere to the gravity or motion that the Earth will enforce. The motion restricts the static bicycle to one allocated position that the Earth supplies to the bicycle. When the bicycle starts to move the bicycle gains a cosmic independence. The gravity effecting the redirecting of the Earth gravity response comes about as the result of additional motion that is introduced to the bicycle, This is the very same process that the aircraft need to get air born because it replaces or repositions the singularity the Earth holds to the singularity the bicycle develop in motion. The aircraft only takes the change in direction of what the gravity is insisting on through changing direction in motion through phase one and into phase two. It all is still part of the Coanda effect. With more motion contributing to acceleration the bicycle will become airborne on condition that it is also given the advantage of a set of wings to increase the effect of creating space-time to the advantage of the motion requiring the change in singularity direction.

I specifically chose to use a bicycle in my explaining because the bicycle is the object that relies the most on singularity achieving the required balance in which to operate. It is singularity, which puts space in balance of the time the space uses to duplicate. The singularity create space-time and such space-time results in a balance of space and time $a^3 = T^2k$. It is through the Coanda effect that marries the motion to the space that gives the balance that keeps the bicycle up right while it is singularity that allows the bicycle to move or duplicate the material by relocation through time. It is an act of balancing singularity that gets the bicycle as a machine working properly. It is also the next best thing to illustrate how singularity by motion provides gravity in addition to that which the Earth already produces. In the Coanda principle there are two factors where one is motion and the other is space and the two provide both duplication as well as contraction of space-time.

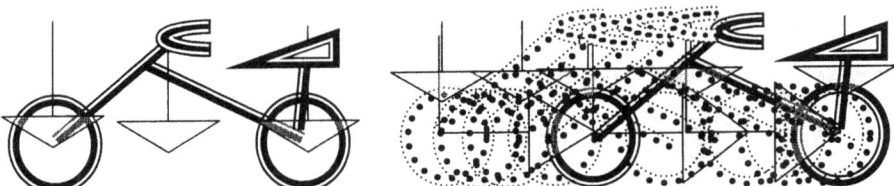

In the normally applying of gravity, we find contracting lines running vertically as the lines connect with the Earth centre. There are numerable lines running from the outer regions to the centre in diverting by 7^0. Motion provides extending of the 7^0 establishing the centre connecting points to the Earth to which it connects.

When it is only the Earth providing the motion there is only one spot of space allocated to the bicycle in the time frame applying at the time. The Earth takes on a specific size by which it duplicates the space it holds in relation to what it renders the bicycle also sharing the space, which the Earth has. By that standard the bicycle also holds an exact relation of space and volumetric size related to its position within the Earth. **This is where motion through duplicating changes the dimension in equation. Motion is the duplicating of existing space from time in the past through time in the present towards in the future to time.**

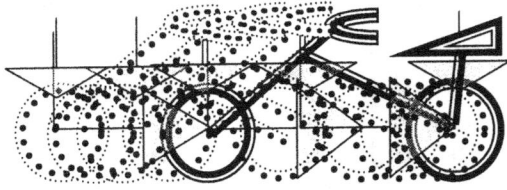

The instant the bicycle starts to move independently it increases its share of space it holds within the Earth. By duplicating the space allocated to the bicycle at a higher premium than the Earth does with the motion the Earth provides, the space the bicycle charge increases in ratio to that which the Earth charge because the bicycle maintain the duplication that the Earth grant but then still add to that space by enforcing more space provided by the motion addition. The motion of the bicycle not only extends the vertical connecting lines and not only changes the direction of the vertical connecting lines, but does both. The value added and the change in direction contributed is what brings about flying and moreover is the cause of the sound barrier.

The Earth takes the position that was before the motion of the independent object came about previously and held by the Sun . By establishing the directional motion in accordance with the k^0, which the Earth then provides instead of as previously provided by the Sun , relevancies replace previous ones.   In normally applying gravity, we find contracting lines running vertically as the lines connect with the Earth centre, which is the gravity we confuse with mass. It is a state of contraction and is the result of space being confirmed in relation to the motion of liquid time. Motion provides extending of the 7^0 establishing the centre connecting points to the Earth to which it connects.

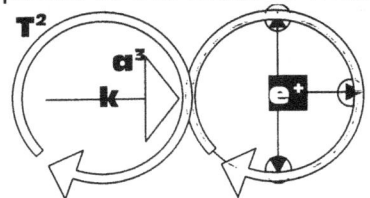

With The bicycle moving the bicycle change associations and where **k** was the factor securing the bicycle as part of the Earth that factor **k** then goes 90^0 in relation as the bicycle in motion then is in association with motion putting it then in terms of what forms a part of the liquid.

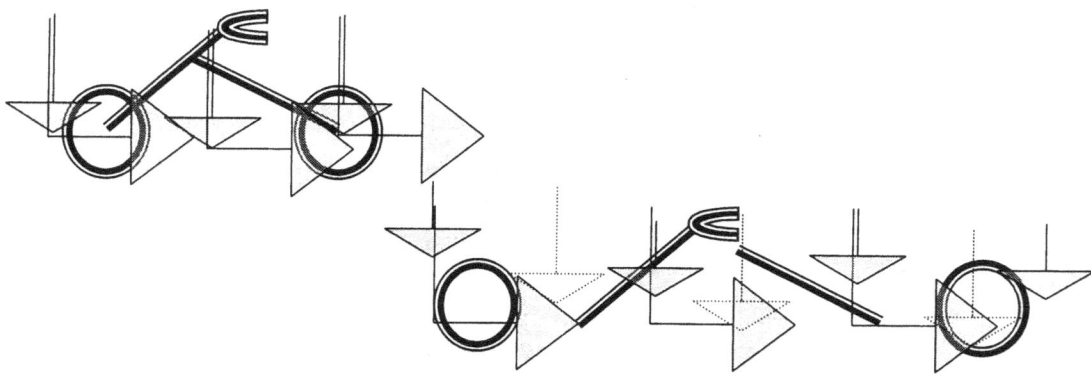

The motion provides the bicycle that is already confined to share space with the

Earth its due in motion. The bicycle can only have independent motion if and when the bicycle has the correct number of atoms filling the space the bicycle holds that will grant the bicycle independent motion. Without the unit being able to concentrate the correct amount of heat that will enable the space to generate motion, such motion under cosmos standards does not exist. In fact under cosmos standards the entirety of the bicycle unit does not exist. Only when the unit forming the bicycle can hold the number of atoms, which will produce the amount of motion through which the required heat will be subtracted from time in order to generate the gravity or motion needed, will the bicycle under cosmic standards gain motion. Before that the cosmos does regard the bicycle as the Earth and only the atoms standing independent from the Earth seems to regard the bicycle not being pert of the Earth. As the bicycle confirm the required and the displacement to launch an independent duplication wills the cosmos regard the bicycle in such a manner. When the atoms forming the bicycle unit are able to condense from time enough heat to give drive in order to provoke independent motion does the cosmos put the emphasis on the unit as a star. But in order to be the carrier of such independence the bicycle will have to grow. On its own as it is there is no way in hell that the bicycle can get fired up and start going as a star cosmos style. If not for life it will go nowhere but be consumed by the Earth to become a part of the Earth in time. In the cosmos the atoms forming the unit provide the motion and only when the total effort of the combined unit atoms manage to move abruptly and with the required confidence can the gravity be generated where the gravity generated is the motion of the entire unit. That means by having motion the fact of having motion grants the bicycle unduly respect from the Earth. The bicycle having motion of any status enlarges the bicycle in atomic space in ratio to what it has when without motion in relation to what the Earth has when only the Earth provide motion and when the earth stands in size in relation to a static bicycle.

When the bicycle is motionless, the bicycle is part of the Earth by gravity applied. As soon as life steps in and brings about separate and artificial motion but still uses the support of the motion that the Earth provides it will inevitably do better than the Earth as long as the motion that life provides is not in conflict with the motion the Earth provides. The bicycle becomes an object with the ability to transform the direction of the Earths domineering motion by redirecting gravity there in find the ability to change the direction of gravity. When gauging what happens we must also admit that it is highly unlikely to find running rocks on Venus or moving craters on Mars. The motion that applies to the bicycle is an extension of the second force in the Universe, which is not part of the Universe and only affects a very small part of the Universe within the Universe. Life giving motion is an alien product and gives an unrealistic adding to the Universe. The motion however has nothing to do with mass but is only extending what was to what will be through what is. It is refurnishing what will be with what was before it now is present.

When the motion of the bicycle accelerates such points that are forming by the increase in motion the forming connections extend to match to motion putting a standard of duplication per time unit extra to what previously was the norm. More space fills in the same period or the period reduce to match the filled space. The motion then contributes by increasing the space factor to keep the commitment with gravity valid. The bicycle breaks its form but because it is structurally bonded. Other aspects concerning gravity have to commit to the breaking of space. When expressed extremely crudely it is put as follows but is very bluntly stated.

Yet, it still is the best way to explain the basics of the sound barrier. The bicycle is the compiling of the independent space within the atmosphere space or time concentrated to be more exact, that is holding the motion in duplication where the motion is continuing from a facet going to the next facet by duplication what was through what is to what is going to be. While the bicycle is filling the space, in motion that is part of the space holding all aspects within by the atmosphere and the atmosphere is holding all that is in it together in the atmosphere of the Earth. That is time performing in ratio to space filled by material. Because the conflict the gravity experiences by having motion within motion, gravity first tries to break the object that is in independent motion.

As the motion continuous, as motion extends the reflex to the situation is then to contain by the breaking to the connecting devices such as the sound waves in the adjoining space. The atmosphere does the breaking of the relevant **k** on behalf of the object in motion since the moving space holding the object in motion as a unit shows much stronger bonding in structure unifying. We experience such breaking of space as the breaking of sound, which is showing motion or gravity differentiation.

The wing holding singularity while maintaining singularity puts the object in a Universe apart from other Universes. The aircraft becomes a separate identity maintaining a singularity and as all singularity does, such a singularity insist on two factors. The one factor is duplicating by motion the production of space-time while the other is the dismissing if space-time by controlling of space-time. The contact with space-time allows the motion to present the wing (and the aircraft) as being much bigger than in reality because it is not only the size of the space-time that maintains the singularity but it is also the contact or relevance which holds the dimensional area of control which stands as a controlling factor. These factors combined make the aircraft become bigger as the capability of motion suddenly allows the relevant size of the aircraft to grow in stature. The contact that the air or space has on one side which is more than the contact that the wing has on the other side where airflow is restricted, makes the side having more airflow larger by motion that the other side has in size by restricting motion. The motion enlarges or restriction reduces the contact and therefore the size per time unit. This alters the size as much as it redefines the balance and that makes the craft fly.

From the allocated position we hold the bicycle seems to be stationary when it is not moving. When we stand still it seems to us that we are standing still when we are not moving. However that is a human conception like mass is and is far adrift from a cosmic reality. We move as fast as the Earth spins. We move as fast as the Earth rotates around the Sun . We move as fast as the Sun rotates around the Milky Way. We move as fast as the Milky Way is rotating around another common centre because there shall be such a common centre that is allocated to order another common centre and this role diversification goes on running up to eternity.

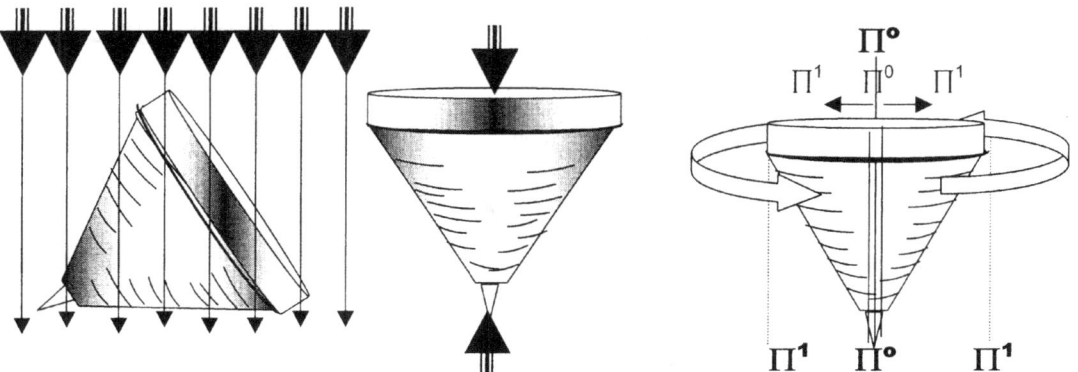

When any object is not moving the object form a part of the object, which holds the first object, captured. The cosmos disregard the existing of the first object in the event of it being stationary. However as soon as motion apply the cosmos grant the object having motion a position of existing by recognising it as an independent Universe that is entitled to all the privileges granted to a Universe.

As soon as the motion enters the equation the bicycle gets cosmic status because the bicycle generate time in the manner of parting singularity with time. The bicycle achieves cosmic independence and Universal recognition as an independent cosmic Universe. The bicycle is keeping upright because it forms a relation between time and singularity. The position is changing because time is changing singularity by the movement through time that gives the bicycle the opportunity to remain upright.

The position the bicycle had in the past taking the bicycle through the present into the future is what is keeping the bicycle in balance because the one position in time is supporting the next position, which is supporting the next position. It is the fact that time has one position, which supports the next position by supporting the next position, and this support forms a line, which brings about the balance by which the bicycle stays in line. Once the motion is no longer present, the bicycle – Universe collapses.

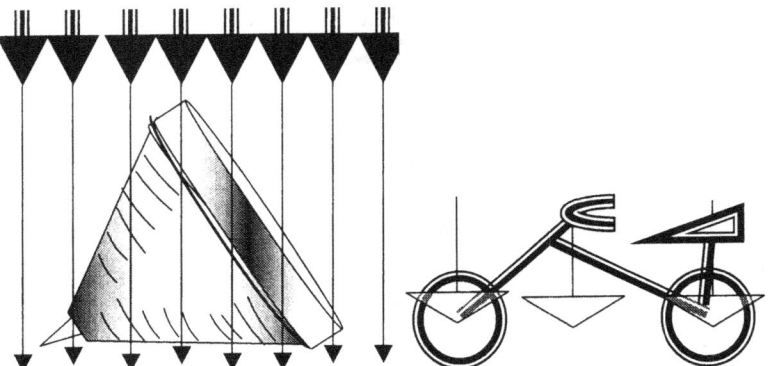

As the bicycle is standing still it is still in motion but holds a position to time where it fills a certain volume of space in that given time in motion. It is Kepler's $a^3 = T^2k$. Let us now forget the fact that life is responsible for the motion and pretend it is all a cosmic affair.

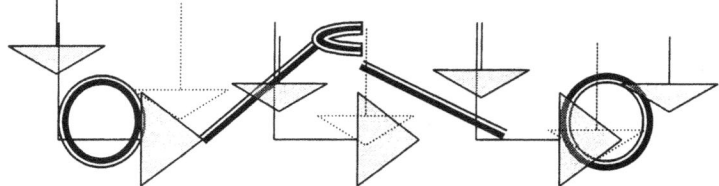

Since the bicycle moved faster than what it did when it was within the Earth motion it now fill more space than what it did before the individual motion commenced. In having individual motion on top of the motion the Earth supply, the bicycle is filling more space than it did before when it was stationary.

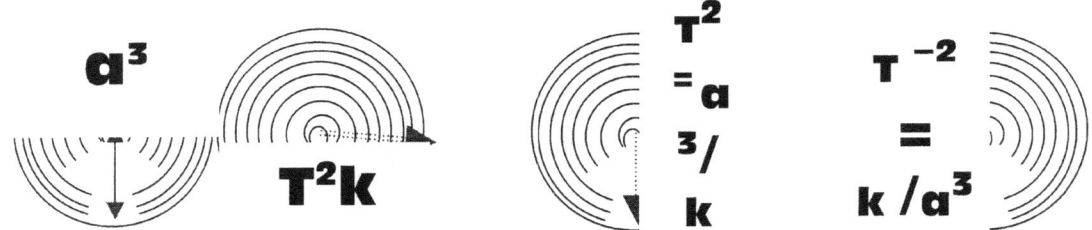

As Kepler stated there are two positions to the Universe unit. There is always space confirmed by the motion thereof in relation to the motion thereof.

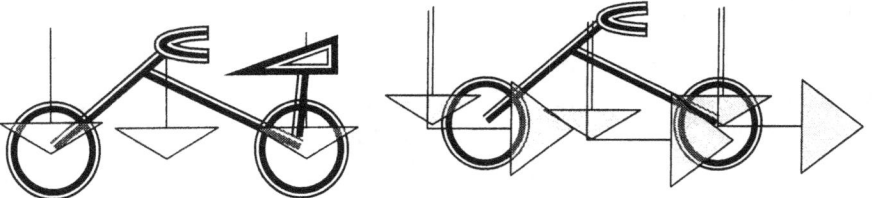

The motion fills space the Bicycle fills but also the bicycle fills space the Earth provides. It is filling more space within the Earth, which is space of the Earth than it did when only the Earth provided such space. Their stand bigger in it's filling of the earth space than it did before it started to move independently. It has more of the gravity going around than it had before when it was only dependant on the Earth to allow gravity going around. The bicycle grew bigger by the same margin that the Earth grew smaller in ratio to each other.

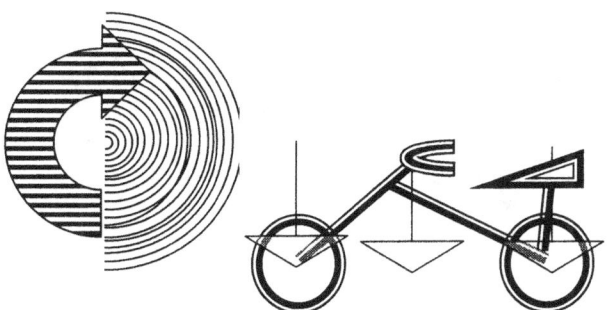

The bicycle without motion forms part of the space, which the liquid space confirms as part of the space. The bicycle then forms part of the other side of the Universe that is part of the solid.

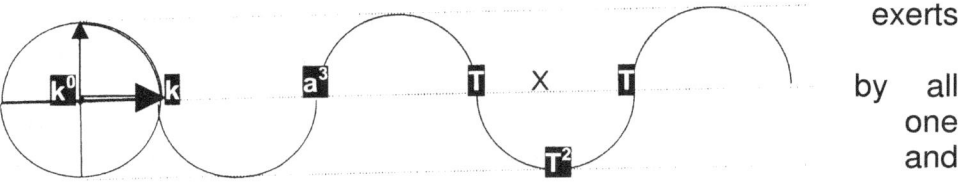

In the motion we have two contributing factors that plat crucial part in the dynamics as to how the material in forming the atom construction reacts on such motion. The atom comprises of spin around the proton by the electron.

Any motion an influence factors on another the exerts by all one and

dynamics is irrefutably connected. Yet at the same time the factors are so small it can never be detected and yet it is so huge it spans across an entire Universe. Remember every atom is a Universe in its individual making.

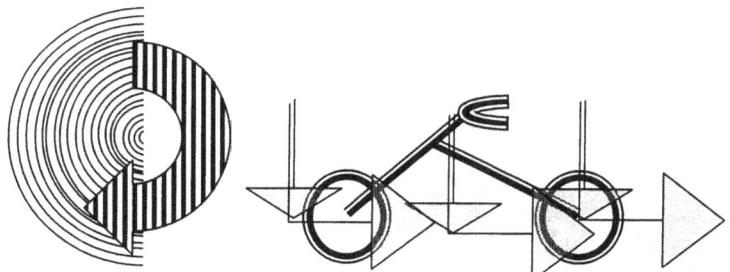

The moment the bicycle move the bicycle switches sides and switches allegiances. The bicycle then becomes part of the liquid that the space confirms not as space but as the liquid that extends the space. It then forms the gravity adding space instead of the gravity confirming space. It becomes $k^{-1} = T^2 / a^3$ instead of $k = a^3 / T^2$. What applied before then does not apply any longer because the bicycle is then on the other side of the Universe. Yet it is much more complicated than that because when stationary the bicycle was part of the Earth extending solid space which is gravity in relation to contracting being $k = a^3 / T^2$. As soon as it moved it became $T^2 = a^3 / k$ which is relative to the old position $T^{-2} = k / a^3$. **This too has no principles that it can share in the mass applying gravity idea. It is about being stationary in relation to moving.**

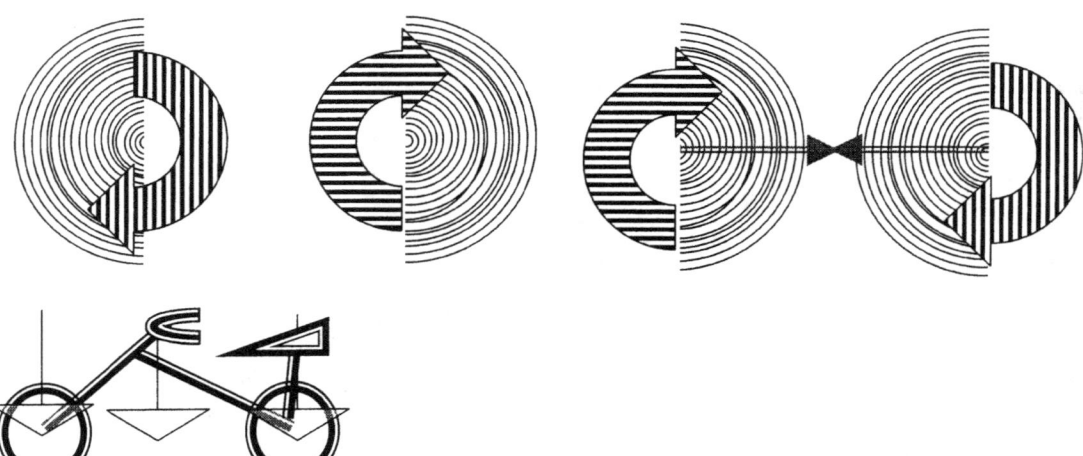

Taken from the Earth perspective there are lines (**k**) running at 90^0 to the rotation of the Earth $\mathbf{T^2}$. The earth gives the bicycle one line in singularity which to confirm its position as far as the space the Earth grant the bicycle to manage. That gives the bicycle a specific space to hold in relation to what the Earth has and in relation to what the Earth offers the bicycle. That is gravity. That however have no principles it shares with mass pulling mass. It is being on the one side of the Universe which space holds motion in relevance $\mathbf{a^3 = T^2k}$. In moving the alliances switches where the bicycle forms an legions with motion by duplicating the space it holds in relation to that which the Earth grants $\mathbf{k = a^3 / T^2}$ and therefore becomes part of motion where motion forms part of time $\mathbf{T^2 = a^3 / k}$.

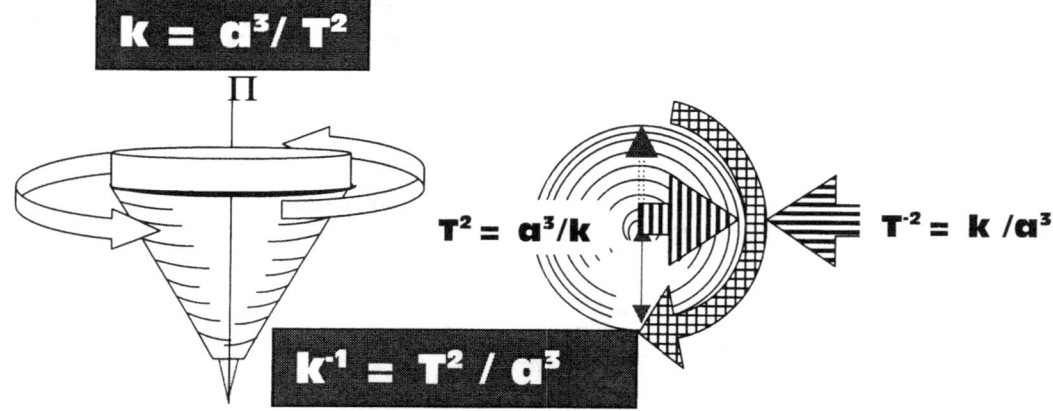

There are always two sides of the same Universe forming one Universe. There is the space extending to confirm the liquid by producing a solid and there is the liquid attaching to the space to extend the space by motion that secures the space. The one stands related to the other by opposing the other.

When the bicycle is part of space the motion confirms it by implicating it as space in the motion, which forms the extending of space. In that case the motion forms $\mathbf{T^2}$ that produces the contracting lines in singularity that runs a reducing and reclining formation into the centre. When the bicycle is apart of the space the line attributes to its space disposition and the space the bicycle holds allocated the bicycle the position it has.

There are always two opposing time lines forming one united space. The one is the line **k** and the other is the half circle T^2 where from those perspectives there then form the triangle a^3. By moving the bicycle then forms the 90^0 cross reference to the allocated position it had before. It is **k = a^3 / T^2 or it is T^2 = a^3 / k.**

The moving of the bicycle involves duplicating the space and the position the bicycle has in relation to the space the Earth allocates and the position the Earth allocates to the bicycle. The faster the bicycle goes is actually the number of times such repositioning of the space the bicycle holds are in response to what the Earth allocate and what the Earth takes in a specific period of motion of the Earth. When faced with the question of how the bicycle manages to stay upright it always comes down to charging the achievement to a balance...but a balance of what? What goes into balance to achieve the upright position? The bicycle repeats its position in relation to the position the Earth grants and when the repositioning of the bicycle is faster than the re-allocating of the earth, the duplicating of the bicycle from one position to the next will sustain that the bicycle can cross the vertical lines faster than the vertical motion will effect the stance of the bicycle. The bicycle firstly crossed its allegiance by no longer forming a partnership with space but becomes a factor of liquid presenting motion.

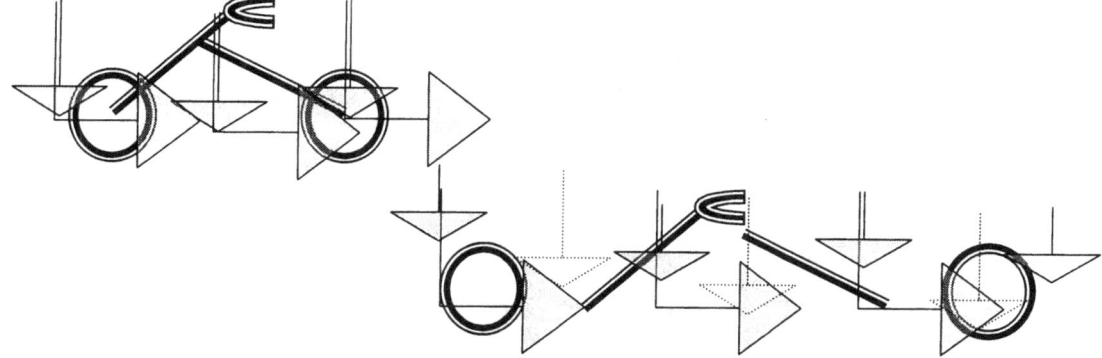

With the motion the bicycle is duplicating in ratio more space than what it had while being part of the Earth. The motion holds more space because it holds less space per time unit and there are more units of space per time unit. Since the bicycle is propelled at a faster pace than what it was when the earth alone supplied the space and forced the time by establishing the duplication tempo in the time contracted. By supplying more motion the bicycle grew larger in ratio to that which it had when the Earth was the sole space-time provider. After accelerating the bicycle then has more space in relation to what the true status is.

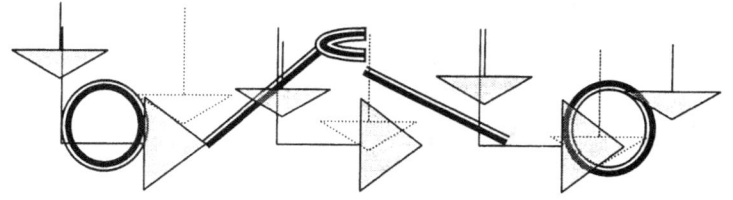

The bicycle holds a larger part of the Earth by which the Earth has to reduce the space it offers the bicycle. The earth had to shrink in order to provide the bicycle with more space.

In the normal relation that the bicycle has with the earth when the bicycle is motionless (or in the manner we think about the status of the bicycle in terms of only its position within the scope the Earth provides) being motionless and then started moving the space increased rapidly as the Earth space decreased rapidly. With the new motion the bicycle finds much more duplicated space and that disturbs the ratio of space shared by the bicycle within the confinement of the Earth. The earth now presumes the bicycle to be much bigger than what it was. By moving the bicycle physically got larger and this is a fact not only by relevance but also by actual annexing and capturing of space in any given period of equal ness. The earth provides a certain value but as the bicycle moves faster the bicycle annexes more of the Earth space and that improves the size of the bicycle in a volumetric and physical measurement. It is a^3 grows bigger therefore the earth a^3 has to compensate by reducing that much actual space. In this comes a problem.

The bicycle does not even contribute a morsel of space when compared to what the Earth delivers. The earth that actually became smaller resents it becoming smaller by the demand of such an outrageous exploiter. To the Earth the bicycle is motion and the motion is liquid and therefore taking space contravenes the being liquid part. Being liquid is also being heat but being heat means becoming hotter when becoming more. To the earth the moving bicycle is liquid motion and being more makes it being hotter and being hotter is therefore is more volatile. Therefore the Earth refuse to become smaller and the bicycle being space

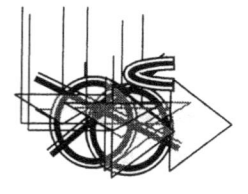

claimed cannot become hotter without destroying its independent molecule unit.

By the size the earth holds the Earth will not allow such a renegade to grow bigger and the Earth sanctions the time in space in accordance with the Earth generated governing singularity. Since the generated singularity the bicycle has still adheres to the dominance that the Earth generate, the bicycle reduces it proportional space in the face of the Earth showing such a strong reluctance to abide by the will of the smaller bicycle. The earth crushes the bicycle in response to the bicycle growing and when the response is more than what the bicycle can withstand, the bicycle crushes buy reducing space. That is what happened to Challenger when it entered the Earth and was liquefies by turning into gas in time.

From the past   In the present   Onto the future

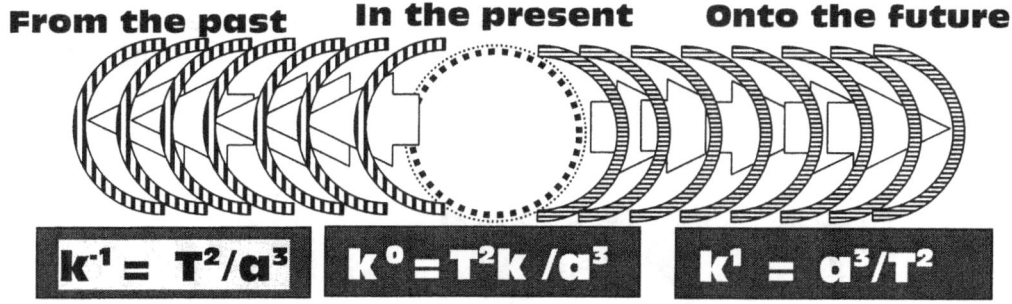

$$k^{-1} = T^2/a^3 \qquad k^0 = T^2k/a^3 \qquad k^1 = a^3/T^2$$

The ratio between space and heat goes into an imbalance where the liquid that is standing as ($k^{-1} = T^2/a^3$) totally dominates the space factor $k = a^3/T^2$. The cosmos informed Kepler of another gravity, which the cosmos applies much more widely and is used by nature all over the Universe. Being with life and being part

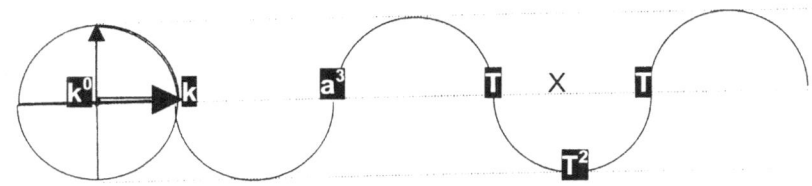

of life we humans take motion in content. Life is the manipulation of space-time by motion and since we can move because life is about motion we generalize motion with contempt. Motion is the most complicated process in the Universe because the Universe is motion. But why would motion occur because the fact of motion proves the presence of a Universe.

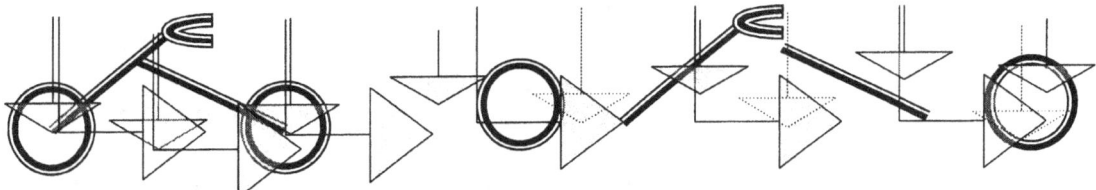

Motion distributes space and therefore decreases heat. By spreading the space over a larger area the heat in the space is reduced because the density of the heat allocated to the space reduces. Motion decreases the heat and therefore the Sun is the coldest place in the solar system while outer space is the hottest part of the Solar system

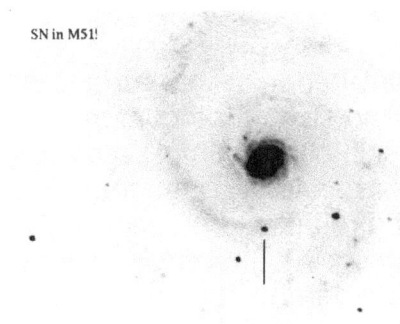

SN in M51!

Supernova stars are stars that overheated. The overheating came as a result of the motion in duplication that was not in ratio with the control in contraction. The expanding of the space the supernova held before had nothing to do with mass. The expanding had nothing to do with Newtonian gravity. It is all a ratio that puts a relation between the space that is taken by the material in concentrated heat (or

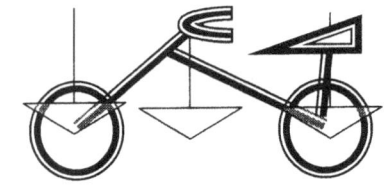

time) and the time in ratio to the duplicating material. If the material forming the motion is not very highly dense the motion is poor. By the same token is the material density high when the motion is volatile. It all depends on the motion that the atoms forming the Unit generate and that motion is forming the gravity. There is no mention of mass except in the imagination of the Newtonian mind.

There is no pulling between particles compressing them into plums of pressure. That is nonsense because a star has no containing wall on the outside to capture the pressure on the inside. It all depend on the duplication of material confirming the containing of the unit in relation to the distributing of the material in ratio with the liquid time with which it is in partnership. It is all about containing heat and distributing heat at the same time. The more motion that is present the more heat is condensed and is contained.

By duplicating at a higher ratio the heat contained is spread over a greater part of time, which reduces the heat contained in the material as it is distributed in a larger ratio to the liquid heat that is maintaining the balance.

There is a defined ratio between liquid heat that is the basis for time and solid heat, which forms the norm for space. The higher the ratio favours the liquid the  colder the solid will be and the lesser the solid is in contact with the liquid the hotter will the solid be. Stars are ice-cold ice cones floating in outer space because of the motion they generate. If we pump air into a compressor the air gets more inside the compressor. The compressor gets hot while the air gets more. The size or the compressor remains the same while the air gets more inside the container and that means the compressor is shrinking while the air is remaining the same because the air cannot get more while the compressor is unaffected.

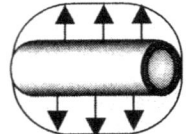 In relation to the heat the heat gets more because the surface of the compressor remains the same while the size of the compressor within the relation shrinks. Because the size of the compressor that shrinks the space outside the compressor has to accommodate the flow of heat because equilibrium has to be re installed. The size will remain reducing up to a point where the compressor is just to small to accommodate the air. The air then will expand. However the air was always expanding from the pumping started because the compressor inside became

smaller as the heat ride to expand the size of the inside of the compressor to match the outside of the air.

 The same goes for material blown by wind to reduce heat. The object has an indicial size to start with. Then we put heat to the object and the heat makes the object increase in size. That is hardly the increase worth noting because the relevancy of heat in the air to the heat in the heating object goes array. The heat has to increase the size of the object in relation to the match it has to find in the space it is within.

With the heat coming into the object the relation the object has with the heat or air outside makes the object that many times bigger because the ratio in the heat balance is disturbed. If we blow air over the object we increase the size of the object by allowing the surface of the object to make contact with much more air in the same period of time, which will bring the size of the object back to normal because in relation and considering the contact with air the object expanded by the motion of the air in contact with the object.

In the normal flow of time the object has a heat to space relation set by the time the Earth dictates. Then we go and increase the heat on the object and in that event we actually increase the size the body has in relation to the heat in the air. By blowing air over the body we increase the air and therefore we increase the size of the body in the same period of time.

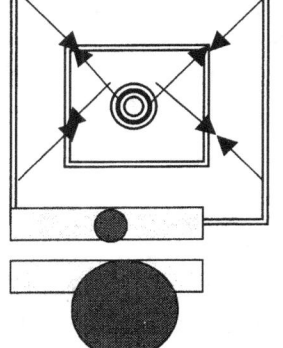 There is now a dispensation of many times the body carrying more heat and contacting many times the heat or air which bring the equilibrium back to normal. There was a body size and by applying heat the balance shifted to the reducing of the body size in relation to the heat. The body then had to expand in heat because the body was too small to incorporate the large heat. Blowing the air over the body increase the size of the body and heating the body decreases the size of the body in comparison with the air it comes in contact with. The body is either expanding or the body is redefining and the balance in heat places the body in relation to either gravity cooling by contraction or expanding by overheating. The very same principle applies in the sound barrier.

In spite of outer space being as expanded as anything can get we still regard outer space as being incredibly cold. Anything expanded to its limits and which can heat no more is as hot as anything will ever get. Outer space is the very edge of expanding of space where heat cannot expand into space any more. On the other hand we fin that concentrating heat is producing cold making anything in the atom in

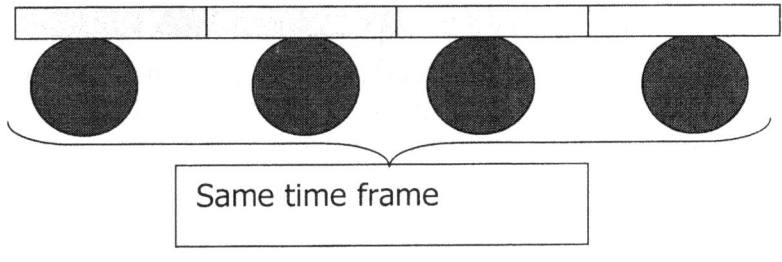

Same time frame

stars as cold a there can be. That heat filling the outer space lacks motion and is

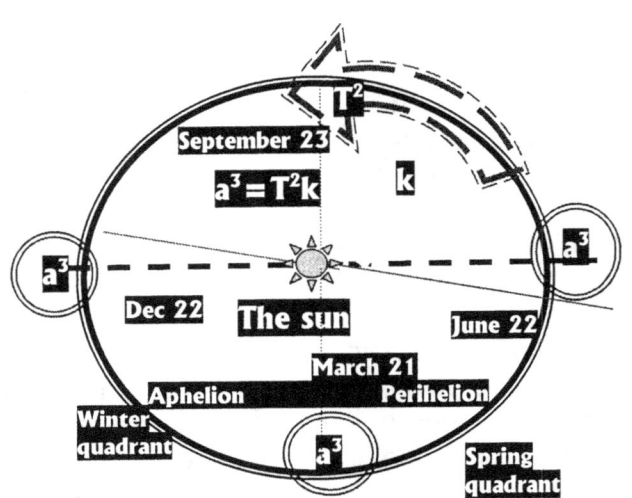

therefore space in another form of material that could conduce by diverting from space to constrain further expanding through motion and therefore was unable to marry the union of space by becoming more space. One must look at outer space and judge outer space from the findings only considering outer space. By motion space duplicates and by space halving it removes heat in space as well as by dismissing space. In the case of material the electron is spinning at the speed of light to contain the heat inside the atom at a higher rate than the speed of light. It is containing heat at a greater motion than what light can travel The atom by motion is the condensing of heat by contraction in relation to expanding or duplicating of heart. The concentration or release of space with heat or space from heat is a direct contribution of the singularity in control of the space-time. The regard of the singularity stipulates the conducing of heat in space or the release of heat to form space by means of bisecting the occupied space. By applying motion the space duplicates what it is from the past through the present and into the future. This is no hypothetical suggestion but is the actual flow of time coming from outer space as a liquid and is incorporated into the spinning atom on the way to confirm singularity. While we are in gravity the manner in which gravity applies in our use of gravity makes us part of the Earth by mass forcing us onto the Earth as a semi unit with all other Earth belongings. Is that which we have truly gravity? By using mathematics, the cosmos spoke to Kepler personally and by the use of mathematics as the medium, it provided Kepler with information about the cosmos coming directly from the cosmos.

The picture we see coming from the Hubble telescope shows why, in the perfect Universe...but can the Universe be perfect when... we see a radius between the Sun and individual planets is not using a regular distance as one would expect of gravity in being a force driven by the mass and in that sense the mass is producing the gravity that always remains even because the mass doesn't alternate. As the mass is never changing on either side, that steady mass has to keep the gravity steady. But in our imperfect understanding of the Universe we find that the radius that should be constant varies considerably proving either that mass somehow adds by measure unnoticed while the structure is in orbit and later allows the same amount of mass to escape undetected; or it's the seasons adding and removing mass at will. This is an absolute contradiction to reality if mass was the factor determining the radius we find between the Sun and the planets. This suggests strongly that we'd better be getting very suspicious about the idea of mass contributing to gravity. But in contrast to this, science is unshaken about their confidence in the perfection about facts they use in terms of correctness. It is well known amongst all persons that science only uses dependable and ultra reliable facts coming from sources beyond doubt. Referring to any work done by any scientist will find a remark about science only accepting facts they use to work with. It is accepted overall by all communities that in science those in science use one hundred percent accurate facts or they use no facts at all. If our view was as perfect as science would lead us to believe it then must be the Universe that is imperfect as it otherwise would not behave so mystifyingly. The unshaken confidence science uses has us believing at first consideration that the drawing of gravity should produce an even diameter positioned between the Sun and the planets because of ever dependable evenly distributed gravity... but I believe there is a perfect Universe and our understanding carries the doubtful suspicions.

Delving deeper uncovers even more contradictions and the level of accuracy contained by our scientific understanding then arouses more suspicion about the correctness of science. Remember Newton changed what the cosmos told Kepler leaving much suspicion as to how far the misdirection takes science. We have to correct the facts we doubt because when correcting the facts they use in science concerning our view about science, such correcting brings along a better understanding and then the Universe has to become ever more perfect as one learns to understand the perfect Universe even better. But it does require an open and clear mind and it needs no culture driven preconception that should confirm interpretations about facts surmised even before they are carefully studied. It becomes obvious that Newton never gave careful attention to Kepler's findings because if he did he would have seen what gravity was. Kepler described gravity without using the name that later was given as 'gravity'. Kepler did not give the name gravity, but Kepler's studies gave Kepler the insight to coin the concept of gravity. Nevertheless it was a name and not the concept that was later named by Newton. The naming was the contribution of the Englishman. The concept that Newton later introduced is totally incompatible to the concept that Kepler introduced. What he (Newton) introduced as the force of gravity, he connected to mass, which diverts totally from Kepler's findings. With giving a name, the Englishman also changed the concept that Kepler introduced. Kepler made no mention of size or mass as part of the phenominon that later was named as gravity, yet it must be gravity that holds the Universe together. The concept the Englishman changed when he introduced what he introduced with the name he introduced. That what he introduced, he corrupted beyond recognition. The concept that accompanied his new name strayed completely from what Kepler introduced. Newton brought in something that was mismatching what Kepler saw in Kepler's view of the phenominon that holds the Universe true to form. The name was dominant but even more dominant and totally inaccurate was the other concept Newton introduced. In truth Newton only gave the world a name of an idea, which he then corrupted as far as cosmic physics are concerned. It is important to admit that as far as cosmology is concerned Newton gave the concept the name but *only* the name and not the concept of gravity. Newton's persuasion on matters of gravity as gravity functions between cosmic structures orbiting one another as we find in outer space is inaccurate. What Kepler saw, Newton saw differently and used the opportunity that Kepler left by not giving any name to the process he (Kepler) and Tycho Brahe worked on for two life spans. Newton did seize the opportunity to name what he, Newton, saw but that what Newton saw did not include that which Kepler uncovered. In Kepler's era the name or title was lacking but Kepler established the concept of gravity and the formulation thereof. The concept came from Kepler even before the name gravity was used by Newton to describe in the concept of whatever we today (after Newton) became accustomed to believing what the concept of gravity is about. With the help of Newton everyone since Newton confused Kepler and Newton on the issue of gravity and this confusion even begins with Newton. Gravity might not have been named but became a proven concept and factor after Kepler formulised it, which is before Newton named it. The concept of gravity that Kepler saw is about the manner in which the structures orbit because there is a space that circles around a centre and this process has kept planets secured, connected and rotating around the Sun which is the same concept that is keeping the Universe secure and comes about with a process Newton later named as 'gravity'. What Kepler saw is not the same as what Newton saw when he saw two objects drawing closer by pulling on each others mass. Then later on Newton named, what he thought he saw as the force that Kepler saw but introduced another completely different concept. Kepler saw cyclic formations keeping the Universe together and never approaching each other. Newton ignored what he wished not to see but he changed as he saw fit and what he thought that should be. His experience as a young man drove him to establish a process he formulated as the process that is keeping the Universe together. In that act he corrupted as much as ignored the work of Kepler, which he also named as the same gravity that he saw as a young man. Why he chose to ignore Kepler's findings on gravity we shall never know but why the world still chooses to ignore Kepler's findings about gravity almost four hundred years after the fact I shall never know. My saying this has literally made Academics ignore me as they would avoid the plague. I am not pretending nor do I exaggerate when I say there were those in Academic institutions that questioned my mental development. Some went as far as seeing me as a joker of sorts and I have correspondence to show evidence to that fact. I know by now while Newtonians are reading this letter I have aroused the tempers of every Academic reading this far, therefore let's see what is being ignored by the Academics which I blame to do just that. .

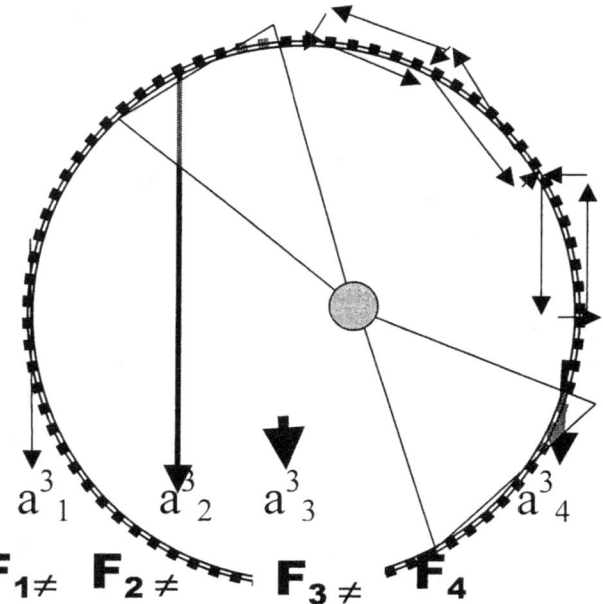

$$F_1 \neq \quad F_2 \neq \quad F_3 \neq \quad F_4$$

We live through seasons which comes from being that at one point, (a^3_1) the distance between the **Sun** and the earth **is less than** at another point we call a^3_3

Let us put a value **of a^3_1 = one** and a^3_3 = three. This means that each year, for the past 4 500 000 000 years the effect of the common gravity between the earth and the Sun has a greater effect than at another point six months later.

That means at one point the earth should be drawn or pulled closer to the Sun   and after another six months interval the earth should stand less effected by the Sun 's gravity, therefore it should move away from the Sun .

Kepler said gravity in space is about the area a^3 that would always keep equilibrium with the time T^2 it takes to travel the distance of the full circle position placed by the indicator **k**, therefore adjusting **k** as the need arrives. With **k** shifting in length a^3 will have to readjust and therefore T^2 will find a new relating value each time. This was the finding of Kepler and came after his intense study of orbiting planets.

Before I attempt any investigation into this matter there must be coherence in our agreeing about what gravity is. If you the reader insist that the falling of objects is the only gravity found, your further reading will convince you little. Anything we do decide upon must support the fact that it is gravity that prevents planets from dislodging from the grip the Sun has on them. Gravity is not about the Sun trying to catch the Earth by attracting the Earth...no, there is so much more to gravity. We must be under no illusions about what gravity is and that being the focus of our discussion and where that gravity is because we have to identify and not confuse the gravity we are looking at. We are now discussing the gravity, which is keeping planets circling around the Sun , and stars around specific galactica centre. In that we do not find one example to use as proof in connection to stars coming tumbling down on galactica centres and crushing into galactica centres. If that is gravity keeping structures in orbit around specific centres we must look at the behaviour of the structures in gravity. We have to find a reason why the planets do not reduce the radius between them as Newton suggested but we must trace the reason why it is gravity, which is keeping them apart because if anything, they are departing as they extend the radius connecting them to the Sun. That is gravity because it applies throughout the Universe. The gravity Kepler found is the general gravity that is keeping structures from colliding and in that the principles are avoiding collision or on the other hand avoiding abandoning each other. It is about confirming respect for one another's independence and clearly staying at a predetermined distance while at the same time both are sharing a common space unit. That then must be the defining of gravity we have to study to find the Universal enticing gravity holding the Universe together. By close investigation one will find three factors in urgent need of investigation. There is firstly a centre that draws the object closer. This gravity is clearly a synonym to what Newton saw as gravity. If it were not drawing the object closer the object would not be orbiting around the centre and applying motion. It will draw and absorb all rotating things in its field of gravity.

The fact it does not draw the object into its ranks is because there is another gravity standing alongside this first mentioned gravity. Our recognising the first gravity forces us to accept the presence of another part of gravity. This forces us to recognise the second gravity. When saying this we are not using Newton's cosmic formula concept $F = G\,(M.m)/r^2$ because that can barely be what is out there happening. What Newton saw was falling. If that what Newton saw is the only gravity then whatever Kepler saw including all other parts of everything out there that are spinning around some centre must come closer to one another and connect in collisions. While that is not happening we must start to look past Newton to new grounds we can investigate. We have to go beyond Newton and admit there is more than that what Newton led us to believe because it is clear that what Newton had us believe…is not happening. That confirms the presence of the second gravity. The fact proves that everything is departing and not arriving. Even the moon is drifting away from the Earth and this information comes about from the most advanced investigation up to date, including a moon visit and the placing of measuring devices there.

Looking at the gravity intensely we find the roving structure travels in a straight line, which repeats another circle around another centre but because of the influence of a centre keeping the roving structure attached to such a centre the motion allows a circle to form by reforming motion from the original straight line to

that of a partial circle. There is a centre; a connecting line travelling between what the two points establishes the specifics of a centre within a circle and the end of the circle. According to Newtonians the centre supposedly draws the rotating object closer. That is half the story.

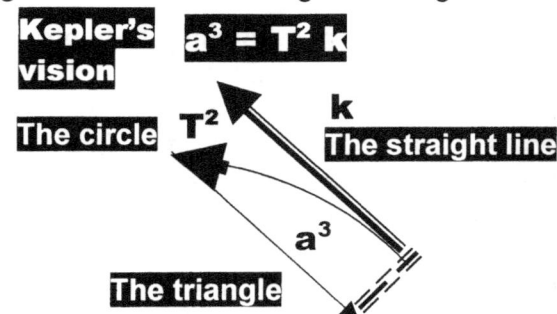

I suggest we do some deliberation and in deliberating may I remind you THAT NEWTON'S OWN LAWS ARE IMPLIED, and again the planets disobey these laws completely!! In the modern age all evidence points away from contracting and favours eternal expanding.

The latest news confirms that the lot is apparently not coming any closer!

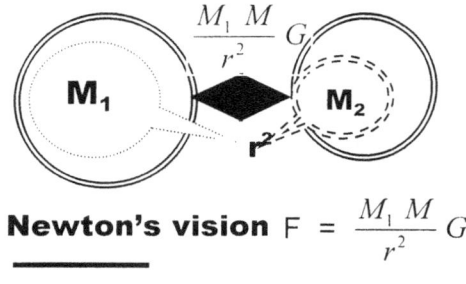

Newton's vision $F = \dfrac{M_1 M}{r^2} G$

Newton saw his apple fall and then went on to blame everything on mass… and you think it is all that plain and simple? Kepler on the other hand said (and let's forget what Newton said about what Kepler said for a bit) that a space of cubic proportions a^3 that will always keep equilibrium with the time T^2 it takes to travel the distance of the full circle at the distance (or relevancy) in ratio or relevance k with an indicator pointing the distance the circle is from a specific centre. That means k is as crucial as T^2 in positioning a^3. In placing the allocated position a^3 requires in determining the sectors (we think of it as seasons) a need arrives to predetermine k in order to measure T^2. Every spot a^3 fills is located at T^2 and is allocated where k indicates. When k shifts relevancy (from Earth to mars or even from season to season) the space in a cube a^3 will

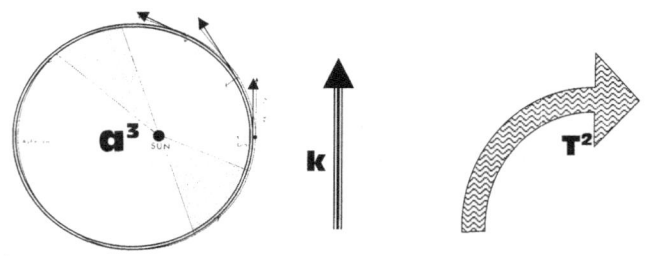

have to relocate as well as readjust and with that T^2 will be redefined. This was the findings of Kepler after annualising the work dome by Tycho Brahe and then later himself. The line forming looks straight from the onset but the line never moves straight but goes bended in relation to the relevancy that k indicates. In what k contributes as a factor it introduce T^2 and from the alliance of the product that T^2k delivers we find that a^3 forms an eternal circle about an eternal centre.

That is gravity if you wish to call it gravity by name. That is the force that is no force that prevents planets escaping from the grip the Sun has on every planet. Gravity is what puts order to what would be the most chaotic arena there can ever be. That gravity has no bearing on mass because with mass pulling that order which gravity then would bring will result in complete destruction. That is not happening. On the other hand if we don't do what Newton did by putting words in Kepler's mouth we find that Kepler gave gravity a completely other meaning. Kepler said (when ignoring Newton's uncalled for interfering and meddling) that

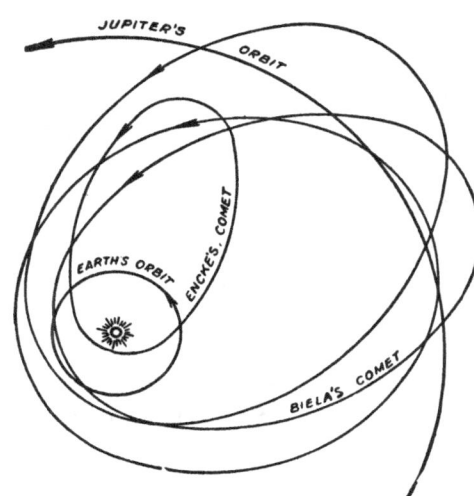

the motion T^2 puts the space a^3 at a distance from a centre and the relevance factor k prevents a^3 to come close or drift further away but stay in the allocated position where T^2 locates a^3. The factor k has it at task to prevent a^3 coming closer and orders T^2 complying with a measured value. By denouncing k everything Kepler said goes array.

Every object in outer space holds as much turning as it commences its run in a straight line. It turns as much as it goes straight. The comet coming towards are no different from the comet going away and by passing a dividing line they change direction from coming towards too going away but in that they remain equal.

A space remains between the comet and the Sun that define the line it crosses when the comet changes from coming to going. That gives the comet a cyclic approach and departure and does not put the comet on any collision course with the mass of the Sun . There is no death defying all destructive route calling for a disastrous end with no chance of avoiding the immanent disaster colliding that is unavoidable predestined to happen as Newton's formula would suggest. As the comet approach it is the relevancy brought on by the changing of k that puts a^3 in a ratio with T^2. It is k that sets the approaching limit as it sets the departing limit and it is k that prevents a collision as it prevents an escaping departure. That is what Kepler said...and that is not what Newton said Kepler said. What Kepler said is quite a different story from what Newton said Kepler said...

> $a^3 = T^2 k$ then $k^3 = k^2 k$ and this is showing that the space k^3 is equal = to the motion $k^2 k$ of the space k^3 seen form one specific point.

In Kepler's formula $k = a^3/T^2$ the smaller k becomes the smaller a^3 becomes and the bigger T^2 then gets. T^2 represents the gravity that positions the space a^3 at distance k from the centre capturing the structure through gravity applying T^2. We all are very aware which star is the mighty gravity producer. So has mass the least say when gravity is generated? It seems most likely to be true.

$k = a^3/T^2$; the distance depends on the position that the orbiting object develop space-time

a³= k T²;the space depends on the distance the space develop from the centre and the speed the space moves around the centre.

T²= a³/ k; the speed the space orbits around the centre depends on the distance of development and the size into which the space developed.

Gravity has two factors influencing space, which are a straight-line **k** and a circle going around the centre **T². a³ = k T²**

Translating Kepler's mathematical expression **a³ = T²k** correctly to the verbal statement in English Kepler said that there is a space **a³** which is equal **=** to the motion in the time duration **T²** thereof between two specific points which holds a relation to a centre wherefrom there forms a straight line **k** and is located on the spot where space begins the circle. Therefore that spot has the least space.

The value of Kepler's space he indicated as a third dimension **a³** does not depend on indicating a structure **a³** that is in rotation **T²** but only needs one position having a constant of some sorts. Any point where **k** may indicate a position one will find a value matching **a³** and the matching location will fit **T²** at that point. That is the relation there is in the solar system between all planets and the Sun . The Sun always indicates the centre and the planets always indicate the rotation. But **a³ = T² k** is only producing a relevancy of three dimensions that is equal to two plus one dimension.

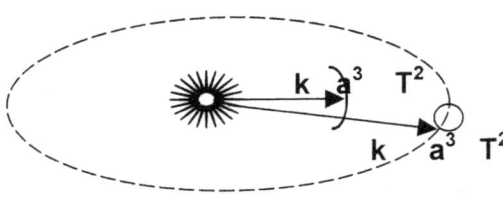

From the Sun there are three points moving between two points from one point to two other points giving six dimensions we find that is forming space. It is space in time or space converting space through the moving of space in time. It is locating a point in the third dimension **a³** that will move according to the second dimension **T²** that will implicate **k** as a reference in the first dimension. It is the duplicating of space by time providing the dimensions to do so.

Let us take it from a point where the Sun provides a centre as one starting edge of **k** then that centre **k** will provide a line from the centre and the line **k** will provide three spots in a formation that produces a structure by the square **T²** of the dimension. Not once did Kepler indicate size as a contributing factor to **a³**. That means every single point that **k** indicates there are three positions **a³** implicating sides of a double dimension. In the same manner is **k** not limited to distance or is **T²** lesser by size. **k = a³ / T²**

There are infinitely more implications in the statement Kepler delivered than what is merely a contribution to motion and only motion as Newton was of the opinion. What is there mathematically not correct in my interpretation of Kepler's manner of translating mathematics to English and why is any changing thereof by Newton or any other person necessary in any way?

We can test any of the following symbolic values in the mathematical expression and also test the principals behind the expression in which Kepler stated them. By such testing we will find that time after time there were never any corrections in the translations required since the translation thereof was never incorrectly presented and in that a case asked for no alterations to secure the correct reporting of the cosmic information being translated. By taking the formula on face value it can change as follows: $a^3 = T^2 k$ can become $k = a^3 / T^2$

When translating Kepler's mathematical expression into English we can see what Kepler said also read as $k = a^3 / T^2$ where k is one point from a centre point that is space a^3 relating to time T^2. From a centre comes space-time. The centre k brings space a^3 in ratio to time T^2, which is space / time a^3 / T^2. Reading this correctly cannot bring any dispute…yet it does…and it's been doing it for centuries on end!

The cosmos spoke to Kepler about space-time coming from singularity. Kepler gave us his findings. Any discomfort that may come when we read what is revealed must be set aside, because we must remember it is not me, or Kepler, but the Cosmos that is doing the revealing and lending us the tools we can use to decipher what the cosmos is trying to make us understand. Kepler translated what the cosmos told him (Kepler) as $a^3 = T^2 k$. Translating Kepler's mathematical expression $a^3 = T^2 k$ correctly to the verbal statement in English Kepler said that there is a space a^3 which is equal = to the motion in the time duration T^2 thereof between two specific points which holds a relation to a centre where from there forms a straight line k. What is there mathematically not correct in Kepler's expression and why is any changing thereof necessary in any way? It says where there is space such space has to move. Test the following symbolic values in the mathematical expression and test the principal behind the expression in which Kepler stated them. Convince yourself about the evidence that Newton saw what Kepler saw where the translation thereof that was done by Kepler is mathematically incorrectly translated by Kepler's interpretation from mathematics to English:

a^3 The fact that any symbol uses a value to the third power indicates space or a volumetric established and separate unit which is serving an under dividable dynamically separate space being within a space. Although being apart the two in space sharing a unit can never be apart but serves as a unit by division of motion. It is space because it is volume using the third dimension. But since the space is smaller than the Universe it must be space being within space, which is within space. There a relevancy is forever present.

T^2 Is an indication of space apart from the surrounding space by granting the independent space by establishing borders through motion, an ability of moving from one point to another point or following a flat distance between two points. It is motion that is taking time in the second dimension.

k^1 Is the symbol used to indicate a straight line between two points with a definite beginning and a specific end position. The two points is valid only by re-aligning an eternal straight line to the figuration of a circle through alternating as well as recognising the control coming from such a centre. It is Pythagoras by the triangle, half the square and the straight line sharing value in the 180^0 they represent. Kepler introduced this absolute basic mathematical principle.

That is what Kepler said. There are three dimensions a^3 between any two points T^2 flowing as time from the centre of the Sun , which is indicated by the line k.

The implication of the relevancy produced by the use of the formula $k = a^3 / T^2$ brings about that when dividing T^2 into a^3 there is k left.

The fact is that a^3 is a three dimension (^{3}) of single k (^{1}) showing one or T^2 is two dimensions of k being the one dimension. It means that k is a part of space a^3 or T^2, which is time. It is the same thing in a double dimension or space being a triple of k then k is one factor and k cannot show a position of zero.

If $k = 0$ then there is no possibility of $k = a^3 / T^2$ because $k = 0$ then $0^3/ 0^2 = 0$. That does not make sense. Mathematically space cannot be zero because those being of the opinion of space being zero or nothing must first prove mathematically that space is zero.

Moreover they then must prove mathematically how does zero grow through the Hubble constant. By translating Newton's vision of the circle in completing a cycle would become zero through rotation…well that does not count the use of the formula a^3. If k cannot be zero then k could not start from zero.

With $k = a^3 / T^2$ no point can be zero because k shows space $a^3 = k T^2$ is no reference to the volumetric mathematical formula used to calculate $a^3 = 4/3 \Pi r^3$.

Nor does it show the use of the circle in the second dimension being $a^2 = \Pi r^2$.

Newton's mathematical vision was the way to calculate the space by using a mathematical formula used as

$$a^3 = 4\pi \ X \ (r^3/3)$$

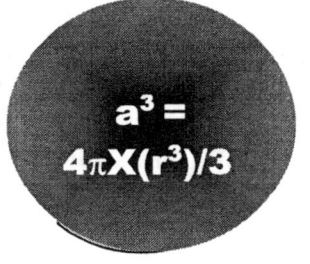

Kepler's cosmic vision was that in the formula $a^3 = k \ T^2$ the space a^3 is equal to the movement $T^2 k$ of the space, which comes about as time T^2 in relation to a distance k

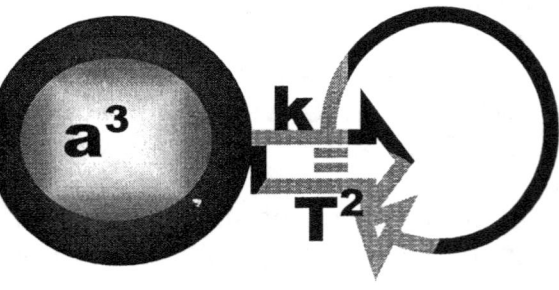

In the case of the Newton formula the circle factor becomes the square as indicated by the duration of the time T^2. The factor standing in for the line which normally would be r and then be the square value is in the case of Kepler not the value indicating the square. That means Kepler never indicated a circle of mathematical procedure but said mathematically the distance of the planet from the Sun k holds space a^3 in relation to time T^2

May I remind you THAT NEWTON'S OWN LAWS ARE IMPLIED, and again the planets disobey these laws completely!! In the modern age all evidence point away from Newton's vision

The lot is more likely moving away from the Sun . The lot is not coming closer!

The lot is more evidently moving further apart

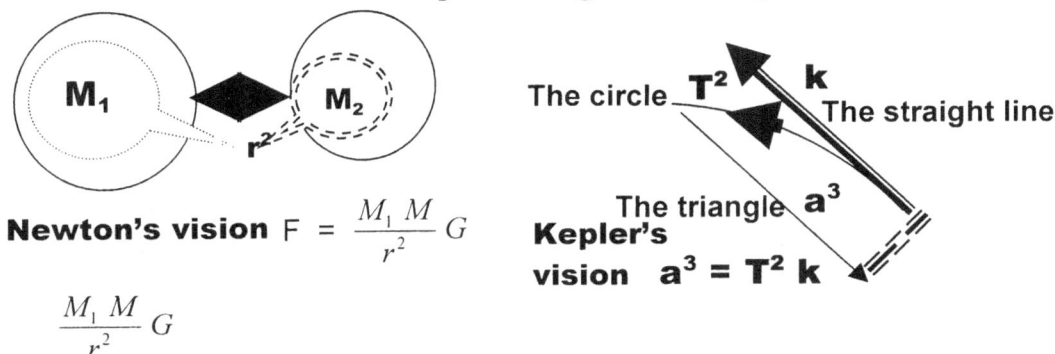

Newton's vision $F = \dfrac{M_1 M}{r^2} G$

The circle T^2 **k** The straight line

The triangle a^3

Kepler's vision $a^3 = T^2 k$

$$\dfrac{M_1 M}{r^2} G$$

$F = \dfrac{M_1 M}{r^2} G$ This is the suggested formula confirming the behaviour of planets used by Newtonian scholars underlining the argument that contraction is coming about between all cosmic objects. What Newton witnessed, if my memory serves me correctly was an apple falling from a tree where both the apple and the tree were part of the Earth and this did not constitute - or lead to - or come as a result of a catastrophic cosmic event happening. In the mathematical sense it does not make sense when Newton's argument is taken out and used in outer space. What Newton saw with his falling apple was a mass influencing another mass to reduce the distance as the influencing involved motion that came about. In outer space there is another gravity where in the case of those cosmic structures in outer space there is no mass pulling each other about or pulling one another onto each other. In the case where there is particles falling from space onto the Earth, that falling also results from gravity, as much as it varies from the cosmic gravity. There is another type or form of gravity different to thes concept Newton introduced. The concept Newton introduced is not the cosmic gravity Kepler formulised. What Kepler introduced is a duel where both objects are clearly in an eternal compromise therefore neither party relents its position. Newton saw just the opposite...Newton saw both compromising their individual as well as each other's position. But since the mass in both cases is unchanged and the mass is the factor that is establishing the force that is used by the circle to hold the radius steady and in place, these facts point to a balance that formed bringing about the above-mentioned steadiness. In the view of science however it is the mass that either draws the orbiting objects closer or is keeping them apart. The mass does not change and since that mass of both produces the radius between both, the logic is that there has to be an even and steady radius that develops. The radius has to be equal all the time since the mass never changed throughout the rotation. The radius must be the same from any and all given points that form the rotating circle which must keep the radius equal from every angle...yet we know that Kepler proved this not to be the case even before Newton's naming and changing of Kepler's work came about. What we see is that there is one factor that is trying to run away being a lesser space within the pulling powers of a larger space (the second factor) trying to capture and control and a referee (the third factor) is seeing to it that the even-handedness is at all times applying in the fight.

That gravity which I am familiar with and know is there). In some part but not in all out representing all the gravity there might be because I cannot see the jerking, as much as I do not feel it. That is then most probably another gravity I can see and which is Kepler's gravity which $a^3=T^2k$ represents. We have a motion of pulling...yes and that is what Newton saw...but then there is another motion of establishing a motion trying to depart, leaving the centre by tearing away from the centre and thirdly there is a motion that sees to it that the balance evolves as rotation. That is what Kepler said when he saw all three factors whereas Newton saw but one of the three. The one space is filling the next space as the space duplicates the position it had in the next moving moment that brings about the next position through motion. This eventually will have confined the next point by using a circle motion, which at first was intended to be a straight line, which is stopped by another straight line. The quest in this book is to find out why the other two factors apply in outer space as only one of the factors comes about on Earth under normal applying conditions.

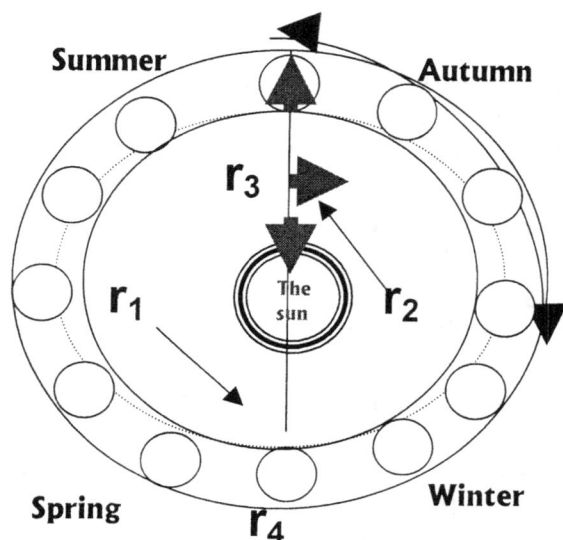

As the two factors are in a motion directional dispute there is obviously one of the two factors or strengths fighting to cut loose from the other one's grip and run off. If there were not such a force trying to escape, the first force would have a quick and decisive victory by reeling in the loser just as Newton predicted. The fleeing object and its matching fighting partner has a third party referee that allows the fight to go in a specific direction as long as there is no decisive victor.

This book, which I produced in the form of an open letter, is on a quest to find the missing two factors and I can declare with some delight and with even more certainty that I found the missing factors. By Newton's introducing gravity as a force with the formula $F = G\ (M_1.M_2)/r^2$ a precedent was set of gravity being a contracting force forcing distances supposedly to grow smaller. Apply Newton's view to comet behaviour. Newton insists that the Sun has gravity reducing distance between the objects and while lecturers are teaching this during the day, at night they all witness how the comet follows this principle in detail showing Newton as a prophet. No sooner does the final conclusion draw near by orchestrating the final demise of the distance separating the two cosmic components when the opposite changes all concepts taught by institutions of science, the next minute out of the blue with no pre warning of the comet changing its mind, the comet defies all logic in scientific circles that apparently even included defying Newton and his logic. Because at the very point you'd think there is no chance of any return where gravity supposedly should peak because the comet is so close to the Sun and due to that fact makes the collision unavoidable…then the comet chooses that very point to dart away into the blackness of outer space, missing the definite collision by miles. By the time the collision is truly unavoidable with the radius between the Sun and the comet being as small as it realistically can be the comet starts gaining on the radius distance in spite of Newtonian denial of any possibility that such an event can in fact take place. The radius that should be shrinking further is instead enlarging. The radius that now begins to stretch proves Newton incorrect and it even depicts Newton as possibly being a fraud. The gravity applied that focussed on the comet reducing the radius between it and the Sun was not acting predictably by maintaining the reducing of the distance until collisions come about as Newton insisted on. In our reading the Newton formula in English it says that $F = G (M_1.M_2)/r^2$ which when one translates that which is said in mathematics to a verbally spoken linguistic dialect, the translation then suggests that a force is committing the material that forms the factors involved, and forcing the material into a path that is leading to a collision. It says that the two will eventually collide because of the non-retractable mass inside each one that enforces the pulling which by the mass in each case is creating the force. The unchangeable ability of the mass and the unavoidable pulling each mass creates would bring about such a collision. The mass contributes a force making a collision imminently unavoidable. The collision is beyond any attempts of diverting any oncoming objects away from the inevitable possibility of contact. The force that mass contributes is ruling out all possible evading each other or avoiding the destruction. (By enforcing a mass it created force that removes all chances from diverting away from the collision that is about to occur). Such a force then removes all possibilities of avoiding the oncoming collision. The force will not allow any attempt to try and bring into the equation other possibilities in as much as rerouting the approaching object and changing the course in the imminent collision that is due and in due course will come about between the comet and the Sun . That which I explained is what Newton mathematically suggested with the formula. That is not what Kepler said notwithstanding so many arguments with Academics that I had in the past who tried to prove to me that the two visionaries views were equal and the same. Well…it's not the same because when we go onto translate Kepler to the verbal English the letters that come out do not even spell the same words.

It is conducive to remember that there is another part of the two relevancies applying where one is a^3 that is relevant to k but also there is the point where k has a duty to place a relation to a^3

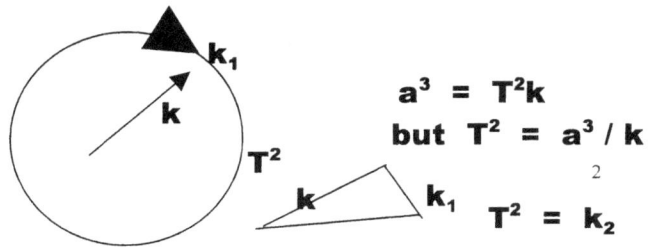

$$a^3 = T^2k$$
$$\text{but } T^2 = a^3/k$$
$$T^2 = k_2$$

The correct Translation of Kepler's mathematical expression will be $a^3 = T^2k$ which proves that Kepler said that there is a space a^3 which is equal = to the motion in the **time duration T^2** thereof between two specific points which is a straight line **k** that holds a relation from a centre to an end where the two ends run from the beginning of **k** to connect at the end of **k.** I might not be the smartest boy on the block but I'm not that stupid either. I know how to translate... and I translate as follows:

a^3 must have a volumetric interpretation because the third dimension is sure evidence of multiple conjunctions of dimensions put together in three sides opposing three sides having the third dimension in place. The fact that any symbol uses a value to the **third power** a^3 indicates **space** or a volumetric established and separate unit. Using a cube by three dimensions symbolises a cube, a room, a space to be filled, a unit able to hold other ingredients on the inside when empty or partly filled. It is space because it is volume using the third dimension.

T^2 is an indication of something having a cubic nature other than the square forming motion that is provided by the motion the square indicates, which is where the moving object is representing a third dimensional object that is moving from point to point and it is this point to point that multiplies into the square. The space is moving as a unit from one point to another point and the moving between the points are represented by a flat square or following a flat distance between two points. The cubic space was in one instant in one place and then the second instant in the other and because time can never stand still or become single dimensional (this I am about to prove as the letter unfolds) insisting that time must always support the motion it consists of or time cannot be. It is motion that is taking time, which is motion in the second dimension moving the space in the cube.

k^1 is the symbol used to indicate a straight line between two points with a definite beginning and a specific end position. It is the location where the cube is holding space and where the space was and where the cube in space is going to be in very the next split instant that follows. That will then in multiplying form the square that indicates the time the journey took to move the cube of space from one point where **k** is indicating the location of the space to where the next indicating of **k** will shift the space being the cube pointing at the end of **k.** Since time represents the square and with **k** being the distance that proves that the **k** represents the distance the space representing the cube went to take the time represented by the square through the motion. It is the distance moving space in the cube to complete time in duration in the square of motion; therefore **k** is permitted to be in the single dimension.

There are infinitely more implications in the statement Kepler delivered than what is merely a contribution to motion and only motion as Newton was of the opinion. What is there mathematically not correct in my interpretation of Kepler's manner of translating mathematics to English and why is any changing thereof by Newton or any other person necessary in any way?

We can test any of the following symbolic values in the mathematical expression and also test the principals behind the expression in which Kepler stated them. By such testing we will find that time after time there were never any corrections in the translations required since the translation thereof was never incorrectly presented and in that a case asked for no alterations to secure the correct reporting of the cosmic information being translated. By taking the formula on face value it can change as follows: $a^3 = T^2 k$ can become $k = a^3 / T^2$

 That proves that the establishing of distance **k** will produce space **a³** and set space **a³** in motion **T²** where such motion is in opposition to singularity, which means gravity or contraction is the deliberate opposite of expanding **a³ /k = T²** . In the beginning the expanding then also involved three more points all just outside the border of singularity but within the atom exclusivity. It extends **k** while it introduce a returning relevancy back to singularity **k⁰** by creating motion in spin and duplicating space by reducing space.

With this mathematical reality what then later formed the grounds for any individual to develop any need to change Kepler's translations from the cosmic given to mathematics and then from mathematics to English while the guilty party is renowned for his superior skills in mathematics?

Kepler translated what he found to be the cosmic given to mathematics which we humans are able to interpret from the mathematical expressed to the verbally pronounced and written but Newton still saw a need to change what the cosmos said about how the cosmos is presented and by no one less than by its own interpretation of its self structured composition.

When viewing my interpreting of what Kepler said I might have asked myself countless times what did I not translate correctly from the mathematical expressed to English after encountering a battery of Academic onslaught and resentment on my Newtonian views because after all it is directly diverting strongly from the teachings presented by Mainstream science and the diverting is not coming in a small way.

In truth from my diverting I came across very new ideas I am able to prove. By my translating Kepler's work correctly I came upon answers not yet uncovered by Mainstream Science

Kepler gave the World mathematically translated cosmic answers he received from the cosmos that Kepler uncovered long before Newton, Einstein and others got wise about cosmology...and later the wise came up with old news (old views as far as Kepler expressed their views before they, the wise were born with the purpose of coming to the conclusion that those wise men eventually did) and where the conclusions that the wise concluded brought much surprise to the world with the originality of the later Masters' initiative while Kepler said the same thing ages before…!)

Such is the advantage of recollecting Kepler facts that it does answer many questions, which went unnoticed and therefore not spoken about up to now and some were previously never even thought about.

Newton said a sphere is a³ = 4/3 Π r³, which is mathematically correct, however

Kepler said the cosmos told him a cosmic sphere is a³ = k T² There are the two distinct possibilities which Newton saw and which Kepler saw and both are most valid. Between the two concepts there is literally one Universal difference and the two can never be mistaken as promoting the same principles. 'Ever try to answer facts about the Universe in as much as…what brings about the expanding? Kepler said the Universe plus its entire content is expanding centuries before Edwin Hubble realised what he was seeing through his telescope.

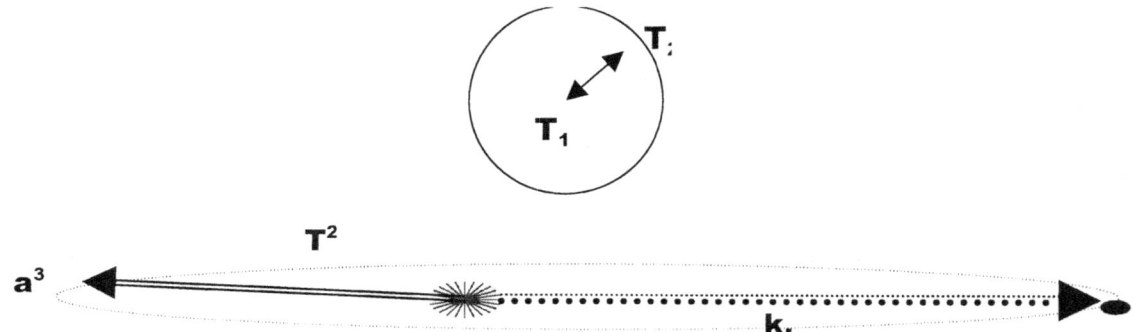

Kepler was the very first person to mathematically introduce **space a^3 centre k** and **time T^2**. Not only did he introduce **space-time a^3 / T^2** but he also placed **space a^3** and **time T^2** in a relevancy long before Einstein did and placed **gravity in space-time a^3 / T^2** even before Newton named gravity. He showed that space **k** is growing in the measure of what means the Universe attend to by promoting space-time as a^3 / T^2 = k^1. Kepler was the person who placed gravity as the ingredient in the Universe that determines **space a^3** and **time T^2** and much more. Kepler was the first one that said that gravity comprises of two factors being **k** or linear gravity and **circular gravity or T^2** as gravity keeps space in form while all is staying together.

Although not one Academic has ever openly admitted to me that they as members and part of Mainstream science are more aware than I am of all the facts and doubts I point out to them, such evidence then becomes clear whenever I mention the matter to them I get more than the impression it does not come as a surprise to them and hit them like a brick between the eyes. The lack of surprise and initial doubt they should show at first when they discover the incorrectness of evidence in their theory is a telltale sign confirming my suspicions about their evidently knowing all this information all along. They clearly seem very agitated about every detail I show when I bring the mistakes and double talk to their attention in the hope that they may confirm my doubts.

Never is there a whisper of a surprise or a hint of a suggestion that would initiate an argument carried on by the bewilderment or the astonishing surprise they should feel confirming my arguments because there is a mild complacency in their voices. My jumping them total unexpectedly about matters they never contemplated in the least leaves them unturned. The rush in blood pressure that should be a factor on their part and part of the instant where total surprise will bring about some confusing thoughts that will inspire the unleashing of an argument in defending their holy grail should at least carry a surprise in an attempt to save what they believe as being the Gospel in science and with that defending their honour. They lack embarrassment, which they should have in their disputing of my claim as they fight off my allegations with a countering of denial claiming foul on my part as they are in shock when finding out about any doubts. A lack of true emotion on their part is a telling sign that they also may have some serious thoughts on the quiet about any inclination presenting a flawed view about what they always thought they knew to be true.

There is only that eerie dismissing of the seriousness and the lack they show in excitement that would deny or support my credibility as I present my findings. If they know about the inconsequential facts in science why is it not generally acknowledged and pronounced as a matter of fact? Why is there the covering up and hiding facts that we associate with some professional criminals such as politicians. The fact that Academics are aware of this evidence in general terms about the misinformation and doubting evidence about Newton's cosmic vision but moreover underlying this is their total denial of knowing about it and that is what is so seriously unforgivable. The fact that all Academics are aware of my evidence even before my presenting them with such evidence is beyond doubt. If that is the case then why are they forever trying to kill my viewpoint and forever try to silence me where I am only the messenger because I bring the solution and the answer? Please note that the answer and the solution are unbelievably simple and unsophisticated. It lacks all the splendour and grandeur expected by all Academics concerned. It is because it is so simple that it went amiss for four hundred years. It is because it is so simple that it misses the grandeur that will entice them. Instead every academic accuses me of not understanding Newton while they can't show me what part it is that I can't understand and I on the other hand can't see what there is not to understand..

Newton said that it is the reducing of the distance between the objects that would bring about the un-reversible reducing that will end in a total demolishing of the radius that is between the cosmos structures, but instead we find the gravity applying in outer space is one of the instances where gravity provides an orbit circle that gravity seems never to completed as the orbiting objects follow from closing any circle that is leading into a following circle up to where the circle is completed in cyclic precision. That is not the gravity that Newton identified although Newton admitted that there is a presence of a centre forming a point in the middle between the two objects. He was unable to know what caused or even the presence of the Coanda principle, which forms so critical a part of my theory. The formula concerning cosmic balanced gravity however leaves no room for the admitting of such a point and by not leaving a possible inclusion of such a point in his formula Newton did by such gesture in principle repeal his admission of such a centre. This had me cast doubt on what is taught at institutions of learning. It motivated me to venture back to an era before Newton came to influence science. I came to acknowledge Kepler as I came to understand Kepler. The accepting of that what I understand in Kepler involves much more reading into what Kepler said by finding what Kepler did not say in the way that he did say what he said than the reading about what Kepler said as it is written in the precise detail and to the letter used in his statements. He never directly stated what he said. Again I must stress this point: when I refer to what Kepler said it most likely means reading into the part that is being a part of the part that he did not say when he was saying what he said but I accept that he meant to say what I am reading and translating from Kepler as part of what he did not say but meant to say. I have to read more with my mind than with my eyes. This comes as a result of interpreting Mathematics to the verbally expressed. I had to learn to read with my mind and not my eyes and I found that that is the manner in which one has to approach cosmology. From the first time I discovered what manner one should use if one wished to read into Kepler's findings I saw Kepler was all about uncovering the unknown. Realising that, the conclusions I drew by reading in such a way cemented my better understanding of Kepler's work, which then helped me improve my insight into Kepler's work as it increased my understanding about cosmology several fold. This helped me to realise what implications were to be found underneath Kepler's discoveries. From my realising what approach I should use, it helped me to improve my cosmic realising by using the method of reading Kepler and from that I could come to appreciate what Kepler introduced.

Only then did it bring insight and proof to me as a student of Kepler and this proof I found by dissecting what Kepler **did not** say instead of what he **did** say, which I now present to you with this letter, you being a superior intellectual person. Kepler said $a^3 = T^2 k$ and that correctly translates to a mathematical expression $k^0 = a^3 / T^2 k$ which in the verbal statement in English translates that Kepler said that there is a **space a^3** which is **equal =** to the motion in **the time duration T^2** thereof between two specific points which holds a relation onto a centre k^0 where from there forms **a straight line k** that is centred on the spot where space begins from k^0 **that produces k** as well as producing the circle therefore that spot $k^0 = a^3 / T^2 k$ has hold k^0 at a value of having the least space. The line **k** is centred onto a spot where space begins specifically at k^0. This point not only produces the line k^0 but represents also the space that forms the eventual circle T^2. Therefore from the centre holding k^0, k^0 leads to **k** that forms the roving space a^3, which is rotating at a distance **k** where T^2 forms the outer limit of k^0. Mathematically $a^3 = T^2 k$ will be $k^0 = a^3 / (T^2 k)$ because $k^0 = 1$. But $k^0 = 1$ also present the single dimension where all factors are a product of one. If one can locate k^0 one will find singularity. That is where gravity is because gravity is strongest where space is least. Then that suggests that gravity is strongest at k^0 because space is least. That is gravity because that is what keeps the orbiting object in orbit but also that is what Newton completely missed when he changed Kepler's work. Newton failed to recognise gravity as the only ingredient in Kepler's formula. He admitted he missed this because he admitted he did not know what gravity is while Kepler explicitly showed what gravity is. Gravity is what keeps the orbiting object orbiting. $k = a^3 / T^2$ is **distance^1** = **space ^{3}/ time^2** forming from a pivoting centre k^0. That is a cycle and moreover it is a cycle formed **by space/time**. What Kepler said is that space is a^3 in motion $T^2 k$.

That says **space^3 (a^3)** relates directly to **time^2** that uses the symbol T^2. This is also what I refer to when I say one has to read what Kepler did **not** say when one wishes to see what he **meant** to say. Kepler introduced space^3 –time^2 long before Einstein's date of birth appeared on any calendar although Einstein is credited with the formulating of the concept of space-time and giving it a name. Going even further Kepler stated that the space a^3 is on the move T^2 around in a circle at a distance **k**. That is what that comet we are discussing is doing. The space^3 (Comet) is circling the Sun using a radius **k** to establish the cyclic time^2 as a period of continuous motion and continuous motion is gravity. That reads much more correctly and closer to the truth than what Newton predicted what according to him (Newton) was happening in space. Remember in this statement I am separating cosmic principles applying from the way that gravitational principles apply on Earth. I distinguish that which is the rule in the cosmos from what we find ourselves trapped in on Earth. The two just don't mix. I am removing cosmic physics from normally accepted physics because the gravity concerned is not the same.

The proof I bring is real however simple it may seem. It has none of the mind-blowing complexities normally associated in the presenting of investigative analyses of Astronomy. I realise the information in this book carries the arguments in a childlike manner which are very simple to follow, and for that in the past I have been blamed over and over again as being unprofessional. In my answer to that I can only reply by using another question: Are only professionals adequately equipped with minds that make them (the professionals) the only ones able to think? We being part of the human race are all thinkers. Everyone as a human being can think. Every person on Earth is a thinking thinker that uses his brainpower by exploring thoughts mainly and normally to his or her personal benefit. It is what we think about that produces the results of our efforts by which we accomplish what ever we are thinking about. I have met professional Academics that I found foolish as much as there are other cases where the so-called amateurs can credit themselves with much more wisdom and insight. Albert Einstein as a patent clerk was that much but to name one. Please understand that I do not compare my achievements or myself in any way, shape or form with the likes of a Master such as Einstein although I speak my mind when not being totally in agreement with some of his or other views. My unsophisticated retracing of Mainstream physics concerning the Big Bang in detail helps to reinvestigate established principles and moreover investigate proof in the light of modern evidence. In principle I distinguish between Kepler and Newton in that Newton is one hundred percent correct concerning gravity on Earth but as far as outer space forms gravity the conclusions of Kepler and Newton do not match and they had totally different ideas about what they saw in gravity. I am in disagreement with some basic principles that science acknowledges and I divert strongly from all accepted roads Mainstream physics follow. My doing that prompted those who are considered and accepted as self-proclaimed members of Mainstream Physics have categorised my views in the past as incoherent. That I do not accept. I admit that my line of thought is extraordinary and controversial but only to Mainstream science and not to the standards lay down by nature. Since the concepts I follow start at the beginning, and I take Kepler at the point where modern cosmology began and in that mindset I re-evaluate Kepler's work. I start by tracing a new approach as to what I see Kepler found. The main condition of my investigation is to establish a divorce between what Kepler said and what Newton thought to add to what Kepler said. It is this divorce I create that Mainstream science finds repugnant or even in some persons' opinion repulsive. I believe the repugnancy does not come from or is not manifested in any part of my work to the letter as such, but rather what my work suggests and who is doing the suggesting. To my view in cosmology such adding to Kepler by Newton was unnecessary and it diverts Kepler's work away from cosmology. But as the generations moved on Newton became religiosity in the mind of science wherever science was taught. To students there is little or no choice in the matter since the only choice left to them is one of understanding by forcefully accepting or die an academic death since Newton is academically accepted without asking questions or raising an opinion. For the second choice, the less accepting students are greeted with a

Dear John good-bye letter sending them off into the unknown Sun set that such a future outside physics will bring them. That is brain washing.

From studying Kepler I saw that we have to gauge what we find in the Universe. What we find is not that what we realise with our eyes but that what we observe by using our minds to translate from visions coming from our eyes to our minds. We have to test the part that we are seeing much more than merely accept what there is to see on face value. We have to not only see what other life beings blessed with much less insight most probably also should see. We must stop using our eyes in the same manner as animals do and start seeing with our mind, as humans should do. By being the superior evolved species that we are, it gives us the ability to read into that which only we can see and that we only can see by using our intellectual mindset. By seeing with an intellectual understanding what there is to see when we see what we can observe, we should therefore have the ability to be in understanding by looking at what we can see but moreover understand that which we cannot see. It is the same as playing chess. See what there should be moved instead of noticing an object not having an ability to move by own initiative. This I first found to be true about Kepler's work and when I started projecting this method of observing what the Universe is, as it scattered most previous perceptions I found that using the new method brought along answers so fast I could sometimes hardly keep up with the interpreting thereof. But as is the case with Kepler so is the case with the entire study of cosmology: One should see what there is about the cosmos, which is unseen to us and then we may find so much more in the cosmos unseen to us representing that which we cannot see and that which we cannot read because we have to learn to read what is not written in light. Armed with this realising I then proceeded from that point by further arguing and debating the full implication of Kepler's contribution. Kepler placed cosmic structures in relevance to one another and so does the Big Bang Theory. The backbone of the Big Bang is that relevancies apply in dynamics and such dynamics are placing all structures without any reservations independent from each other. As the Big Bang progresses all filling the Universe to the inside is in the same Universe that was then at the time of the Big bang just as much as it will always be and the lot remain the same, however the relations that the elements comply to bring across new relevancies with new positions to fill. The father of the Big Bang concept is a person by the name of Father LE MAÎTRE, GEORGE ÉDOUARD (1894-1966) who was a Belgian priest and cosmologist. He was the first person to embrace the fact that the Universe expanded from an infant stage. His model of an expanding Universe (1927) was superior to that of W. de Sitter in that it took into account mass, gravitation and the curvature of space. Similar models were proposed in the early 1920s by the Russian mathematician Alexander Alexandrovich Friedmann (1888-1925) but Friedman compiled various such possibilities. Lemaître argued further (1931) that the quantum theory supported an origin in the explosion of a 'primeval atom' or 'cosmic egg' into which was originally concentrated all mass and energy. As modified by A.S. Eddington, Lemaître's model provided the springboard for G. Gamow's Big Bang theory. In the wider picture of science in general a lot changed to just allow such turnabout in thought since the day of Isaac Newton. From Newton's attraction and contraction many things came into place that allowed change in the most hardened minds. Accepting facts about the Big Bang concept is quite radical. By promoting expansion the Big Bang theory contradicts gravity and our accepting of the Big Bang has to change all other concepts. By accepting the Big Bang other changes are also involved.

KEPLER, JOHANNES (1571-1630)
The German mathematician and astronomer KEPLER, JOHANNES (1571-1630) became Tycho Brahe's assistant in Prague in 1600 A. D. where he undertook to complete the tables of planetary motion Tycho had begun. Kepler first calculated the orbit of Mars. He spent much time trying to reconcile Tycho' s accurate observations of the planet with a circular orbit, but concluded (in Astronomia nova, published in 1609) that Mars moved instead in an elliptical orbit. Thus, he established the first of his laws of planetary motion. A theory that the Sun controlled the planets by a magnetic force led him to the second and third of his laws, which were published as part of his treatise on theoretical astronomy, Epitome astronomiae Coernicanae (1618-21). The Rudolphine Tables (named after Tycho's patron, the Holy Roman Emperor Rudolph II) of planetary motion appeared in 1627 and were still in use in the 18th century. Kepler also wrote De Stella nova, on the supernova of 1604 and Diptirce on optics and the theory of the telescope. The overall view followed in this book **an open letter To Selected Academics ISBN 0-9584410-9-X** places the true significance of his work in true contents. In KEPLER'S EQUATION is the equation that relates the eccentric anomaly of a body in an elliptical orbit to its mean anomaly. The equation is $E - e \sin E = M.$, where E is the eccentric anomaly, M the mean anomaly, and e the eccentricity of the orbit. It is important as one of the mathematical relations enabling the position of a planet about the Sun , or a satellite about is planet, to be calculated from the orbital elements for any time. However this only relates to the solar system, and KEPLER'S LAWS only apply in the contents of the solar system. The three laws governing the orbital motions of the planets, discovered by J. Kepler is as follows: The first law states that the orbit of a planet is an ellipse with the Sun at one focus of the ellipse. The second law states that the radius vector joining planets to the Sun sweeps out equal areas in equal times. The third law states that the square of the orbital period of each planet in years is proportional to the cube of the semi major axis of the planet's orbit. The first law gives the shape of the planet's orbit; the second describes how the planet must continuously vary its speed as it follows its orbit, moving fastest at perihelion and slowest at aphelion. The third law gives the relationship between the planets' average distances from the Sun and their periods of revolution.

Instead of studying the true value and contribution of to Kepler's laws an Englishman going by the name of I. Newton placed his own interpretation to Kepler's laws, and in doing this, he wilfully destroyed the principle working of the Creation. Saying this I hear the alarming hooters announce Newtonian dismay. In the past my experience was that all the revered Academics lost their appetite for any further investigation of my work. That is sad as much as it is regrettable. Through Newton's tunnel vision, he applied his own misinterpretations to the correct presumptions of Kepler and through the Newtonian tunnel vision Academics did not move an inch away from repeating the same procedure. In the past it was this that had Academics shying away from me because at the point where I raise criticism of the Newtonian viewpoint I am rejected. The point where I declare my suspicions concerning they're accuracy and the correctness about their theories, which is where I should then be raising their doubts about their way of thinking is the point where in stead I raise their suspicions about my way of thinking. That is what caused the rejection of my criticism about Academic Newtonian science and evoked their criticism in the past about my views instead of them following the logic by investigating what I said. Their rejection of self-investigating got me and my work rejected to a point where the applecart lost its wheels on every occasion. It is where Academics read my remarks and what brings (seemingly in an instant) wrath to Academics. I say this because I realise that reading my remarks or hearing me remarking about this notion brought much resentment on their part and if the reader at the present moment is a Newtonian, boiling his/her blood. It is blood boiling because I believe they see my remarks as belittling that which they feel they have accomplished. This is not the case but still my remarks have the same effect on the Academic as pouring icy cold water down the back of his shirt. I mention this because I know it has happened many times before and if possible I wish to avoid this response. Therefore I ask you kindly to please be warned about the negativity you must feel towards me where you are the Newtonian and I am not. Before you lose interest in reading this letter any further please allow me to finish. In the past Academics thought me to be presumptuous and that normally became the point where all the Academics find their interest vanishes. That should not be because if Newton's work is as utterly accurate as those with faith in his work believe it is, then every aspect about Newton should stand above any and all reprimanding or any form of doubt causing a notion to reprimand. The testing of Newton's work should withstand all testing notwithstanding the person or the prominence of such a person's social or academic standing in the Academic society or even the prominence that such testing will deliver. From what I see about Kepler's work it is a flow of circumstances that lead to Academics neglecting Kepler's work and the realising of the theory I suggest is not forthcoming due to my personal brilliance. I do not consider myself to be the brilliant in any way as to be the one that can remove the verbal splinter from the eye of the Academic. Yet...if there is a splinter what else should I then do...Newton reduced the implication that Kepler findings hold by introducing to the law of gravitation. He then went about and changed it to three laws of motion. It is clear that while he formulated the laws on motion he missed the way Kepler introduced gravity as space a^3 coming about through motion T^2 and that gravity is space a^3 within space k within motion T^2. Newton also missed the fact that gravity is at its strongest where motion and space cease to be. This is most important to recognise about gravity in one of the two forms it has. I. Newton generalized Kepler's first law, verified the second law, and showed that the third law should be amended to the form; $4 \pi^2 a^3 / T^2 = G (m + m_p)$. In this, the value of "T" and "a" are the period of revolution and semi major axis of the orbit of a planet of mass m_p about the Sun of mass m, and G is the gravitational constant.

It should be clear to any person investigating Johannes Kepler and his work that Isaac Newton hijacked Kepler's work and any time there is the slightest referring to Kepler about the research Tycho Brahe and Johannes Kepler did such referring to Kepler always lead to and always include the mentioning of Isaac Newton changing the work of Johannes Kepler. It is as if the World never could acknowledge Johannes Kepler because the work of Johannes Kepler would be completely wrong and misleading if it were not for the intervention of Isaac Newton saving the skin of the less admirable Johannes Kepler. This comes in the midst of every one realising that Kepler used the information he received directly from the cosmos. I do stress this on many occasions throughout the letter because the embarrassing part is that Newton changed the work of The Universe and not of the man called Kepler. Should you reading the letter entertain the opinion of Newton and feel any urge to defend Newton you should ask the question as to who is standing corrected, is it Kepler or is it the cosmos that gave Kepler the information he concluded? The cosmos supplied all the information by using mathematics, which Kepler then had to translate. But Newton destroyed the accuracy by altering what the cosmos said and directly by adding to that what he (Newton by name) thought that the cosmos left out. This set a precedent by Newton in cosmology and also set a trend, which was retained in all future cosmological development and it lasted in cosmology for three hundred and fifty years. In this book you are reading, I am about to show that such practise should no longer be accepted in cosmology. In the process the world of Mathematics developed by the world of cosmology stood still for almost four hundred years. Faculties contributing to cosmology and feeding off cosmology improved as much as they developed, but when cosmologists see the Roche limit in action in the lens of the Hubble telescope and refer to the event as "stars blowing bubbles" being the ultimate response coming from those persons who are supposedly the Masters of cosmology affairs, then the truth of what I just said comes down on you like a ton of bricks. Everyone having any remote interest in cosmology will find they are being very disillusioned by such "official" testimony about the evidence the Ultra Wise report about. This book is about showing how great Johannes Kepler was and how enormous his work was. It will show he preceded all ideas of everyone that came later and officially introduced the novelty of such ideas. Back during the time Kepler was introducing his work the stature and the magnitude of his work was beyond any person's understanding (including Isaac Newton) and this prevailed for most of half a millennium. I do not say I am the brilliant one to uncover Kepler in the face of everyone failing that came before me, but as I am not a Newtonian such bias was not part of my repertoire and denying me the fortune of being a Newtonian added to my fortune of realising Kepler. Yet as you will notice, the work I contribute is much below the sophisticated norm of modern investigative research and the levels that modern research accomplishment demands to better the effort of the understanding ability in the splendour that investigative research work should deliver in view of our modern times. It is only pure neglect in science circles that moved science past Kepler. Not seeing and therefore not investigating through almost half a millennium has paved a road past the inferior levels that the researching of Kepler's work holds because it was rocket science four centuries ago but the brilliance of it has faded since then. My contribution holds no astonishing flair that may add to science in general. Only failure to notice what I see on the part of those truly brilliant can explain my being able to present my contribution about my work in investigating Kepler. Only by their passing such degrading levels of the Academic establishment in the past and the present can bring the blame for such an obvious discrepancy because any involvement in the work at such an inferior level as that which I bring cannot interest and excite a salted Academic and when thinking about it, the idea is totally unthinkable. This letter, although it is on this inferior level is about correcting this tendency and has in mind the effort to put in writing what would place Kepler in the greatness and glory he deserves. As I already said, if Kepler was wrong then the cosmos was wrong about facts and applying relevancies and tendencies in the cosmos. I yet again wish to reiterate we should never for one moment forget that Kepler received his information directly from studying the cosmos so how could the cosmos stand corrected? In spite of all the brilliance attributed to Newton nonetheless if Newton had the mind to change Kepler's work and my saying this includes all persons agreeing with such changing by Newton of the work of Kepler those persons admit that he or she or Newton never took any time to really and truly investigate what the cosmos told Kepler. Through understanding the work of Kepler I prove gravity, the Titius Bode law, singularity, space-time, space-time relevancy, the Lagrangian system, the Coanda effect and the Roche principle, the sound barrier, the principle behind the Black hole. The precondition for my ability in doing so is that I have to remove Newton's opinion about Kepler's work from Kepler's work. Whenever cosmology comes into question and all the phenomena, which I mentioned just now remains unexplained and by that token alone it shows to what degree did cosmology remain undeveloped. Whenever there is any mention of Newton, Kepler is never mentioned. But the reverse is always applying. Mainstream physics holds the opinion that Kepler may only have an opinion if Newton can change the opinion. Kepler gave space-time, gave gravity, gave singularity, gave the Plank theory, gave the theory on relativity but no one ever found Kepler's work deserving enough to launch any investigation such as I did. I belabour this because of what revulsion my rejection of Newton unleashed. That is one barrier much unnecessary but it has been an insurmountable barrier this far.

NEWTON, ISAAC (1642-1727) and NEWTON'S LAWS OF MOTION
An English physicist and mathematician who developed his principal theories about gravitation, optics and mathematics between 1665 and 1666. In 1668, he made the first working reflecting telescope. Most of his work remained unpublished for long periods, partly because of criticisms by c. Huygens and the English scientist Robert Hooke (1635-1703) of his early work on the corpuscular theory of light. However, in 1684 E. Halley persuaded him to organize his work on the celestial mechanics of the Solar System, which was published as the Principia. Newton's other major work, Opticks, was not published until 1704. It contains his corpuscular theory of light, and the theory of the telescope. His greatest mathematical achievement was his invention of calculus, independently of the German mathematician Gottfried Wilhelm Leibniz (1646-1716). His profound influence on physics and astronomy is reflected in the phrase 'Newtonian revolution'. Three laws published in 1687 by I. Newton concerning the motion of bodies.

1. A body continues in a state of uniform rest of motion unless acted upon by an external force.
2. The acceleration produced when a force acts is directly proportional to the force and takes place in the direction in which the force acts.
3. To every action there is an equal and opposite reaction.
4. However there is one more law on motion that went undetected by Newton...This book is not about trying to disprove Newton...it is about adding too science more than there now is available without removing any that science already accumulated.

In this book I use Kepler's formula to either prove or to disprove the following accepted principals in cosmology and if any person in the past gave only the slightest attention to Kepler's work, many statements would have come much sooner delivered by someone else or may never have come at all. By applying Kepler's formula correctly in this letter I can either agree with or in other cases deny the following principles.

It began with NICOLAUS COPERNICUS who changed the status quo. COPERNICUS, NICOLAUS (1473-1543) was, according to the Anglo Americans, a Polish churchman and astronomer although this is just more politically inspired propaganda because his parents were both German (in Polish, Mikolaj Kopernigk). While he was completing his studies, he had realized that the Earth revolves around the Sun and not vice versa. Such a view was in that time, held to be heretical. As I pointed out in the first few articles, the Church regarded the geocentric world-view of Ptolemy as consistent with its doctrines. Copernicus set down his basic ideas around 1510 in the Commentariolus, which he circulated anonymously, because of the Islam link. In 1512-- 29 he conducted his study and concluded the observations that he needed to support his theory, while carrying out ecclesiastic and local administrative duties. In this time, he had to defend his mother in court on charges of witchcraft. In 1539, the Austrian astronomer and mathematician Georg Joachim von Lauchen (1514-74), known as Rheticus, became a pupil of Copernicus and began to spread his ideas. The published work was openly spread as the Copernican system, in spite of the life-threatening dangers connected with such a "crime", in 1543 in the book De revolutionibus orbium coelestium. However, the reality of a heliocentric Solar System was only commonly accepted, after the work of Galileo and J. Kepler. The ideas introduced developed along and proved to be correct until such a time it met a solid wall with the investigation of Max Planck.

PLANCK CONSTANT
(Symbol h) A constant that relates the energy of a photon to its frequency. It has the value 6.62076×10^{-34} Js. It is named after the German physicist Max Karl Ernst Ludwig Planck (1858 – 1947). PLANCK ERA. In the Big Bang theory, the fleeting period between the Big Bang itself and the so-called Planck time when the Universe was 10^{-43} s old and the temperature were 10^{34}K. In this period, quantum gravitational effects are thought to have dominated. Theoretical understanding of this phase is virtually non-existent. It is named after Max Planck (1858-1947).

PLANCK'S LAW
A mathematical description of the energy radiated at different wavelengths by a black body: $E = hf$, where E is the energy of a photon and f its frequency. It was formulated in 1900 by Max Planck (1858-1947), who realized that energy is radiated in discrete packets, which he called quanta, and it formed the basis of quantum theory. The quantum of light is a photon, the energy of which depends on its wavelength.

There is one rule which is well established and which Mainstream science agrees about. It is one aspect, which forms the very principle that holds the theory about the cosmic starting together under the covering of a verbal blanket. All in science agree that it all started with singularity but I manage to go one step further where I prove that it is also where it ends, as singularity reunites space-time, which is from where Creation split in the very beginning.

Singularity is as follows: Singularity: a mathematical point at which certain physical quantities reach infinite values, for example, according to the general relativity, the curvature of space-time becomes infinite in a black hole. In the big bang theory the Universe was born from singularity in which the density and temperature of matter were infinite. From singularity flows space-time.

Space-time is as follows: Space-time is a four dimensional position of the Universe where the position of an object is specified by three coordinates in space and one position in time. According to the theory of special relativity there is no absolute time, which can be measured independently of the observer, so events that are simultaneous as seen from one observer occur at different times when seen from a different place. Time must therefore be measured in a relative manner as are positions in three-dimensional Euclidean space, and this is achieved through the concept of space-time. The trajectory of an object in space-time is called world line. General relativity relates to curvature of space-time to the positions and motions of particles of matter.

SPECIAL THEORY ON RELATIVITY
A theory proposed by A. Einstein in 1905, based on the proposition that the speed of light in a vacuum is constant throughout the Universe, and is independent of the motion of the observer and the emitting body. A consequence of this proposition is that three things happen as an object's velocity approaches the speed of light: its mass goes up, its length shortens in the direction of motion, and time slows down. Hence, according to special relativity, no object can ever reach the speed of light because its mass would then become infinite, its length would become zero, and time would stand still. In addition, Einstein concluded that the mass of a body is a measure of its energy content, according to the famous equation $E = MC^2$, where c is the speed of light. This equation describes the conversion of mass into energy in nuclear reactions within stars.

GRAVITATIONAL COLLAPSE
The collapse of a body that is unable to support itself against its own gravity. Gaseous bodies undergo such collapse if they are not hot enough for their gas pressure to balance gravity. This can happen in the early stages of star formation, or when nuclear burning ceases in a star's core. The time taken for such collapse decreases rapidly with increasing density, varying from about 100 000 years for the birth of a new star to less than a second for the formation of a neutron star. Star clusters may undergo a similar collapse if the random motions of their constituent stars are insufficient to offset gravitational effects, either during their formation or at an advanced stage of their evolution.

GRAVITON
A hypothetical particle or quantum of gravitational energy, predicted by the general theory of relativity.   gravity - motions have not been observed but are predicted to travel at the speed of light and to have zero rest mass and charge. A graviton is the gravitational equivalent of a photon. It is this anti-photon-being-a-gravity - motion by just merely swapping direction and all is proved that I find not very indigestible in modern science. One of the main issues that I wish to protest by my writing of this is my argument that if the Universe can be compressed back to the size it had at the point of 10^{-38} seconds after the Big Bang the daily outdoor temperatures of 10^{27} K will also come about once more. The expansion was the

result of compressed space, which then formed into heat and in turn resulted in finding a Universe with all the insufficiency of space less ness prevailing throughout and wherever space was needed. By that it forced space-time to come into being. Space-time came about at the time of endless time duration without space availability, which brought about the period of the Big Bang wherein space growth was the converting of such heat to space. If the Universe was in a vacuum as big as being available now then what was the temperature of the vacuum while it was empty before material filled it later. Then I presume the vacuum was there present as it is now in this present day. If the Universe then employed the space of say one atom, the impression comes through that from edge to edge and from Universal border to border the space occupied was the same as one atom will claim in our present day and age. Normal gravity started at 10^{-43} seconds. The Universe was the size of a neutron or somewhere in that vicinity. The Big Bang began and GUT, or the grand unified theory, produced the attempt to describe the strong and weak nuclear forces and electromagnetism in one single mathematical theory. Somewhere before 10^{-12} seconds of counting the Universe cooled to about 10^{15} K the electromagnetic and the weak interactions acted as one single physical force. Science reckons that unification may come about at temperatures of 10^{27} K, which was the temperature of the day at 10^{-38} seconds after the Big Bang. This statement echoes my viewpoint but one has to look carefully for that to surface.

In the suggestion the presumption claims that all the space that the Universe made available at that time was the total space one atom might take up today. If that might be the case then where was the rest of the space that now fills the Universe? Or was the rest of the space we now find in the Universe and what is now explained away as the vacuum, also available back then. Did the Universe only have that one tiny hot spot it filled with huge volumes of heat? Was the rest of the space vacant being out there all along during all the time running to the present date but filled with emptiness standing around as a big vacuum with no better to do than sucking on the Universe while the Universe was exploding at the speed of light. Then that statement suggests that in this hot Universe there were light-years upon light-years of vacuum waiting to be filled by the intense heat soaring in the smallest spot. If that is the case then why did the vacuum not fill in the blink of an eye by all the exploding expanding material growing at the speed of light? Was the Universe overall bitterly cold where the vacant space was locked in with one spot of the vacuum filled with temperatures so hot we can only produce it in numbers suggesting a value but never claim to be able to digest the reality thereof in the human mind? If so what happened to the natural consequence that heat flows in the direction of cold and equalise between hot and cold. Was the space being available at present available then or was the hot space the only space available at the time. If so what prevented the heat from instantaneously filling the eternally cold vacuum because with the rules controlling vacuum in affect, it should have filled in such a manner in less than a heartbeat? I believe that singularity formed space-time and space-time developed from the overflowing thereof at the time it was extending. With time marching onwards and outwards to this day space-time developed. Space-time developed another product that everything in the cosmos has to have. It must be in such large quantities everything imaginable in the Universe has to have it and that is space using time to move about. I suggest that it is space that is holding heat in a quantity providing density and ratio to space available and in relevance to the space being available to quantify the presence of the heat and which then proves to form the time factor. The container and contained all together mixed by motion. From that very first separating of heat and space, which is what formed from singularity to produce space-time. The Universe was full... It was overflowing by the speed of light in the beginning...so where and when did vacuum or nothing enter the Universe as a factor if and when the Universe was so full.

The answer to that is absolutely crucial because how did the Universe decide to fill some parts with a variety of something and decide to fill some parts within the in-between with nothing? If that is true why did gravity not prevent the vacuum filling because no gravity that came about since can beat the force that gravity had back then? This leads to another question following the previous one in asking why did gravity at the time when it was so strong with r^2 so much compromised not fill the nothing immediately as it entered with something that could absorb the nothing. At the very beginning the mass that was pulling on the mass by force was immeasurable and none quantifiable. Even more to the point is the question to be asked in how big was the radius between the materials with the immeasurable mass placed in such a little space. This is all the more important in the light that the smaller the radius is the bigger the force will become from the immeasurable mass pulling.

With the immeasurable mass that was producing the first gravity between the particles divided by an almost non-existing radius the gravity produced had to be in gigantic proportional quantities and with the separation of the radii being in the infinite measure that it was at that point then how did the Universe establish the chance to expand. It did expand, as we all are witnesses to in spite of this contraction of gravity that had to have been compromising the expanding factors. Still the expanding filled the unknown part of the unoccupied Universe, which at the time was there or was not there and where it was not there it was then filled with nothing. If the nothing was not "nothing" then the nothing that was not being nothing was also filling the rest of the vacant Universe that was or that was not because if it was it was filled with nothing and if it was not then it was nothing.

This is then taking into account that then all the reducing that is resulting from Newtonian contraction and that was going about in the space available at that time was something filled with nothing and surrounded by more nothing. With everything in the Universe being that much crowded and crammed where and how did nothing enter the Universe and fill the rest that was unfilled? What factors introduced nothing into the picture since the entire Newtonian concept finds its base on the principle that matter reduces using gravity by force which then brings about reducing or the removing of the many nothing between particles, which will then lead to nothing that has to vanish even before nothing can enter the space. This question may seem small-minded belonging to the mentality of a child or to that of the mentally impaired with not much factual appreciation developed yet. Please do not see it that way. If you think in those lines it will be because you do not have an answer to challenge these silly questions. Beware, silly as they are they represent official backing by the Wise-and-Informed. If the space is nothing and if the space was as large as it is at present then there was no need for such a small area to fill with something leaving only the rest filled with nothing at first since all the space we know about was there present and by being present it was there then for the taking. What ever filled the Universe had to start at the centre of the Universe and fill the entire Universe all over from a centre as it moved outwards filling from the inside outwards. This is a natural human instinctive realisation but is beyond proving by using Accepted Scientific policy. But that leaves Newtonian science with a massive unsolved problem: where is such a centre at the present time and where does the centre produce the limits or border it apparently has to form as it expands.

By expanding there is an additional contribution too that which was when that was, it was receiving more than there was before the addition increased that which was and by then becoming more than there previously was it had to be improving the border from where it must have been before the adding took place to where it was after what that was added was added. When that was less than it became when it was added, it was at the limit that was there before it was added too and that limit there was, was a limit that is the limit that I am referring to as a border being there. The cosmos is filled with unrecognised borders. The expanding has to be an ongoing filling that is at the same time expanding from the inside towards the outer limits of the Universe. Since nothing can enter from the outside where nothing is, the filling of nothing as a substance that would take up vast quantities of room had to fill from the very centre spot where all other filling came from. This filling of nothing with material has to be well mixed. The truth about cosmology is that space forms no borders but by using any Newtonian centre from where mass is attracting we must find a point where there has to be the ultimate Universal centre which is the cardinal point in the entire Universe and it is the first, the prime position to locate coming before any other concept one wish to put forward because all concepts has to start with locating that cardinal centre. There has to be the ultimate r^2 radii located precisely between the ultimate mass drawing the other ultimate mass closer. If there was a Big Bang then there has to be the spot where from the Big Bang developed therefore there has to be such a centre connecting the past to that ultimate centre with the line of development flowing onwards to this day. The fact that science is Newtonian proves that in the meantime Mainstream science is still of the opinion that there was the specific centre in the Universe that is nowhere to be found as it was filling the unknown with nothing coming from nowhere, but which somehow is still somewhere in the centre of all of that which is something. On the opposite side of nowhere there is an outer border in space producing a limit to nothing and serves nothing with a specific point to stop being nothing because that point is precisely where nothing ends and forms a beginning of a Universal border or a Universal end. How one will stop vacuum being no longer nothing was a question everyone comfortably missed to ask therefore no one ever seemed to deliver any form of answer.

One night some years ago very close friend of mine had a meal at his restaurant and as the conversation progresses he asked me about space and where it must end. I tried to explain to him what I believed in comparing to what Mainstream physics believed, but soon saw I was not gaining in his understanding. Then I decided to jot it down on paper and he could read it at his leisure as he saw fit. That led to the first book written by me (in Afrikaans my native language). What I tried to explain to Johan Boonzaier that night, is that if the Universe was the size of say even a tennis ball with only the size of a tennis ball being the very all of space there is available, then yes, it must take time to expand from that having the excessive heat there was back then in all the space we have at present. It then is converting heat into space bringing about the expansion. But one will find most expanding within the atoms, as the atom must grow since the Universe in all was the size of what one atom is today. The space in the atom pushed the space outside the atom but there must be plenty more to the growth. Something outside the atom contributed in it own right because there is more expanding than there can be blamed on coming from the atom. But the space then also developed as the Universe developed and if space developed then it cannot be total vacuum filled with nothing because "nothing" cannot develop. You the reader must judge whom is correct between my view that space developed with the Universe as part of the Universe and reject the official view about space being nothing or otherwise you the reader must then decide that I am wrong, but should you do that, then find a reason why the Big Bang started out small and filled all the available vacuum or what is contemplated as vacuum that we have with the motion of time. When Mainstream science accepted the Big Bang as the principle that will take science into the future the view about such a Big Bang concept unlocks a different door to another view on the cosmos from birth to end. It calls for the revision of all aspects of the entire history on cosmology and change that what is dead wood and that, which needs to be chucked out. Most of all it was my following the lead I got from Kepler that unlocked the doors I now present to you. I claim there is no gravity - motion as there is no gravity forming weight or forming mass. I hope the sketch contributes to my explaining effort:

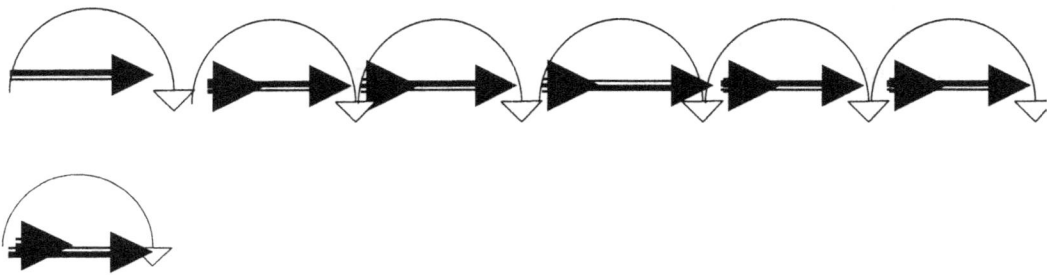

The duplicating frequency the Earth shows as k_1

The frequency of motion duplicating my body maintains as k_2.

k_1 minus k_2

The frequency of motion difference my body has minus to what the Earth has where that difference in motion becomes my mass. It is the sum total of the reducing of the motion that my body has in comparison with the motion capability of the Earth that is the mass value.

In k_1 as well as k_1 the symbol represents motion, however in the case where k_1 minus k_2 that shows an incapability of motion, which is motion, frustrated or a more commonly used name would be <u>mass</u> is created

The new k that is applying the relevance, must link the space a^3 being equal to the motion T^2 to singularity k^0 in order satisfy k^0. The flying of the aircraft is then unequal to the motion in the previous relation that was in place

where it was part of the Sun and the Earth motion relation and the new motion will bring a correcting in the relevant distance **k** to put the motion in balance with space. Space will always demand a correct establishing of the miss-interpretation of the equilibrium that is needed to sustain the effected singularity because of the space-time factor.

There is a point where the two points forming the relevancy unite in shared singularity. It comes because of shared motion. In all space-time, one finds at least two relevancies where one is at the centre. That part Newton saw and formulised.
He missed another part. Crossing a limit of inclusion is the limit of division and such limits are in distinction by motion producing the gravity, which is parting the two objects.
Motion brings about a relevancy where two positions no longer share a common point in singularity. That is what Newton missed. That is the gravity aspect Newton and all other Newtonians miss.

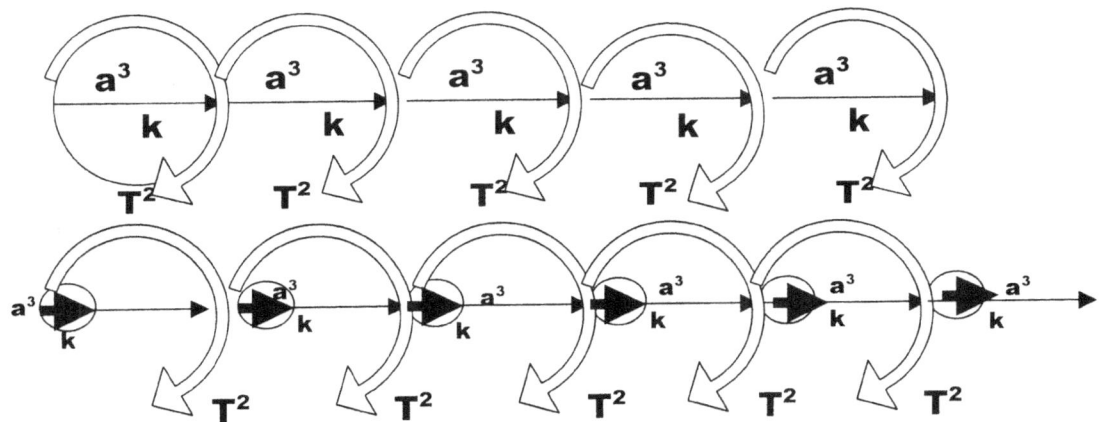

Two objects of substantial size differences are travelling at the same time but one has a space, which it has to move when it travels that is considerably different from the larger space. The larger space will produce an extending line equal to the space it moves while the smaller space will also produce a line in ratio to fit the space it holds relevant and which it has to move.

Mass has precious little to do with the whole affair except to be an obstacle intended to restrain the motion of the hosting space. The difference in size between the one in circular motion and the space in contracting motion must bring about that the smaller object has to move about a circle much closer to the centre because the larger space form the centre hosts it. However there is no large or small in the cosmos but only those better developed or those poorer developed. By duplicating there is more to duplicate in the better developed than in the lesser developed. When the lesser-developed space is duplicating the less developed space would hold a lesser extending from point to point forming a shortfall by distance in comparison. The motion being extended needs less extending and should therefore be closer to the centre in relation to what the better developed space would need in extending by a duplicating effort. This is the principle we find behind the sound barrier. The motion the aircraft produces forms an increase in the duplication of the aircraft, which extends the duplication of the aircraft splitting the Earth and the duplication that is producing an extension of the aircraft. The splitting does not align gravity lines with the Earth as it did before. The aircraft is reproducing more in a shorter time duration by duplicating and extending space filled by material that goes beyond the attempt of the Earth's extended of duplication by such motion.

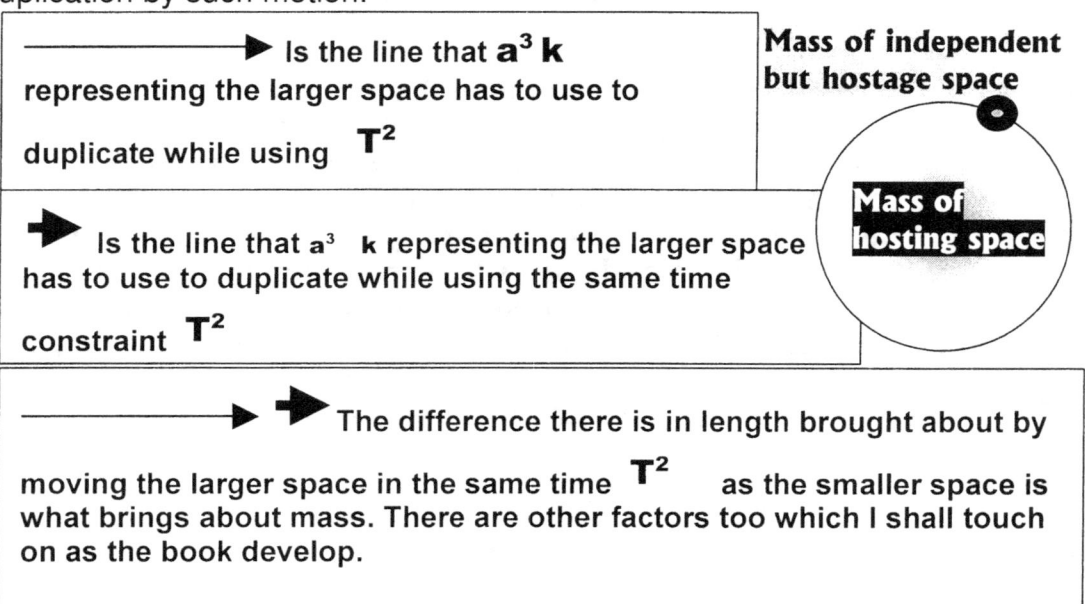

Is the line that a^3 **k** representing the larger space has to use to duplicate while using T^2

Mass of independent but hostage space

Mass of hosting space

Is the line that a^3 **k** representing the larger space has to use to duplicate while using the same time constraint T^2

The difference there is in length brought about by moving the larger space in the same time T^2 as the smaller space is what brings about mass. There are other factors too which I shall touch on as the book develop.

I know this may sound barely believable but please hear me out. While we use gravity, the use of gravity as such makes us part of the Earth. We see gravity as some influence or force producing mass and that mass is forcing us down on the solid ness and onto the Earth. By having the mass we become a semi unit with the Earth. That is how we on Earth see gravity but when investigating gravity in outer space we must come to a basic question: Is that what we experience as gravity on Earth truly gravity? Much of the proof about gravity is part of our perception about gravity because we experience certain conditions with gravity while we find ourselves bogged down on Mother Earth. But are our perceptions about gravity truly correct? We experience mass but are the mass the result of gravity or are the mass the product of gravity. We only experience gravity, as a factor from the position we have on Earth and the conclusions we form is a product of a perception we formed while we are being forced to be part of Earth. It's as if we are upside down and have to decide on which route we should follow. I want to make a suggestion, which I aim to prove in the following pages. My personal being on the ground and having mass that is keeping me on the ground comes about because of the speed that I travel through space being the very same as that which the Earth has.

By me not applying a speed difference I then inherit the speed the Earth places on me. But my space which I use $a^3 = T^2k$ to travel and the space which I use tot travel through is much smaller than that which the Earth burdened with to move and to move through. By me having a smaller space to move $a^3 = T^2k$ the space a^3 being moved **k** in the time it would take to move T^2 will produce less space a^3 to shift **k** and therefore a smaller distance **k** to replace all the space a^3 that is moved in the time T^2 the space a^3 needs to enable it to move **k**. To duplicate by motion the smaller space requires a smaller distance to shift the space but the motion will take up, as much time to complete than would the larger space take to complete though the space the larger space has to duplicate will require a longer distance to complete the total duplication of the larger space. A large space a^3 will produce a large extending **k** when using a^3 the same time durationT^2 when using the same time factor as that which the smaller space is required to use when under obligation to use same time constrain. Behind this is the most basic principle hiding which allow us the fortune to be able to fly using a flying machine. It is all about motion supplying relevance and forcing on time constraints.

Because my body that I have is travelling so much slower than the Earth is travelling due to my size in relation to the size the Earth has and although I am using the same time as the Earth does to move, such a speed difference is not in the time differences it takes to complete but in the space differences that has to be completed in the same time but is unable to fill and the space is trying to crush me into the Earth where I am forced toward the centre. If I were able to penetrate the soil solidness I would reach a point where my speed as zero would equal my space I occupy.

The space I duplicate by moving from one position and placing the space I hold in the next position while keeping my space I move as it is identical in the next spot but located in the next position. Such moving by duplicating takes a certain time to move from one spot to the following spot and it will use a certain frequency that will have the same ratio in bridging the gap from one point to the next point as that which the Earth has. My speed of duplicating by motion has to be even in frequency because I am within the duplicating space, which the Earth is duplicating and as part of the space that the Earth is duplicating but the duplicating of my space I do myself. But in size there is a massive difference between the space I hold and the space the Earth holds but to duplicate will take me as long as it takes the Earth. Notwithstanding this common factor the Earth has to use equal time in duplicating its massive space, as I have to duplicate my small space when we both have to share a frequency that will keep us duplicating evenly. Therefore the frequency of duplicating using the same time period will be a lot different to my much shorter frequency of duplicating space.

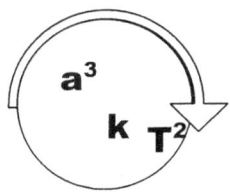

The difference is between me being in mass and me being in the correct position in the space-line the Earth has will place me in the correct position but the heat that then will surround me will fry me into non-existence. Fortunately, for life, the soil forms a barrier through which I cannot fall any further as to correct my location. Being where my position would have no mass would allow me to float there in that location in the same manner as I would float in water. I would be buoyant. It is because I do not harmonise the displacing frequency that I should that I have mass.

My having weight is what Mainstream Physics use to give me my gravity. Science purposely switch my having mass and confusing my mass with my having weight to explain what is beyond explaining. It is said that while I float in outer space in state of suspending hanging above the Earth in the weightlessness I still have all the mass that I had on Earth. But in order to prove that those in science will give me a mass even in outer space whether I deserve it or not. By that token science first has to cheat all logic by reasoning in some bizarre way that I take my mass up there to where there is only micro gravity. They firstly claim that all of a sudden I take my mass to outer space and in their next argument they say I have micro gravity in outer space since my body is floating as if it is in the sea. But if I stop floating and start falling to the Earth my body and I did not gain any mass. My falling then comes as the result of my motion being much smaller in relation to the space I claim and my motion then is being less than what is required to keep me in the position I have which in I maintain my orbit up there. By moving to slow I fall. I do not fall because my mass grew. But science has been proven wrong by their work without any of them aver admitting to such a defeat. All the satellites fall if the satellite motions are not reset. The satellites do not gain or lose mass. They gain or lose motion. By amplifying either my space (using a Hot Air balloon) or by accelerating my motion that I have in relation to that which the Earth forces me to have, I will break free from my weight or mass. I shall become airborne and float as if I am in outer space. By pretending my mass can be multiplied many times over in using a process, which then is called not gravity but momentum. But motion and gravity is all the same because motion is gravity that is redirected, which then forms another part of gravity where gravity again is also only motion applying. Science maintain the argument that when I am in outer space and am no longer part of the Earth I then will only have mass. But since there is only micro gravity I will be in a state of weightlessness. My mass is what gives me gravity and while being up there I take my mass long with me. But with my mass up there I will only have micro gravity. I am floating with my mass and it is my mass that is responsible for my gravity and I am floating above the mass of the Earth, which is right down below me, but still I have micro gravity. That is true if I wish to incorporate the dubious use of double standards by separating mass from weight. The mass my body will have in a Black hole will be a billion times (at least) more than what it is on Earth. With that the Black hole destroys the fact propagated in science that my mass will be the same everywhere. That is more than permitting double standards. Because our motion is much slower than the Earth is spinning, we place a breaking effort on the velocity the Earth has and that breaking effort we accept as the mass we have. The truth is that my mass comes about from the lack of motion I have in relation to the space I occupy and has nothing to do with any gravity - motions pulling me down. If I increase the motion I have there shall come a point where my motion will be sufficient to pull me into the air, as I then will have the required velocity to lift from the ground. That motion being in excess of what I have and is complimenting the motion that I receive from the Earth counteracts the motion of gravity that is containing me. The motion I adopt then release me from the motion containing me and if motion can release me by only becoming more, then gravity is my motion not being enough in the first place to keep me onto the Earth. Nowhere and at no time does my mass ever gain by having more protons that will get me back to the ground as if I am bigger or carrying more material or does my mass reduce to get me into the air as if I am smaller or carrying less material. Please note that this is my way of explaining to you about the fact of bodies having weight or mass. It is not mass or the lack thereof or any means to measure occupied space within the atmosphere of a larger body that pins me onto the ground. My body is claiming space by motion in space. Gravity is the result of motion because it is in the motion that bodies have that gravity affects them. This is proved because by adding motion the mass does get more but the body never gets bigger or hold more material, and in defiance of that statement by increasing the motion my body lifts and flies. The reality is that my body in motion has more mass being momentum but still my body lifts when motion allows my body to lift. This statement confirms Kepler that a^3 becomes more (massive) when motion T^2k becomes more (moving).

Mainstream physics admits all along that nobody, human or otherwise knows what gravity is. While investigating Kepler's work with employing much motivation and detail in order to give his work the much duly credit it deserves it will also serve a valiant purpose when by the same token we try to establish what gravity is, because I believe Kepler possibly answered that mystery. We have to start with the person that introduced gravity or so does everybody acknowledge. Newton saw an apple fall from a tree and he subsequently realised there is some force pulling the apple to the Earth.

Although he still was a student he announced his findings and became a genius on the spot. The concept he introduced as gravity gave him instant admiration from which he became the legend he is today and that reputation he gained there at that moment would last him from that day he instantaneously unveiled his mastermind, and that same genius still serves him in his honour to this day long after his death. He found that this force has to have some thing to do with the weight and the mass of that particular object and the mass of the Earth. There is some force pulling that apple as much as the force is pushing the apple and the same goes for the Earth because the mass the Earth has is doing the same to the apple. Between the two objects facing gravity there is a force that develops where such a force is pulling the apple on a constant basis towards the Earth even after the apple is already in a steady state on the Earth. That forms the mass and the mass forms gravity. He concluded that the mass is responsible for the pulling. Remember this observation came three point five centuries ago when knowledge and brilliance carried a much different defining than what such defining of brilliance is worth today. He realised the pulling on that apple brings about weight that brings about mass because the apple departs from its location and arrives at its end location when the falling is completed. Then he went out convinced all that was in line of finding the needed convincing because no body before Newton thought of what Newton thought quite in the way that Newton thought about gravity.

Newton succeeded because he found a way in presenting science with the fact that objects move closer because of some force. He went one step further and named the force he fathered as gravity. But there it stopped! Any and all other further defining the matter or going into any possible observations of whatever magnitude concerning the topic never realised any motive to go further. Inspiration to further commitment just flew out the window as the essence to do so immediately expired as far as the rest of science is concerned. What might he have missed if he missed anything? We all fall down when we are unstable and out of balance. He never realised that balance is more crucial than brutal gravity because that part is the defining part about gravity. No one ever gave a thought about the balance part even centuries later even as we grew into all the sophistication we now enjoy. What brought about the balance that secured objects in an upright stance and supplied some form of control over the managing of a position? Any other position than being flat on the floor would have a better defining than being just at the mercy of the force gravity. Standing tall is a stance that defies gravity so there is another force other than the pulling of gravity. Admit tingly the force would first and foremost have to aspire to the rules of gravity and then comply with other demands. True enough is the fact that that position would ultimately and firstly by all accounts have to satisfy gravity before any further motion could commence. Yes but then by balance motion defies gravity by changing gravity's force of pulling everything straight down towards a visionary centre between the objects. In affect this means somehow there is control over gravity and gravity does not leave objects beyond outside control. Gravity is manageable and can be controlled; we just have to find a way…

Years later some one came up with the novelty of hot air ballooning. Ballooning proved that there is antigravity but that part was missed by all even to this day. Some people speak of antigravity as if that is some mystifying mysterious concept that is so well hidden in the secret annals of the hidden Universe that only Ali Baba and his magic words can reach it. Please consider the following statement. If gravity was bringing the object down, because of the affect of gravity which is that what we experience as the gravitational sensation and that is what we interpreted as gravity by our sensation and observation, then that is only coming about by our bodies that is in a state of being dragged down. The dragging down of the body is in the direction of the Earth centre. That sensation of being firmly locked onto the ground constitutes to what we believe we experience as gravity.

When some influence brings about the very opposite effect, which then results in establishing the opposite result, it deserves to be anti. In example we feel dragged down but anti will be the lifting of the body into the air. Anti will be going in an opposing direction of the motion that gravity inflicts. It will counter the influence that gravity applies. Such motion has to indicate antigravity. The counter acting of the mass dragging us down must be anti gravity pulling us up into the air above the ground. Antigravity must come from such an opposing influence that will bring about the lifting of my body. If hot air ballooning gave the object an opportunity lift, then ballooning must be antigravity. The balloonist and the entire balloon found a manner to counteract the pulling of gravity enforcing weight. The balloon can lift what gravity depresses and if Newton said gravity is the falling then later Newtonians must agree that the opposite of falling is flying or lifting. A balloon is lifting-and- flying. If gravity is pulling down objects in the direction of the centre of the Earth then flying is antigravity. Moving away from the Earth by means of motion and in particular flying is using whatever means to defy gravity where the lifting can also be the hoisting of a body by a crane. Lifting by ballooning in a hot air such balloons escape from gravity where the balloon constitutes to bring about the effect of establishing antigravity. Climbing up mountains must fall into the antigravity department because parachuting down the mountain definitely falls in the gravity department. Nevertheless it still does not answer the question of what gravity is.

Let us look at antigravity because the antigravity is releasing the object from the gravity that controls the object by an Earth fed force. The balloon starts flying when the confined space of the balloon is veraciously and violently heated in access. The balloonist shows us that in order to overcome gravity we have to introduce heat. That is the only manner in which we can defeat gravity. Even by an engine driving an aeroplane such flying can only result if an engine combust solid fuel by creating motion as the fuel mixture is turned into heat. It is heat that makes the difference. That is the very thing that Kepler said. Expand the space a^3

and the motion T^2 will move further increasing **k**. Blowing hot air into the balloon is increasing space within the balloon a^3 which then results in providing the balloon with a larger distance **k** from the Earth centre k^0 that still holds time with in the Earth atmosphere with the Earth T^2 within the space of the Earth **k**. Using Kepler provides us with insight and the ability to see what gravity is by showing us what antigravity is (a^3 gets bigger and that will bring in a larger **k**). But moreover the larger space in enough compensation to bring about extra motion that will defeat gravity by the extending of **k.** If that is not antigravity then we can forget about Ali Baba and his magic rhymes too.

The balloon assists us to escape the Earth's hold on our body, because there has to be the force producing motion countering the motion of the Earth gravity. The balloon shows that releasing enormous quantities of heat into an inclusive area excluding space such as that which the balloon canvas provides, which is establishing the release from the gravitated containing force on the body giving the body a means to escape by floating about above the ground. The motion is at that point breaking free from the containing gravity by moving in a specific direction, other than the direction the Earth gravity inclines the body to travel. By concentrating the releasing of heat into the balloon, the direction of motion starts to contradict the enlisting of the Earth gravity and the heat breaks the balloons confining properties while the balloon is released from the Earth as the balloon and us lift up into the air and away from our confining to the Earth.

At the point of explaining we arrive at the point where we can say what we think the difference is between the balloon floating in the air above the Earth and a body suspended in outer space floating above the Earth's atmosphere. The difference is the heat that is in the confined air per volumetric ratio favouring the heat being more in the space than what the heat is outside the confined space. If we had any method to put the required heat we need to escape from the limits of the Earth to outer space into the canvass of the balloon there was no canvass left to contain the heat. The heat is available to do the job but the means to do the job with the tools in hand is unavailable as far awe can use the balloon. By having more heat in the one area than there is in the other area beats of the pulling of gravity. Obviously it is antigravity that keeps the balloon in the air and what keeps the balloon in the air is having a larger volume of heat per space unit than what is in the atmosphere. The balloonist shows us that by applying more heat we can defeat gravity more. Someone took the advice, because the next minute the Germans had rockets. The launching of rockets brought about the ultimate defeat of gravity but it involves almost the ultimate releasing of heat.

In antigravity we find heat more concentrated in one definitive area than the heat concentration is elsewhere. The more the heat is that we release into space the more the antigravity is that we achieve and the more release such antigravity can produce. But what connection can gravity have with heat and if there were any connection between heat concentrated and gravity, what would such connection be? The history behind Carl Benz should bring the answer but more so would be the story behind James Watt and steam although the James Watt story may not be that thought provoking because it is much less filled with the ever popular cheap thrill only sensational gossip can provide…Still both stories cover the same principles. In the Carl Benz story a housewife leaves a pot of benzene fuel on a coal stove. The pot with benzene heats up where the pot with benzene becomes hot and under pressure. This performing of heat increase, such increasing expands as space and releases the heat as newly creates space, which then removes the housewife with her house from the neighbourhood she used to regularly frequent as her residential address. Afterwards almost the entire neighbourhood is not there to tell the tale or ask why...

It was a stupid tragedy that brought about the end of steam and the rise of the internal combustion engine and on Earth billions on billions of human souls are in torment not to please or suffer for the advantage of coal Barons any longer but now they are dying and suffering in agony to please the wishes and desires of oil Barons. How much did the world not change...While it is no longer the coal Barons shackling us in chains and telling us democracy broke our burden of slavery, we have now the pleasure of the oil Barons enslaving us with democracy and telling to be happy because we are the fortunate slaves, there are others circumstances in which they can enslave us that will leave us worst off. All this came just because the pot of fuel created a houseful of space that was enough to remove the house from the address the house previously enjoyed. But Mainstream science neglects to appreciate this. They see the heat, they see the antigravity but they fail to add the heat, the anti gravity and the space that no longer housed the house of the naive and rather impractical thoughtless housewife. They call the tragedy an explosion but then again everything that expands while using a noise during the expanding is an explosion. Adding of new space to the space holding the house at first altered everything that was previously proportional positioned in the space where the house was. Such exchanging of heat to accumulate and introduce more space in the process referred to as an explosion was bringing in more space that came directly as a consequence from the explosion which was producing more space where the increase in space brought disorder because the well organised material distribution and placing was before the event filling just enough of the required space arrangement that was holding every object in a prearranged order of tidiness.

Then suddenly out of the blue the space which held the house in a tidy arrangement had to accommodate more space therefore the ratio of material per space volume increased dramatically many times over in the favour of the space in the balance. That part no one ever acknowledges. However the losing of the house was not much surprising to Mainstream science back then and even today because who cares about old news. All of Mainstream science was at the time, as they are today, very familiar with all explosions because of wars and bombing that leads to maiming and killing and all the unspeakable monstrosities we associate with war so that the dirt poor can suffer and die to leave the disgustingly rich even richer. The poor has not the means to pay science to be clever and devise methods to save their lives, so the rich does the poor the favour of paying science to find methods whereby more poor could be killed as long as the rich saw it as a good investment with great capital gain on the part of the rich. Therefore science is well established in the method of creating more elaborate and destructive explosions that the rich pay them to invent. In the explosion caused by our housewife no one put up money to investigate what happed during the explosion but money went to why the explosion happened.

That inspired an investigation in connection with the fact of the finding more about what takes place during the carnage as more money goes to finding means to create more carnage per money unit spent. At least that is why the poor were invented and that is why wars are invented. It is invented so that no money goes wasted on saving the poor people except if the poor has the money ready and available to pay the rich for medicine to enable the poor to stay alive. So science goes out and develops more fuel for carnage but fails to find out why the housewife and her house are no longer part of the neighbourhood she used to frequent. With the loss of the presence of the ignorant housewife with her house her neighbourhood and all were a normal way of leaving us with a new way of tapping and harvesting energy and untold riches which was born with the death of the absent minded housewife. But according to the mindset of science they saw not what the incident presented in space producing for to their view nothing new came about since it was just another exploding of fuel...so no body bothered as to finding out how. What they missed was the part the coal stove played in the whole tragedy. Without the intervention of the coal stove producing the heat that turned the liquid fuel to liquid heat liquefying the space that turned the liquid space into a gaseous space where the liquid space revealed its true incentive in nature by turning out as space and the newly created space that was in fact liquid space that went on to become more space, well that space was providing the one main factor in space-time relevancy. The stove's heat was producing space by transferring the heat from the stove to the pot filled with volatile fuel and the transfer of the heat to the fuel brought about the expanding of the space that the fuel claim to need in the pot filled with the heating and volatile fuel. The fuel space requirements became more as the heat filled the space that was filled with fuel and that took up more space, which the pot could not cope with since the pot had no room to allow such an increase on the demand for more space. The fuel expanding as such was claiming new space to sustain and accommodate the growing requirement for more space to be created in order so that the volumetric increase of heat added to the fuel could be accommodated. At one point the asking for space became a claim on new space, which we see as an explosion. The heat transformed to new space by an exhilaration of breathtaking increase in heat forming space and this increase of newly formed space was transforming all other surrounding space within the room, the house and the neighbourhood in general. By increasing the volumetric quantity of the space it rearranged all space, which included some of the space held by material that was in a solid form and scattered the rearrangements as fragments in all other designated places far away from each other. This meant there was excessive and all around rearranging and relocating as well as re-allocating of space in general. This to science sees as shock waves resulting from an explosion but is merely heat expanding space to set new required standards. It is rearranging every aspect that contains space or that space contains. It will bring a much different looking end. Everything about this concept is missing from Newtonian science because Newtonian science failed to investigate Kepler. Kepler said space a^3 is equal to the motion T^2k thereof and then that says without Kepler directly saying it, it says that if space a^3 goes bigger as a result of the explosion then such increasing in space will constitute to more space a^3 which has to produce an increase in motion T^2k where more motion T^2k will bring about faster displacing space. This is one small fact that Newton robbed the world of realising with his ignoring of Kepler's work.

From every point there may form on the outer circle line of every part of the circle

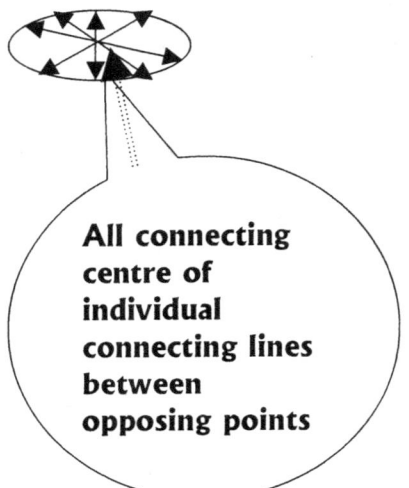

All connecting centre of individual connecting lines between opposing points

structure and all structural positions of the circle in all circles, all circles refer to the centre in perfect aligning. Every point wherever located on the sphere has a matching and equal but an opposing point on the other side of the circle but in equal position on the other side of the circle. Between the two controlling points runs a precise straight line connecting the two opposing points in counter balancing. When drawing the connecting line between the two controlling points and connecting such points on further edges of the circle by lines formed the lines will all cross the centre? From wherever a line may cross and from every point forming a line to the other side of the circle rim holding the connecting points there has to be a counter point located on the very opposing side that when connected by a line, such a line crosses in the centre. In the middle the centre spot bonds all sides coming from any and every direction there can possibly be. The line will run to an equal point on the other side across the same distance from such a centre and that then has to be where the strongest gravity can be located.

We are now serving time in the twenty first century. One Professor once told me I must realise that Newtonian science took man to the moon and back several times and in such a view I am rather annoying presumptuous to criticize Newton. The Professor missed the point. I criticize Newton on what he did not give us, which he gave us as incorrect by his own admitting that it is mostly guesswork on his (Newton's) part and his guessing about the facts where later that guesswork became institutionalised facts believed by all concerned to be correct and to be proven to a degree of correctness that is far beyond doubt. Newton gave us gravity but Newton never gave us the explanation about gravity. At the time Newton met strict opposition from his colleges and piers because others felt his introducing of an unexplained force was taking Science back in time, which of course it did. Many scientists at the time accused Newton by name of dragging science back in the wrong direction of progress by introducing unexplained forces acting in a superstitious and mediaeval manner. I went one step further by asking myself the question: If space becomes more when heat becomes uncontrolled

$$k^{-1} = T^2/a^3$$

why can space not become heat when space is under control? If space becomes more as we see with every explosion of every kind and such heat forming space releases energy, then why would space being managed not form heat being under control and produce energy. We only have to see what Kepler said gravity is. Motion gives us energy.

$$k = a^3/T^2$$

In the normal flow of events there is a certain ratio of liquid time that stands related to a specific gravity, which is generated by the motion of the liquid in relation to the solid that binds the liquid to the solid. That uses symbol representing the factor **k** in the Kepler formula $a^3 = T^2 k$. This symbol is standard on both sides but due to the inherent nature of rotation qualities, it opposes that what it was by crossing the divide there is. The liquid attach to the solid by diminishing the relative facto bring about that **k** goes negative $k^{-1} = T^2/a^3$.

The solid on the other hand confirms the liquid attaching to the space by the motion thereof and withthat the solid recognises the shift in limits where there forms a new boundery and a new location where space that include the attaching liquid then ends.

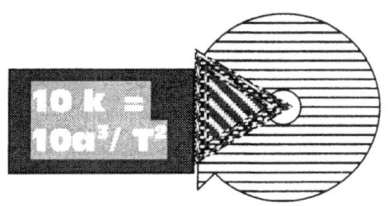 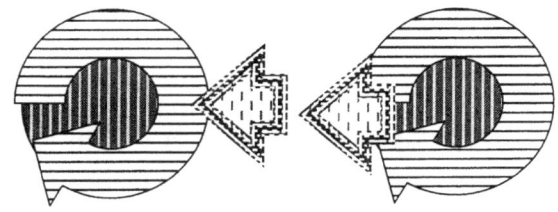 When an explosion occur the solid space melts and the lot in the space becomes liquid. That is a direct result from the expanding of the limiting or relative factor **k** that reposition the boundaries of the liquid turning space into an altogether new dynamic. As the solid melts it amplify the space held by the heat many fold by tarnishing the spin T^2 that increase the space a^3 and reduce the density of the heat in the space a^3 where the velocity of the motion T^2 is reduces as the factor **k** is increased in distance. By increasing the claim on space as the spin velocity reduces and the density intensity decreases the liquid part becomes more since the overheating destroyed the solid part and that which we know as gravity reduces because that we know as concentrated space reduces in intensity. The expanding is due to more overheating than what the governing singularity can control.

Since there is an excessive demand on space through the overheating which reduces the density of the solid and turns the solid into liquid the reduction in heat density by lost of singularity control. The increase in liquids demands more space 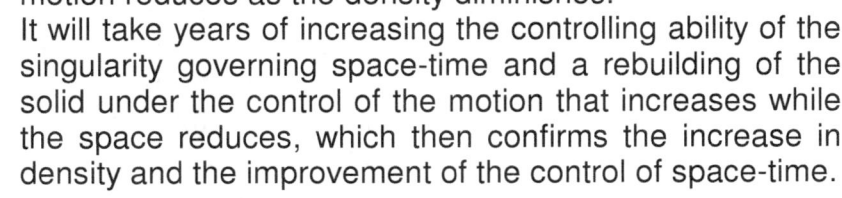 in the light of the decrease in liquid density and the motion reduces as the density diminishes.

It will take years of increasing the controlling ability of the singularity governing space-time and a rebuilding of the solid under the control of the motion that increases while the space reduces, which then confirms the increase in density and the improvement of the control of space-time.

Where space is the least, which is in the centre of the circle, gravity is the strongest. The gravity located in the circle's space less centre holds not only the sphere together but all that is in the surrounding of the sphere outside the sphere as well. It is from there in a giro action that gravity bonds all atoms forming the structure of the sphere as one unit together in a unit as well as distributing a specific alliance in shape and form. How the atoms manage that we will get to in a while, but there is a law allowing for that to take place.

Gravity is the strongest in all cosmic structures holding the form of the sphere and gravity controls all around from that very centre where space is the least, therefore the more material there is to generate motion within a star the more secure would the generated centre be in any star where such centre produces gravity. It is not the material but the motion the material accomplishes that becomes the factor of gravity. The smaller the star is as far as volumetric occupation goes, the stronger the gravity is that is coming from such a centre. The less the space there is the less the motion is and therefore the stronger and more deliberate the motion is evoking gravity. From the centre in the middle where space is absolutely at a premium the gravity grows stronger as it draws all material.

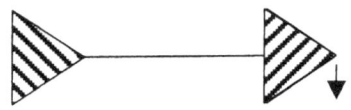 If outer space was nothing then the crossing of light in outer space was impossible because of the total destruction nothing leaves when contacting anything (1X0=0). Light crosses space at so many points holding infinity that it becomes the speed of light or the ability of light to convey space by motion thereof.

The motion is one of confining the space to a centre by the moving or trying to move the flow of space and whatever is in the space into the centre where the space is least. Take the Neutron star and the Black Hole as an example and compare that with the Sun and the answer is simple. I claim that gravity is all about reducing space and not attracting matter but that I explain a little later on. Therefore the matrix of gravity must be permanently located in the location where space is the least. Looking at a sphere we find that what holds the sphere true to form is placed in the centre of the sphere, which then has to be the most intense point of gravity. Gravity is confirming the round shape without favouring any specific point. Such evenness of gravity comes from what is applying at such a centre and is in control of the surroundings. The centre that secures all of the space and material in the space holding the specific form has to be round if it is anything. That shows that in the sphere one can see that the sphere as a form is dominated or controlled from one specific location in the centre. The explanation about the reason there is control coming from the centre, has a very childlike simple answer.

The Big Bang was where gravity held the Universe in the least space there ever was. To find the original gravity we therefore have to reduce the sphere to the circle and reduce the circle from there narrowing the circle down to as far as one can go. The Universe is a magnitude of spheres constructed by a complexity of circles. This is because everything sprouted from singularity. To narrow any circle down will be the same as narrowing down the Universe. In our reducing of the Universe we must first acknowledge that the Universe constitutes many spheres, which is giving the Universe gravity as a combining unifying part which is the part of the sphere giving the sphere form (or gravity) and that confirms that the sphere is a circle in many times over multiplying the positions from where gravity secures form. If we wish to go back in time by taking the Universe back down the same route and at the same time maintain some coherency we must concentrate on a single circle because a sphere is a circle by millions of possibilities linked together by just a name that changes the concept. When one takes this accepted route in thinking that by reducing the connecting line to the connecting circle point in the centre of the lot, it must take us back in time at the same time as the circle reduces to the time during the Big Bang.

During the Big Bang where all circles were as small as they can get, we run into an unknown substance we came to know as antimatter. This theory is propagated according to Mainstream science but what is most surprisingly is that I do agree with this part of the statement. All material produces gravity. I go one step further and say all material applies motion where some motion may be to contain by using gravity attributing to the contracting that leads to the reducing of their space. Then as everything in Creation has an opposing to restore and maintain balance, there had to form another or other material that did not by our lamentable standards produce gravity because those material produce antigravity, a concept beyond human discernment. Antigravity must be the expanding in counteracting contracting. A counter action to contracting is where expanding provides growth of space to that which has then reduces the gravity effects. Forming pappy provides more space by losing density to the advancing of their space. Material either have gravity by solidifying or concentrating the space they hold in ratio to the material within the space they hold whereas others lose their solidness by entertaining more space within the ratio of material to space where such material becomes liquid and in more extreme cases they become gas. Being a gas they float which gives that material a high degree of antigravity being airborne. It is however not clear if antimatter produced gravity as it did when it went to lunch on and ate up all material in the immediate surrounding. It was cannibalistic but the unanswered question is this: was it a gravity producing predator or a non gravity-producing carnivore. Did material find a comrade in their gravity forming of form or did the gravity it produced bring on the demise that subsequently followed the event as is reported by the highly informed.

The Accepted statement on antimatter reads that matter composed of anti particles where such subatomic particles that have identical rest mass to corresponding particles of ordinary matter but opposing charge and are opposing in other fundamental properties. One example given is that an electron would have a positron, which then functions as the anti particle and has a positive charge compared to the electron's negative charge. That is put bluntly in its utmost simplistic form. Unanswered and tough questions arise from such a statement. What kept the electron bonded to the atom since the protons must by implication produce expanding or by definition be repelling the atom and surroundings instead of the normal contracting or confirming of form. What is a positive compared to a negative charge, because it is human concepts that put the directional qualities of material into a positive or a negative contexts as we did with hot and cold. It is human standards that humans brought about to make all human inadequacy by lamented human understanding better but it is not applied cosmos principle. If there is extracting electrons performing in the capacity as antimatter, then there better be protons by other name in service to the anti electrons, which then of course serves the anti electron in the capacity of an anti proton with an equal but negative charge to that of the proton. When matter and anti matter meet, the two opposing particles annihilate each other until one vanishes from the Universe. I have to add that at the time this theory was devised the first computer games became a crazy fashion played by young and old, those wise and those foolish all alike. This game was called the packman and the packman ate up all the skulls and after eating left nothing as evidence. The theory about antimatter has some very striking similarities to that packman game. It still does not answer the most ardent questions: What makes a positive electron different from a electron in the working place each has and can any person show such an object found in nature.

Can people take a positive electron to an investigative bureau and are acclaimed for bringing about such evidence? It is unwise to substitute nature with human concepts just to further mathematical equations. This was apparently presented as normal as nature was when nature developed with the Big Bang and nature then did behave this oddly just after the Big Bang came about. But one huge misgiving in this argument is declaring that everything the antimatter had as a meal vanished and even moreover then antimatter went and vanished too. Where could the combination that was produced when the matter and antimatter collided go after it disappeared and did it form the by-product of antimatter science is talking about, which since then apparently vanished too. What a bloody none-intellectual fairytale that is on the in addition as well as one of those made–up-as-they-go-along stories, which is told by persons that supposedly should know of better. Since there is no place other to find a location to be within than being in a place inside the Universe it is hardly possible to vanish from the Universe except in fairy tales because for one simple fact: there is no other home to have but the home we have which we call by the name the Universe and we have no where to escape to but within the walls that the Universe provide for such a purpose.

There is one Universe containing all and preserving the lot. Mainstream physics is accepting this fact. But then by the same margin they accept a principle that allows property that once was part of the Universe to leave the Universe and go somewhere outside the only Universe. They create a loophole whenever it suits them to misplace what they cannot explain readily and logically. In Creation to their and my thinking there can be no hiding of anything but in the Created Universe. This they admit and confirm although with the same breath those very same intellectuals also admit that there is another place outside of what we are able to find in the Universe. When someone comes up with the marvel where such a person can declare in all honesty that the product of antimatter or singularity escaped from the Universe to God knows where that person should leave the field of science and go for fantasy writing such as fairy tales or reporting about politicians inner deepest chastity and integrity. That is what we can find outside the spectrum of what the Universe can deliver. With such a statement on anti matter or the loss of any Universal product disappearing from the Universe then alarm bells should go off in the mind of the trained and professional Scientist working with such matters.

Yet those in charge do not once belabour the question on the validity of a statement that involves stating losses occurring of substance being in the Universe going lost or being removes from the Universe. Their surprise of pressruns stating the loss of factors and declaring the possibility that there was a possibility of such losses where those factors now are outside of what once was part of the only place there ever can be. They can read mathematical calculations and agree on an outside the Universe without stating it in an explanation what happened to the lost and found or they're ability of introducing the concept as a reality, which they claim it is. That such factor can go outside the Universe and leave the Universe by causing a Houdini vanishing act of never –to-be-repeated-again status. Science would have us to believe this antimatter went into hiding in a manner that is out of the Universe.

They applaud this thoughtless presumption while fully knowing that at the time they do this acknowledging that there is no other place for anything wishing for a place to be within then having to be in another place other than inside the Universe. If it was ever anywhere it still is within the Universe merely because there is no other place to go than to be inside and part of the known Universe! There cannot be some factor and then misplace it as if a valid factor calculate the value can prove the disappearance and by disappearing it no longer is. If it was in the Universe it must still be in the Universe somewhere. Then we better start looking for it.

Another big issue is that what ever the Big Bang produced must be in equal terms everywhere. The Big Bang was a process that had the Universe act as a high-speed cocktail mixer of no repeating ever again. Whatever the Big Bang was of all that it was, the most it was in the beginning was that it was one massive mixer mixing everything in it at the speed of light. The relevancies might change slightly and balances may change favouring opposing ends…yes and known appearances did change…yes. But in the end all the factors must always be present everywhere through out the Universe. By this lacking of a fundamental explanation about what antimatter will look like when found Mainstream is incredibly poorly judged by scientific standards. Those mathematicians calculating physics suggest that science should take antimatter as a cosmic fact and then in disregard of other realities they dispose the truth by discarding its properties onto the unknown.

That hardly suggests plausible science by any one's admitting. In the cosmos is, was and will be all the material there can ever possibly be. Our concepts we put forwards can be faulty but nature cannot ever be at fault. Our arrangement of our ideas can be at fault, but we cannot pull a vanishing act on certain cosmic products and in doing that then dismiss the existing of such a factor or factors, which we then claim, have vanished in the further developed Universe. Our concepts of what they became may be at fault and by changing some basic principals such changing may produce a better understanding about what we think we read into mathematics. Mathematics is purely a language and mathematicians are purely translators. Mathematicians translate from the language they read to the verbal equivalent they speak. As in all translations made, certain concepts may become misinterpreted.

The terminology used to explain this is "lost in translation" Mathematicians must see what there is in the translation and try to incorporate what there is available in the cosmos to what the Mathematician sees in his mathematical calculations. The Universe was full of heat and it was full of material but it was not full of free space. If that is the case then where did the heat come from and where did the heat go? Hiroshima and Nagasaki taught us many things about the horror of human nature but most of all it taught us that material is heat secured in atoms and atoms are heat tightly wrapped in a cocoon, which we named the atom. Heat in any form cannot have anti in another form. The package holding heat wrapped can unwrap as it does with nuclear atomic demise. But the anti to heat is cold and cold is space.

The undeniable fact about the Big Bang theory is the accepting of a growing state in which the entire cosmos seems to be in. With all the expansion that went on we came to the point where we now are at and in such growth all aspects in the Universe must grow in relation to quantifiable progress in all different aspects, which takes us to that which is seen and that is unseen and which came along as products in the Universe where everything took everything on a growing spruce by unveiling space. That is where we now are. Such expansion include all there is including everything and not just with outer space growing. The dynamics of outer space alone cannot grow by leaving the growth of material behind. Should we wish to see where we came from we have to reduce that which we now see in our surroundings to apply to the measures that once applied in all aspects of the cosmos. Mainstream physics is over pronouncing the growth of space and with that suppresses the part matter must play in such growth by simply ignoring the issue. That is the reason why they prefer to ignore the evidence that material is growing notwithstanding that material is growing or that their disbelief about the matter of material growing do not change that material is growing in any case. Because they cannot find any reason why material should grow they refuse to admit that material does grow.

This is hiding from the truth by hiding the truth. If space grows and the Universe is getting bigger then all space grows to allow the Universe to get bigger. That includes matter and space not in matter. Space can only grow if materials that also hold space also grow within the space that is growing with the growing space. It means that stars get bigger by the cosmos growing from the Big Bang onwards and outwards to the moment in which we are at the present. But if stars grow then the atoms forming the stars are doing all the growing as they secure more space within the space they claim.

If Hubble saw space grow, the growth of space must include the growth of space holding material as well. In studying the Hubble's expanding theory we come across evidence that makes it clear that all material expand in a manner as if the expanding comes from the centre of each and all particles within the expanding space and the expanding grows outwards from every particle centre. It is using every star centre to grow from in all directions proportionally in all directions evenly. This leads one to believe that gravity is this securing of space in the material just as Kepler showed it to the world. It proves a connection with deliberate implications coming from every as well as in every specific centre. It proves that the centre $k^0 = a^3 / T^2 k$. It becomes apparent if and when separating Kepler from what Newton thought about the work of Kepler which Newton accepted as being inferior and all incorrect.

That is making the Universe small and as man grows man allows the Universe also to grow in relation and corresponding to man's ability to comprehend. We see the cosmos as a circle and we accept the circle because the circle is what gravity implement when the choice of form is coming from material that has all options to freely choose from. By taking the circle back one will follow will trace the rout of the cosmos to where it then started.

All stars are many circles in many dimensions, which form when all circles join into what we call a sphere, but that leaves us only with the circles in the plural.

Taking the cosmos back can only lead to one point and that Kepler told us we will find singularity $a^3=T^2k$ which is $k^0 = a^3/T^2$ k. We can only reach $k^0 = a^3/T^2k$ if we repeat $1/k = T^2/a^3$ in a continuing manner indefinitely. When one does the effort of reading this correctly, it says that when distance **k** brakes from singularity $1=k^0$ that is then $(k^0 =1) /k = T^2/a^3$ where the space a^3 produced a time T^2 equal to singularity k^0 and singularity k^0 is equal to eternity which was where all was equal to a never changing cosmos that was holding the single form into one dimensional space that included all the filled and vacant material filling in from all sides.

This is one way of looking at the issue and by doing that I am about to prove that singularity is Π. I am about to prove that not only is the planets adhering to the Titius Bode rule of seven over ten and ten over seven in relation to the Roche limit but that the Roche limit explains the very, very first instant the Universe experienced outside eternity. The atoms relates to space in the very same manner of seven singularity positions to ten points and from this motion of material interacting with space is securing material on the inside as well as on the outside. By that motion gravity comes about finding the value of Π^2. Gravity uses the relation of the Titius Bode seven on ten and ten on seven as well as the Roche factor to form gravity and gravity is always Π^2.

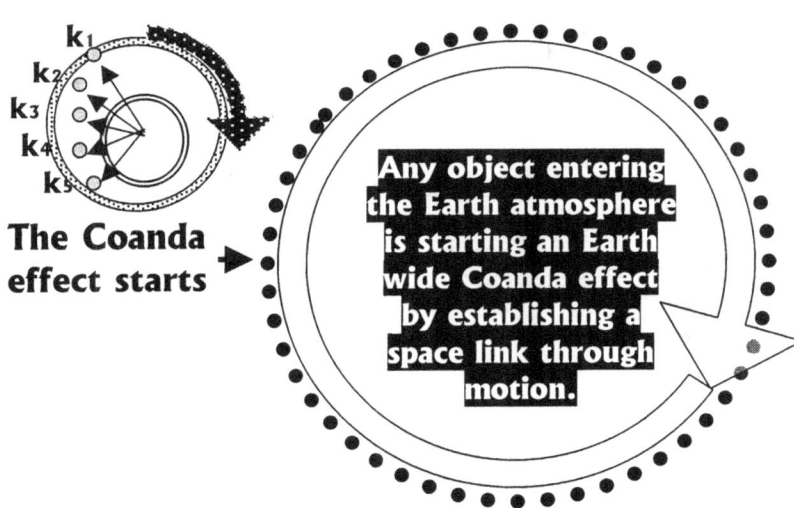

The Coanda effect starts →

Any object entering the Earth atmosphere is starting an Earth wide Coanda effect by establishing a space link through motion.

This I see by reading Kepler's work as Kepler produced the work and introduced the work as $a^3 = T^2$ k. With this formula $k^0 =a^3/ (T^2k)$ must also be true because $a^3 = T^2$ k is a relevancy that has to be in relation to singularity and therefore singularity must be $k^0 = 1$. Where will we find $k^0 = 1$? When an object is within the singularity alignment of the Earth, meaning it is either stationary, or free falling on pure "gravitational" momentum it holds the 4 X Π^2 in relevancy to Π.

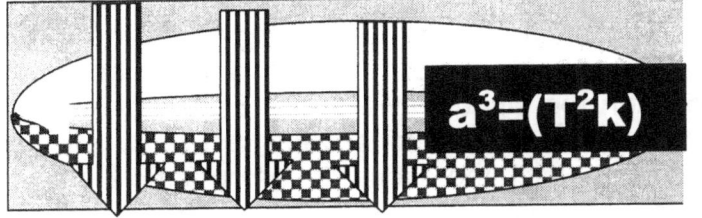

$$a^3=(T^2k)$$

What I try to say by that is that the Earth holds all objects within the atmosphere to a displacement value of $3\Pi^2$ within the seven are where it is using 10 as the liquid in space. Because the aircraft does not leave the Earth's atmosphere the 6^2 and the 10^0 does not come into effect.

Mass is the refusing of any object to dismiss the form it has and to join the Earth solid structure. Mass cannot and does not contribute to the establishing of gravity except by depleting space through motion and such numbers of the protons in a space forming an exclusive unit.

$$k^0 = a^3 = (T^2k)$$

$$k^0 = a^3 = (T^2k)$$

It is when the motion exceeds the mass the aircraft has the ability to break the sound barrier. Galileo proved that no mass is present in falling, which is also matter in the process of flight and because of that can the sound barrier become some form of constant.

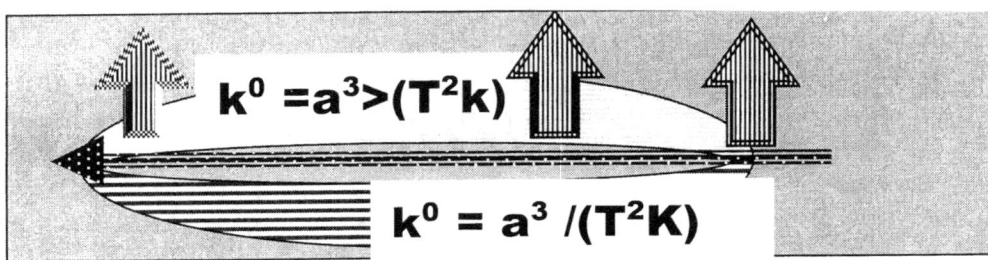

$$k^0 = a^3 > (T^2k)$$

$$k^0 = a^3 /(T^2K)$$

The establishing of independent motion of the craft secures an individual gravity and such individuality leads to the breaking of the sound barrier because the one gravity can no longer subdue the smaller motion, which is producing gravity.

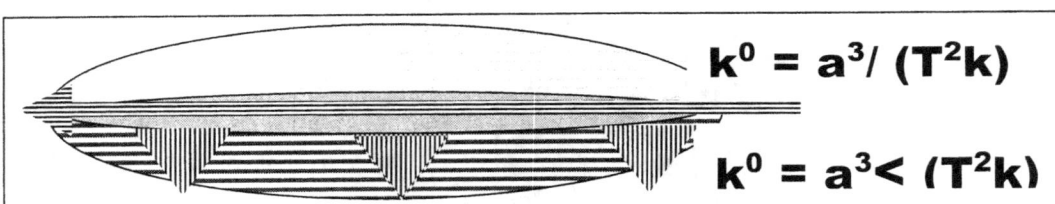

$$k^0 = a^3/ (T^2k)$$

$$k^0 = a^3 < (T^2k)$$

At a height of 31000 km above the Earth the mass of the wing becomes compensated only by a motion of a relevancy that comes about at 2500 km per hour. In that case the craft has to apply motion at a rate of 2500 km / hour just to create the required velocity to keep the aircraft in motion in the sky. Motion creates gravity just as Kepler said when he said gravity is about $a^3 = T^2 k$, which translates to the dismissing of space and the motion, duplication establishes a centre that controls the balance that the newly secured singularity will provide. When the aircraft stands still the Sun provides such a pivoting centre but when independent motion comes about the point shifts from the Sun to the Earth centre where there is a line contact between the singularity that the Earth holds which then forms a new relation in respect to the singularity activated by the independent motion of the moving body which the aircraft takes on a trip in motion and with it a position that the relevant singularity is claiming which are released as part of the minor space.

$7 (3\Pi^2)\Pi^0 = 207.2$ km/h. $\mathbf{R^2/T} = 7 (3\Pi^2)\Pi$ and $\mathbf{R/T} = \Pi^0$

The Roche limit also changes to accommodate this change and becomes either Π^0, Π, or $(\Pi^2/2)$. Falling "free" will then mean that the object holds a position of 7 $(3\Pi^2)\Pi^0$ to singularity. In the half circle applying two of the quarters in time, the position is in the triangle of singularity placing the half circle in the first quarter. Any further linear movement will follow the triangle second line by multiplying the line by the value translated as a number of Π^0.

$$7\ (3\Pi^2)\Pi^0 = 207.2 \text{ km/h or } (3\Pi^2)\ 2\Pi^0 = 414.5 \text{ km/h.}$$

$R^2/T=$ 7 $(3\Pi^2)$ and **R/T = $2\Pi^0$.** This is the first barrier but we only see the linear part as a value and peaks at $5\Pi^0$

7 $(3\Pi^2)(\Pi)=$ 651.13. In this **$R^2/T=$** 7 $(3\Pi^2)$ and **R/T = Π** where then the linear component Π becomes the second line in the triangle of singularity. The linear component or negative displacement can go as high as 2Π but then another barrier would come about and in this we star to locate the principle behind the breaking of the sound barrier. There are definitely TWO barriers to comply with when breaking the sound barrier

$(\Pi^2/2)$
7 $(3\Pi^2)(\Pi^2/2)=$ 1022.79. In this **$R^2/T=$** 7 $(3\Pi^2)$ and **R/T = 2Π** where then the linear component Π becomes the second line in the triangle of singularity.

$2(\Pi^2/2)$
7 $(3\Pi^2)2(\Pi^2/2)=$ 2045.59.

From this point on and above there are a boundary no one can cross because singularity will not allow the crossing of the third quarter $\Pi^0 \Rightarrow 5\Pi^0)$ by any object.

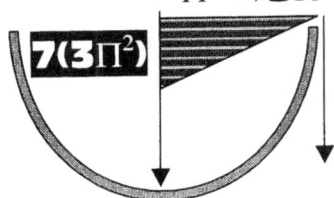

Everything that is in the Universe is heat. There is liquid heat that can flow and there is solid heat that cannot flow. The solid state or liquid state has no bearing on human perception but is in motion. When the body moved in relation to the body that does not move the body that moves are liquid and the body that is relevantly stationary is a solid. Movement is liquid and immobility is solid. '

The sound barrier is in principle Galileo's pendulum arm

The sound barrier is directly related to the swing of Galileo's pendulum arm. By standing still the arm points directly down to a centre point within the Earth. As the motion becomes a factor in the aircraft or any moving object for that matter, the motion will push a diverting of the stationary line as the moving object changes relation to the lines, which the Earth dictates.

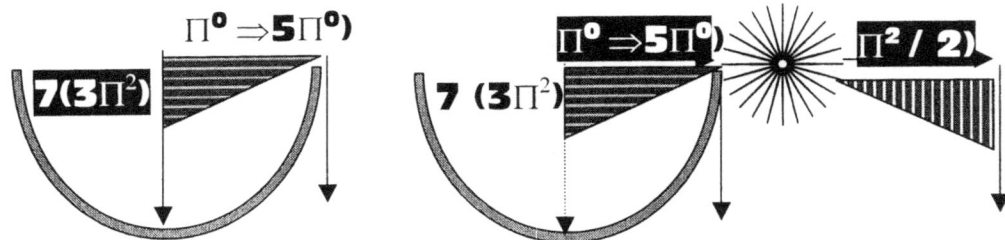

The line holds a velocity of $7(3\Pi^2)$ and that is the motion the space has in relation to the time the Earth dictate by contracting motion of time to the centre of the Earth. Any more motion that will underline the independence of the object in motion will show a diverting from the Earth singularity line by measure of Π^0.

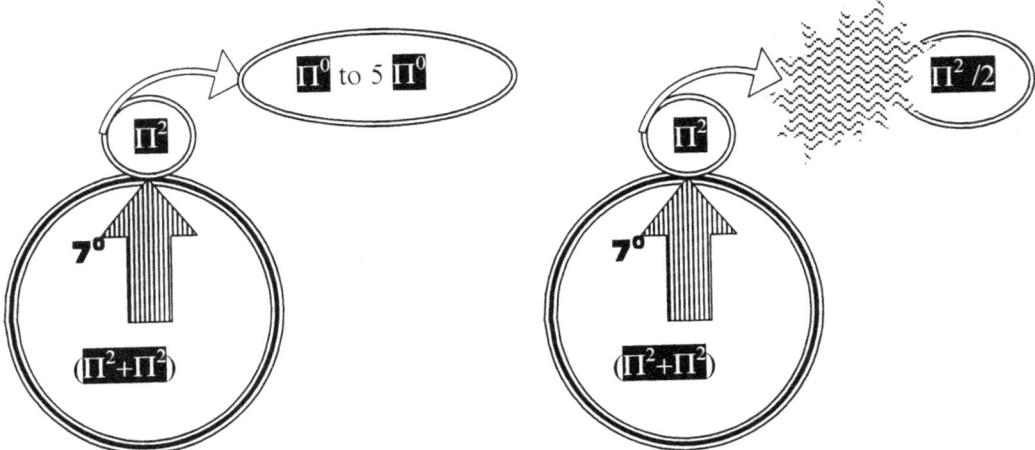

Motion is connected to heat in every principle applying. It takes heat to drive motion and it takes drive to establish independence of singularity. By additional motion brought on by adding heat the motion will show how far the independence divert from the Earth prescribed singularity. The sound barrier comes about when the second form of the Roche limit is crossed at $\Pi^2/2$

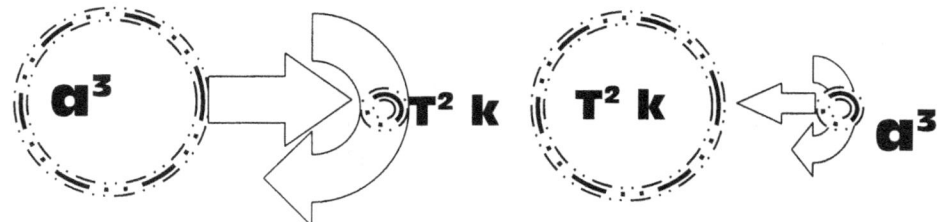

Every aspect however small or however large is serving as an atom that is maintaining singularity. Every atom is a galactica and every galactica is a star. Everything moving is containing heat and everything containing heat is preserving singularity where singularity is the Universe.

THE SPOT THAT'S HOLDING THE LOT.

Every position in the universe either holds singularity in a form, or relates to singularity. There can be no position unrelated to singularity therefore every aspect of the cosmos is space-time in various forms under the provision of singularity connecting. Matter cannot be if not surrounding singularity

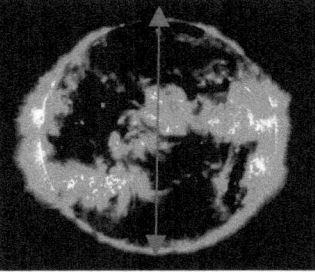

Singularity is as close as any spot can ever come to zero BUT IT CANNOT EVER BE ZERO. From singularity diverts space-time and there cannot be space without time as much as there cannot be time without space, not withstanding the size of space or duration of time.

Through space-time singularity connects as much as relates linking the universe into a network of influences beyond what ever we can ever conduct. There can be no spot that does not participate in the curvature of space-time. From the point of singularity runs space holding time to the prescription singularity dictates.

With singularity connecting singularity will or cannot relieve or release the connecting other than by a method we humans refer too as an explosion, but space-time separating as much as joining singularity dividing can change by applying time too space in a changeable manner, stretching and shrinking the time aspect by changing the density of the occupying heat, which creates the space allowing the spin of the occupying heat creating space setting the time.

There is a Universe that is limitless and without boundaries as much as there is a Universe that is within limits, neither one can ever be in the position we think they are. Our Universe is limited between the Titius Bode law of 7/10 and 10 / 7/. Once the limits of 7 / 10 and 10/7 are crossed and that region in time is passed either way our Universe stops. It is not the Universe that stops but it is only our Universe that stops.

While singularity is the ultimate endless space parted from the infinitive spinelessness singularity presents that which is between the two forms the Universe we have.

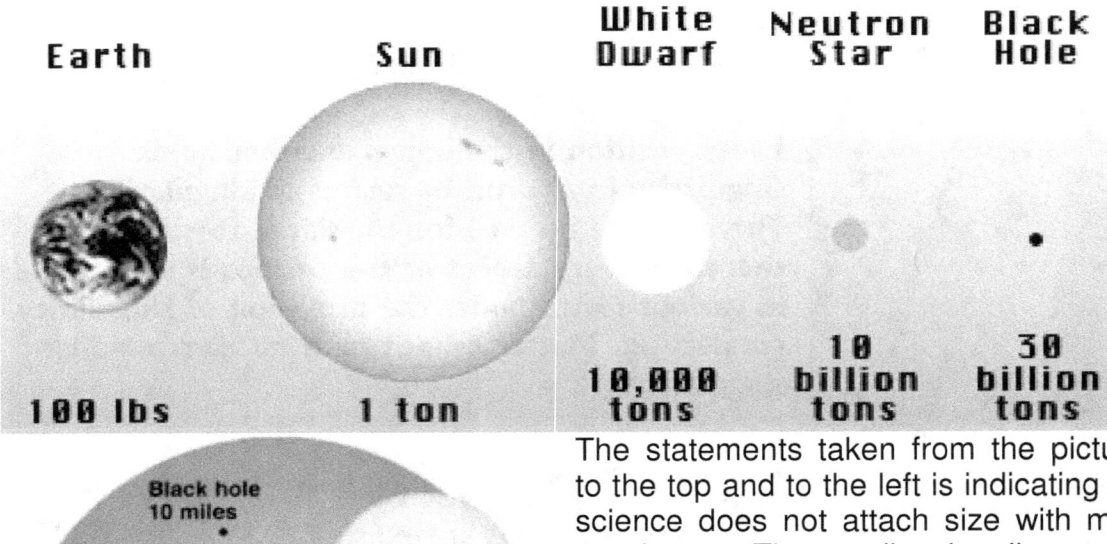

Earth   **Sun**   **White Dwarf**   **Neutron Star**   **Black Hole**

100 lbs   1 ton   10,000 tons   10 billion tons   30 billion tons

Black hole 10 miles

Neutron star 12 miles

38 yards

1 foot

Sun 875,000 miles

White dwarf 10,000 miles

22 miles

Betelguese 875,000,000 miles

The statements taken from the pictures to the top and to the left is indicating that science does not attach size with mass any longer. The smaller the diameter of the structure is, the larger science think the mass is.

The spinning top is all the evidence any one needs to come to such a conclusion. In the past when I have acknowledged the fact that I have no academic background in the field of cosmology, this admitting brought about discontent, or rather dismay and I might even add some scorn from academics. In my opinion this trend of behaviour is uncalled for as every person is entitled to be opinionated. Every person has an opinion and it is the opinion that has validity, not the persons social standing. It is the way one absorb and reflect a personal view about matters that are important and not the manner in which the person arrived at such conclusions but it is the validity of the conclusions that should carry importance. The cosmos contains matter, space and time in heavens or dimensions standing in relevancy to one another. Whether you refer to dimensions as dimensions or heavens it is no different because it is the same thing. Singularity brings about heavens as it brings about dimensions and singularity is a mathematical fact. A straight line cannot start at zero and still be a straight line because zero extending to wherever brings about a full zero. A straight line starts at the point where the pen point meets paper. That point may be any distance from infinity to a measurable dot, but it cannot be zero.

Any straight line is also half a square because the line forming the square cannot start at zero for the reasons I just mentioned. That is singularity pointing an eternal direction from a point of infinity and that is the basis of the cosmos as much as that is the basis of mathematics. To escape from nothing one has to become something and by doing that one could not have been in nothing in the first place. If one holds a point in nothing one cannot become something because of the nothing value.

$$180^0 \ X \ 2 \ = \ 360^0$$

To back this argument that no line can ever start at zero is to ask the simple question: what will the length of the shortest possible line be. It must be a line where the starting point is so close to the ending point the distance parting the two is incalculable yet there is the line therefore the end and the start is apart still sharing the same spot.

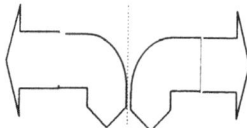 To this end any shortest line will start run and end in infinity Should a straight line start from infinity and never be able to reach a point of zero (because that will bring in the factor of no line ($0 \times 1 = 0$)) that means the line dips into infinity and it has to come out on the other side leading in the opposite direction of the first line in an attempt of avoiding zero.

The length cannot be zero because zero means no line. The starting point and the ending point may be inseparably the same point with virtually no space between the two points but neither of the three points can be zero simply because there is a line (be it infinitely small it is there). If the point is zero, then the line will be shorter than the shortest possible line. If there is a line and the two points starting the shortest possible line and being the continuing of the line it may still be the same point and even by sharing the spot as the point ending the shortest line possible the line must be there and the line holding the start and finish is next to zero but can never be zero otherwise there is no line. If that is the case then the one side of the square also presents the point of singularity to the other line holding the other dimension. That then makes the straight line not 180^0 because it is half that of the complete straight line and indicating the point of singularity of the other line and that results in any form holding only one aspect of the full form. But because the two lines are in a relevancy to a common starting point, both will enjoy the remaining distance of flow still holding the relevancy in comparison to each other and that brings about a triangle that also has 180^0 in number. That means any straight line is also a half square which is a triangle bringing about 180^0.

The co-ordinates of referral will only hold three references as surface contact areas pointing in one direction of the possible six surface areas available. That is mathematics and mathematics cannot be bias or lie. Every object in the Universe holds three points of face value to any direction possible. You can only see three sides of any six-sided object. Because the lines has a starting point in infinity that starting point has to represent singularity outside the sphere because at some point the lines cross the border of being the shortest lines possible to being normal lines. That means every starting point represents singularity by measure of r instead of Π. Every point is also the point indicating singularity at another point pointing in the direction of a point holding singularity as the pointing line as well as the supporting line.

I explain r later on but r does not in any way refer to radius, as does the r that I use for that purpose. By the same measurement, I also provide the third line with a singularity as reference and a singularity of direction. The prominence of this will become most apparent when explaining the Titius Bode law in relation to matter relating to space and matter relating to matter claiming space, but as usual, I am getting ahead of myself. What I have just indicated is pure mathematics moreover the most basic mathematics there is. There are always two lines running in supporting but opposing directions claiming space from that one point of singularity and a third line confirming (controlling) space in a third direction.

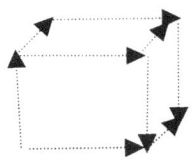 Every line is as much part of a square than it is pointing in another direction relating to a position holding singularity. That brings about my conclusion in changing Kepler's formula of $a^3 = T^2 k$ to my own which is $R^2 / T \times R / T = 1$ where any of the three components can form a square in relation to a mutual point of singularity where each one is the starting point on that lines individual singularity

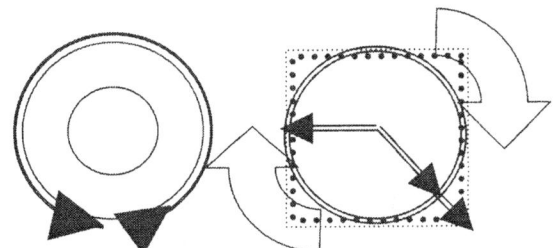 Squares rotating will find a point where it goes into self-destruction through faster motion. This action is imbalance but in principle it has a point where Π forms a meeting with r and Π at that point forms r.

That is space-time holding one square of space in time and directing time in space in a specific direction of flow. You may not agree with me on this issue at this point but my reasons for such a claim comes in other parts of the book and there I substantiate the claim with much more explicate mathematical detail. Space-time is matter-claiming space in time while directing space in that same duration of time from any point forming singularity. Now one arrives at the question of positioning singularity. Singularity cannot be in space because space is claimed from the point of singularity and space is confirmed from that point of singularity. If space is claimed and controlled Einstein must be at fault looking for singularity in outer space.

To find such a point one should once again look at the possible dimensions available. A square is six sides holding three to any direction. A circle is a square without edges. That is what we are looking for. A square without edges can only be a circle with a definite un-varying point of singularity indicating a specific location. A square can place singularity at any position because space holds no specifics where as a circle holds specifics at a defiant point. Space provides no specifics because an astronaut can and does drift in any direction when not being attached to a sizable object. That throws outer space out as a possibility.

Take the top with some astronaut to outer space and ask him to spin the top in outer space. We do not even have to bother such a busy person because we know the answer. The spinning top will not be able to turn because there is no stability factor in outer space. The top needs support from the needle running upwards, therefore the top will then have thrust running downwards, and the accumulative effort will bring about a point where the opposing points will allow a spin to occur. In outer space there is no such a point therefore the point we seek is within matter. A top cannot spin in outer space. We already established the fact about singularity being the position where two points originate in a square and the third will show direction. The top is a circle.

The difference between the circle and the square is in the direction the indicator follows and a square cannot spin, but through contact with a sphere as a circle cannot be motionless The factor of Π indicates eternal motion and NOT zero motion. There is a massive difference in that concept. If no line can have a zero point to start with where will the circle get the zero to indicate motion! This principle is the most basic mathematic rule and even I the ILL EDUCATED can see this. When the end of the rotation arrives the end rotation also announces the beginning of another rotation and not nullifying of the previous rotation because the rotation will have a line showing the effort it made and as it forms a wave, the wave will be there forever. The pitch may decline to a straight line, but the line remains. When calculating the motion a triangle does the honours.

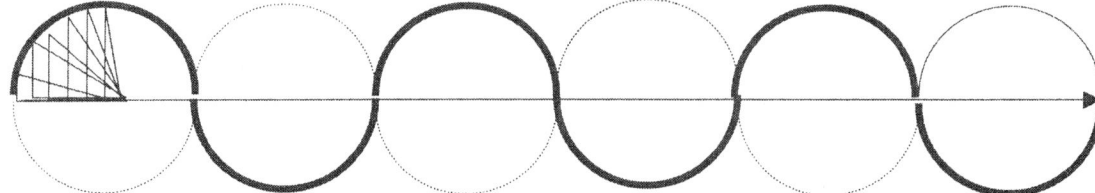

The wave confirms rotating directions followed by the circle as it spins. By stating that a wheel has a relevancy of zero by completion of a rotation such a claim denies the wave its entitlement of existing. The wave going flat, as it becomes a straight line also has an indication to singularity. All spinning matter has the point where the spin is still there but the radius is to small to measure by any means. That point is standing still in relation to the rest of the spin. In relation to that logic I do not accept that science holding the radius of a spinning object unaccountable in the spin, whether the spin is applying or not. It is in the very fact that the spin places the reference factor in the graph that makes the graph that astonishing accurate. It is mainly mathematics through the graph, which affect day and night and not the Earths standing in relation to the Sun that brings about climatically and weather changes and Earth development conditions.

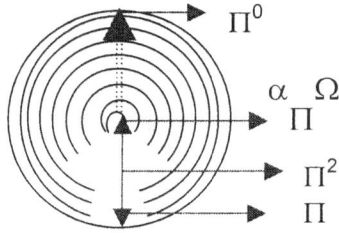

In the centre there is the singularity from which all stems. That is the centre of the Universe carrying the value of Π^0. By rotation Π^2 a line forms Π to Π with the value of 2Π. The forming of the line by initiating Π^2 is the realisation of gravity and the constituting of the Universe, which releases the Universe from Π^0 to Π by Π^2. The motion spawns the gravity because the motion is the gravity. To block the motion is to resist the gravity and such resisting forms the mass aspect that counteracts gravity.

Nothing said so far is high tech or mind bending complicated. All the above arguments from the first page to this point reached are simple and there's only ordinary primary school mathematics involved that every scholar should know. One does not need a brain fitting Einstein to come to these conclusions but just thinking about everyday issues.

All circles have something in common with squares. It has a surface but where the square holds a surface pointy the circle holds no points. To calculate a square there has to be a point where the line starts and that line will hold a value of at least infinity running from that point in two opposing directions.   Arithmetic presents the possibility of zero and mathematics excludes such possibility. That is the difference there is between arithmetic and mathematical science. It is where mathematics departs from arithmetic and is as basic as counting is in arithmetic. There is nothing outrages about that which I have mentioned. Neither does one need the brainpower of a person like Einstein to come to such basic conclusions nor yet it completely destroys the claims.

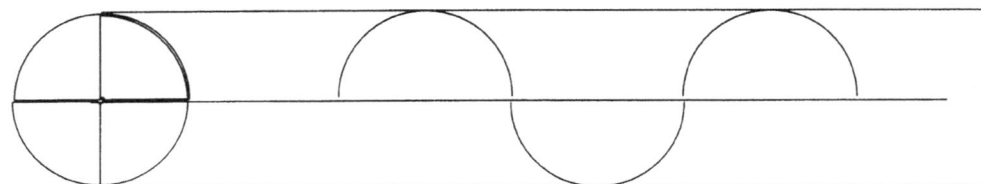

From the graph one can establish the link in the circle's rotation around a conforming unit being singularity.

Being a circle means the thing must be round and spinning. In that case, let us take an example well known to all, the spinning top. The top spins on the thinnest

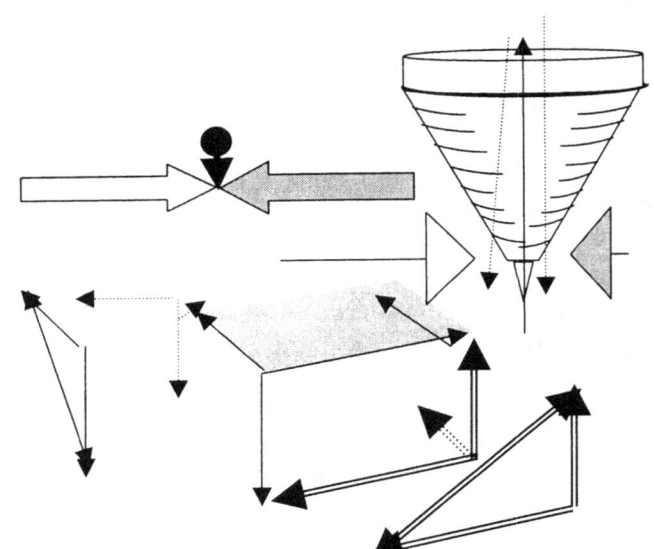

of points, and still maintains a balance.

All circles have something in common with squares. It has a surface but where the square holds a surface pointy the circle holds no points. To calculate a square there has to be a point where the line starts and that line will hold a value of at least infinity running from that point in two opposing directions.

In the scenario depicting the square there is always some three pointing triangle everywhere and the common denominating figure is therefore three.

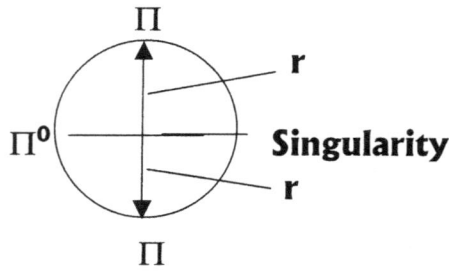

In the circle, there are also two lines where each line holds one point to singularity. Splitting the line in the two opposing directions, so in that way it is the same as a square, but the third line indicating direction brings about a difference that distinguish the circle from the square. The circle direction indicator is always Π placing the pointers at $r \times r = r^2$

Applying the second law $F = ma$ one arrive at the formula $G(Mxm) / r^2 = m (\omega^2 r)$ claiming a zero influence between the radius and the orbit of planets. I shall later again return to this issue but firstly I would like to take your mind to one other thought that seems to have escaped every body. It concerns the medium science rely most on for the gain of facts and information about the Universe. It is the influence of light.

When realising the error of science in accepting a value as zero to be legitimate in mathematics, one can establish from that that the circle does not employ zero as a value after the completion of one rotation therefore $F = G (M_1 \times M_2)/r^2$ is invalid, one has to return to Kepler's $a^3 = T^2 k$ and establish a value from that.
From the graph one can establish the link in the circle's rotation around a conforming unit being singularity.

Saying that one therefore has to admit that the smallest spot has to hold space because the most insignificant dot can transmit light and being able to accomplish that, one must accept it to carry a value of something. If that spot had the value of nothing, it means that spot was not there to begin with. Holding space-time one should return to the original formula indicating space-time in as much as $a^3 = T^2 k$ where a = R and T = T. Being time it has to alternate positions and that can therefore only apply to **k** where **k** will indicate a relation to the space-time in question or the relevancy to singularity being $k^0 = 1$. This reality has serious implications on the speed of light we take as a constant.

Have you as you sit reading this part at this minute sat back and gave a thought about the light enabling you to read? Such a thought brings to mind the most simplistic answer one can imagine. The light hits the page bounces from the page and contact the lens of my eye where the lens conveys the photons becoming electricity to a part of the brain that translate the electricity to an understandable message and that makes one read. It is as simple as that! Ever gave a broader thought about light streaming across the night sky, coming from ends of the Universe we do not even realise is there? How does the photons manage to convey one complete picture coming from as far apart and as wide an area as it does? With a few photons connecting the eye or lens no one ever noticed the wonder of light. The photons reflect a view that seems as if coming from all the billions upon billions of stars. But most is coming from darkness covering an area no man can measure. Yet how many photons can actually connect to the lens of the camera or to the eye? Still a few photons coming from a single direction directly ahead eventually tell the entire storey. It is very simple to take the process of seeing by means of photon conducting very lightly and I have never heard one of the Brainy Bunch really in sincerity dissect the process to its potential. It is impossible that light from such an array of assorted sources can simply come together at the eye lens and show a picture of objects spanning across a Universe as wide as our mind can receive where the objects they reflect is beyond human measurement and the quantity is inconceivable many.

Light is much more than the medium science takes it to be. Light connects the Universe in a way we cannot contemplate. Light being far apart originating from regions not in the same time or universal space connects in a way that present us with a picture holding the Universe in an understandable content. From the point we stand and we watch the Universe the significance of what we see surpasses the sense of understanding of what we are experiencing. How can the few photons that our lenses catch coming from such a vast area covered by the night sky cover transmit the complete picture of what we see. Take a few seconds and inspect the picture of the night sky then rethink the picture applying the full content in the picture to what the size of your eyes. Think how big the picture is that your eyes take in and translate that area to the size of your eyeball in an effort to determine a ratio. One will be forgiven if one thinks of the ratio as eternal to nothing. Yet a few pages back I showed that according to mathematics there couldn't be anything as nothing. Consider the path the light followed from the source connecting to light from all other sources where all particles of the other light may come from and bringing a full picture to the lens one use to look through. In your mind connect a line from every atom producing light and connect the lines to your eyeball and see how you can manage to fit all the lines, as small as the lines may be.

Scientists think of outer space as geodesic zero, with nothing in outer space but space. Geodesic zero means the light travels in a straight line from where it originates unhindered all across space to where the light connects the eye. Such an idea by itself is outrageous because the stream of photons reduce in space to such a minute quantity, that taken the area the photons travel and the space in vastness it covers, the chances of one photon coming across many hundreds of light years through billions upon trillions of cubic kilometres of space and selecting my eye to convey the electricity is less than infinite. Yet such conveying takes place every second of every minute. The position of the location of the second singularity, which is the precise duplication of the first singularity but in a diminished capacity, is obvious to miss when one is not applying a detective mentality, as one should in scrutinizing the cosmos. Culture will have us believe that when one sees a colour shining from an object the colour is associated with the object. Logic tells a different tale. A yellow dot is all the colours in the spectrum but yellow because it is disassociating with the yellow. That goes for red blue and all other colours we may visualise. I think the norm accepts this as scientific fact with very little argument or substantiating proof.

If light came as individual streams of photon flurries our visage would translate that as such shown in the fragmented picture above. It would be a picture unconnected bringing across some photons in the manner where every object stands apart not being related in any way and that will be what we see, if it is anything that we see. That we know is not the case but that means geodesic zero is as much rubbish as anything Newtonians regard with simplicity and with careless thought. Geodesic zero means nothing and how can I see nothing as darkness because "nothing" is not darkness, nothing is "nothing" and the darkness I see is darkness showing the darkness as something.

What then about colours that are technically not colours as is the case with black and white? White is simple. By spinning all the colours in the spectrum the colour white shines through. Black is quite another matter. A friend of mine whom is one of the best painters I have ever come across told me that one couldn't paint black but have to make black a dark blue to show shade on the canvass. That apparently is his success in achieving the realism.
 He also went on to explain how many variations of dark blue form the shadows in one simple tree. This remark set my mind in motion. One cannot see black because black has no colour to show, but black is the colour most prevalent in the

universe. One can see only by colour and since black is not a colour we should not see black, but we do.

If the darkness was the representation of "nothing", then that should be exactly what we must see, nothing but the stars. Taken from the top picture some stars and leaving the rest to nothing is what we see in the picture below. A blind person sees nothing but when we look at space, we see something that we think nothing of as we see as space. One cannot have the ability of sight and see nothing except by closing your eyelids and then you see nothing. But in that case you do not see "nothing" in contrast of "something" you see "nothing" without it contrasting to "something".

Nothing is all about not being and not "not seeing".

By the ability to see the darkness renders the darkness something other than nothing and that changes the acquired value of the darkness from nothing to something. There is an eternal difference between something in infinity and nothing.

The arguments introduced up to this part of the introduction prologue only touches the most basic aspects of my work and by no means can such an introduction secure an opinion. Yet, not once through all my long investigation in the past thirty or more years have I found any other person claiming such views that I have brought about even in this skimpy way as I do in the prologue.

The arguments introduced up to this point of the introduction prologue only touches the most basic aspects of my work and by no means can such an introduction secure an opinion. Yet, not once through all my long investigations in the past thirty or more years have I found any other person claiming such views that I have brought about even in this skimpy way as I do in the prologue. As it applies with all things, so it does in this case as well that when delving deeper into any issue. The complexity of the issues truly comes to the fore ground when analysed in more detail. I wish to advise the reader to treat the seven books as seven different works and in that light I have separated each work in volumes of seven separate books with individual I.S.B.N. numbers with adding one part, the one you are reading, with one sole purpose and that is to bring about an academic introduction to clarify a quick perspective. Then the next three parts being of a general introductory nature there are overlapping in some sense but each highlighting issues in a different manner as to clarify facts used in the last three parts bringing conclusion to different cosmic perspectives. Yet the work is seven parts of one thesis and as such it serves.

I WISH TO DEFINE THE CATAGORISING I USE AS PART OF THE BOOK.

I have the utmost admiration for Scientists and I shall never dream of placing me in the same category as academics mainly because of their intellect and achievements. To substantiate this segregation I use some referring to place distinction between the highly schooled super trained academics that spent most if not all of their lives in preparing to further their minds. Because I tip the opposite of the scale and spent as little time in an official capacity on learning and education I have to be on the "other end". From where I stand and admire you, I can only see intellect: being the academic's common denominator. If that is the common denominator on the one side, the joining factor on the other side *"my side"* must then be the class of stupidity. To you and your class such a remark would be an insult but to me and therefore my class it rings truth and that makes it not an insult but a norm we should accept and learn to live with. It is rather a pity that while the SUPER CLASS will never say it to our faces; the SUPER CLASS is strongly of such opinion that we on the one side of the Universe have no minds to think in any way, and it is our duty as much as privilege to accept what you the ones occupying the other side of the Universe inform us to accept and you live by that idea. As I said I have to live with it too and if I am the illiterate, then the SUPER CLASS must be the SUPER–EDUCATED; where I am the class amounting to stupidity the SUPER CLASS must be the Brainy Bunch. It all comes from the fact that there is such a huge differentiation between us. You consider one with merely one Baccalaureate degree a stable boy; think how low your opinion must then be of my type including me that never even came that far. To distinctly point to grouping or class or whatever you wish to consider the division there is between you and me I refer to your side of the Universe by the names I use above. Further more when I refer to mistakes that I do prove to be mistakes in the book as we go along I refer to it as Xepted mistakes to clear another distinction of necessity.

Introducing the book **Matter's Time In Space** written with facts about the creation in mind produces a problem because the complete picture that I introduce has nothing in common with current accepted science.   The issue remains comprehensive even by using it in a very simple form. In spite of this, I shall explain three of the four unrecognised phenomena I use in proving my statements in this very book aiming at a theme of simplistic introduction being "an Academic Letter". Then in the following books I go into extensive detail proving all of the Cosmic Pillars. With my introduction of the phenomena, which I named the four cosmic pillars you will find it obvious why science do not accept them even if it is documented throughout the Universe and is quite commonly found. Applying them totally annihilates Academic's formula of the basis on which science rests in the formula being $F = G (M_1 M_2)/r^2$. The four cosmic pillars are the following:
1.   Roche-Lobe
2.   Titius Bode principal
3.   The Lagrangian principle
4.   The Academics Gravitational Concept forming the atomic relevancy.

The Universe is a combination of many material formations holding positions in space. Some of such material was covered in the blanket of heat, distributing into more spacious surroundings as the material expanded from the centre flowing outwards. Hubble's constant is proof that the space between cosmic structures are departing from many centred positions between such objects and this is a trend being located between all the objects throughout space but also indicating a definite growth in the radius and such radius growth follows a pattern where the growth seems to flow from any such a centre point away from the centre. With out the absolute and undeniable proof coming from the Hubble constant bringing proof beyond any possible doubt in any one's mind that expanding is very much and a very big part of all Cosmic activity. The accepting of the Big Bang would not be in place. $F = G\,(M_1 \times m_2)\,/r^2$ is in essence a big issue about contraction while Hubble showed the space was not dividing. The space was multiplying. The stars are growing apart and so is the galactica. This then brings in the question of space available.

With me not whishing to go into the formation of structures at this point in the book I would like to point to the fact that my following referring to the solar system is actually referring to a similar solar system that is somewhere and is now a part of a galactica we do not know about. I bring this in to disqualify any academic loophole that may come about from an argument about the solar system coming into place at a later stage of the cosmic development and therefore the argument I am about to present that such an argument does not apply. To avoid such a loophole we now use a hypothetical but real solar system in space, which formed as the Big Bang took place. There are those who avoid admitting to inconsistencies by arguing that my argument about growth is invalid because the solar system was not in place at the Big Bang. To them we now present a solar system that is identical to the one we know in precise duplication thereof. But it represents a precise duplication of our solar system and was in place ever since the Big Bang. That means with the solar systems being apart in millions of kilometres there was a time the planets and the Sun were apart by the measure of kilometres. The Big Bang shows a growth in space. Then there must have been a time when the planets were between fifty-nine and fife hundred and ninety kilometres away from the Sun . How did the planets being the size they are at present fit into such a space and still be apart? Material too must be part of the growing. This line of arguing I suppose is much below the Academics pursuit of matters but since I am much lesser in mental standards of development than they are, such reasoning prompted me to go on some investigative journey. Light journeys through out the cosmos and it will be sensible to follow the travelling of lights.

With objects being apart at some distances and light flowing in straight lines between them it must take light a straight line to travel between cosmic objects. The distance the light has to cover depends on the radius there is between the objects and as such the Universe is then about structures claiming space and space setting objects apart being the radius standing between those objects. The objects are circles by dimensions and the space is also dimensions that are crossed by lines travelling through the dimension. With light being a line and the Big Bang coming from a situation that was a lot more cramped for space than at present, the correct path to follow if I wish to trace the steps of the Cosmos back to the Big Bang is to reduce the straight line between the structures and find where such a line will no longer be a line. The same procedure will apply to the material structures all being in a sphere form. A sphere is a lot of circles forming a unit but not repeating any occupation of the space, which any one particular one claimed. Such a circle also applies a straight line only known by another name but still serves the same purpose. Reducing the line will lead us to the beginning of time.

By dividing the radius r by the half of the value that then reduces r to a point 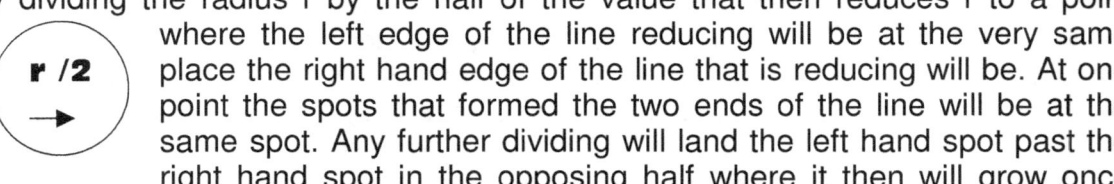 where the left edge of the line reducing will be at the very same place the right hand edge of the line that is reducing will be. At one point the spots that formed the two ends of the line will be at the same spot. Any further dividing will land the left hand spot past the right hand spot in the opposing half where it then will grow once again but in the opposing direction. All possible dividing then ends on one spot where such a one spot shares a location with all other possible sides. The centre then physically is in the single dimension applying as one spot to share a location for all sides. At such a point there is no further dividing possible. On several occasions in the past I have been accused of manipulating the argument to produce none-existing or overrate facts. That is not the case. I am not manipulating facts to create an argument as so many accuse me of. What I am talking about is a mathematical fact that any one can prove by calculating following a very simple procedure. A child is capable of using the two times table and dividing by two every time is the most simple form that mathematics may be used. It is a mathematical fact that a line will reach a point where all sides are at one spot and as such cannot divide any more. At that point all sides share but all sides prevent zero becoming as factor since the sides share on spot. While the different sides are in one place the factor and value is one to all.

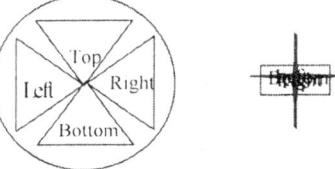

Reducing the radius r from all angles possible throughout the circle will bring about that all possible direction will eventually land on the very same spot with no more dividing possible. Yet zero cannot be a factor since the sides still hold value. A point arrive where more reducing will land the one side on the opposite side of the line but it will not bring about zero in the equation

What this argument further proves is that the circle reducing must then come from all points because the radius might be a line but that line represents a circle through 360^0. Taking that into account it is important to recognise that notwithstanding the size of a line, which any radius of any size is there is another line (or dot) eternally bigger as well as eternally smaller than the line in question. While we are in the third dimension being part of the third dimension then allows that all parts of the third dimension forever can be divided once more until the line in the third dimension is no longer part of the third dimension. When such a line leaves the third dimension it is still dividable because it might not be part of our dimension any more but it can still reduce further as part of the second dimension. By that time it has left our scope by miles. It does not mean that it end there because from our perspective that is where it ends. Yet it can still reduce infinitely more until it has left the second dimension and then at last forms part of the first dimension. Only then when the line reaches the first dimension, no further dividing of that line is longer possible. We can never grasp the size of a line that forms the utmost or the least of possibilities and therefore size belongs to the human mind forming conceptions of big and small, but it has no place in the cosmos at large. This concept not only applies to size, but to all limits and divides we wish to create forming borders that we can appreciate. When looking at the circle in the conventional manner, we persist with errors brought about in culture and not by applying some significant modern logic. Take a circle and reduce such a circle constantly to where it no longer can reduce. Reduce it to a point where only form remains part of the circle because the radius has gone beyond human measure and becomes so small it is not noticeable with what ever tools man may use, then what remains is pi since pi does not indicate size but indicates form, and form is all that then will remain. In any circle or sphere the size only depends on the fluctuation of r, as a component to the circle or sphere but that does not affect the form by indication of Π in any way there may be. The conclusion I drew from following this process is that from this line no start can be at zero because that will be a mathematical impossibility since no line can ever reduce to zero. A line will forever be able to reduce further becoming smaller but it can never reach zero because zero is not on the scale of lines. If a line cannot reduce to zero it then cannot start at zero. A line or spot starting at zero would therefore be shorter than the shortest line possible. For obvious reasons can no line, or any line grow or extend from zero because such a line must then quit zero and become something, thus abandon its original value. That would mean the start of the line has a different value to the end and a line holds conformity through out. When any line is starting from point zero it can never leave zero because of the influence of being zero disqualifies any possibility of growth. If the line then had to grow in all directions at the same pace the line must then become a circle or being three-dimensional, then forms a multi circle we name a sphere. Since the Universe is about circles and lines connecting circles, I came to conclude that flowing from this fact is that in the Universe there can be no zero improvising as a filling ingredient for the space of a point or be unfilled space. In the case of the growing sphere the value of the circle is Π, and that is where creation must have started. That gave me the clue where to start looking for singularity. One would find singularity in the value Π and the value Π will be in all things rotating in a circle. As usual I am again shooting the gun before the hunt started. Lines in mathematics do not start from zero and that is no discovery on my part that was a realisation I came to.

UNIVERSE
Everything that exists, including space, time, and matter. The study of the Universe is known as cosmology. Cosmologists distinguish between the Universe with a capital 'U', meaning the cosmos and all its contents, and Universe with a small 'u' which is usually a mathematical model derived from some physical theory. The real Universe consists mostly of apparently empty space, with matter concentrated into galaxies consisting of stars and gas. The Universe is expanding, so the space between galaxies is gradually stretching, causing a cosmological redshift in the light from distant objects. There is growing evidence that space may be filled with unseen dark matter that may have many times the total mass of the visible galaxies. The most favoured concept of the origin of the Universe is the Big Bang theory, according to which the Universe came into being in a hot, dense fireball about 10-20 billion years ago.

UNIVERSAL TIME (UT)

A worldwide standard time-scale, the same as Greenwich Mean Time. Universal Time is the mean solar time on the meridian of Greenwich. It is defined as the Greenwich hour angle of the mean Sun plus 12 hours, so that the day begins at midnight rather than noon. It is closely linked to Greenwich Mean Sidereal Time (GMST), since the mean sidereal day is a precisely known fraction of the mean solar day. In practice, UT2 is determined by a formula from GMST, which in turn is derived directly from such observations of the meridian transits of stars. The version of UT derived directly form such observations is designated UTO, which is slightly dependent on the observing site. When UTO is corrected for the variation in longitude due to the Chandler wobble, a version of Universal Time, UT1, is derived which has genuine worldwide application. When UT1 is compared with International Atomic Time (TAI), it is found to be losing approximately a second a year against TAI. Broadcast time signals use the time-scale known as Coordinated Universal time (UTC). This is TAI with an offset of a whole number of seconds. The offset is adjusted when necessary by the introduction of a leap second, and UTC is always kept within 0.9 s of UT1. On this issue there is much more to explore than the meagrely mentioned. Time stands related to the position an object holds to a centre such an object refers too while in rotation. Kepler found for instance that T^2, which holds the orbit to a rotation specific, is directly dependent on k to value the space a^3.

Einstein proved that in the presence of a strong gravity time slows down. Surprisingly, with that evidence being around this long, nobody since then in science took those statements and made any further progress from there. It was left in some drawer to dry. Science still sticks to its change that time did not change slightly since the beginning of the time and holds the same pace ever since. With the entire Universe including all the gravity now present and not excluding one Black hole or dust speck pressed in an area possibly the size of a lepton the gravity extending from that must have been beyond what words can ever describe. If the gravity was that high and Einstein already proved gravity slows time down, then there is one logical conclusion and that is that time was n fact standing still. Mathematically it is incorrect to allow gravity to compress the Universe into a spot smaller that an atom and exclude any other factors and relevancies to change. But before coming to the mathematics I would first like to bring your attention to the practical side. I am promoting a theory in which I am able to prove there is as much contraction (moving in the direction of the Big Crunch) taking the cosmic Universe back to the size it had during the Big Bang as there is expansion (moving apart by Hubble's Constant) and the contraction is as much part of the expansion. By contracting the Universe is expanding and everything is based on gravity providing both actions. The Universe rides on a balance and we have to locate such a balance. To prove my theory I firstly had to locate the centre of the Universe. Even admitting to such a notion sounds like madness or in the least a tasteless joke, but please give me a chance to explain in more detail. I realised that my effort to locate the point holding singularity only stood any chance of success if the reducing of the line enabled me to backtrack the exploding Universe to its origins. By applying some basic effort I have located the position from where all movement came and the direction it took moving forward in time...and yes, during my search on locating the centre of the Universe I also stumbled on time as such.

Let us find the smallest possible line first. Reducing the line will eventually leave all sides on the same spot. Such a spot must be round in form. The line being the smallest line will start off as a dot. A line so small it has reached a point not dividable any more will have all sides literally on the preside same spot, and I have located singularity in just such a spot. I came to the conclusion that the spot I found had to be singularity purely on the grounds that that spot holds only one side to serve as a start to the starting point of all directions possible. There in that side is only one spot is only one side applicable and one dimension present. With all the factors given one can only come to one conclusion and that is that there can be only singularity. In such a case more dividing by two will land further positions on the other side of the divide. That point serving as a position for all point and cannot allow further dividing is the smallest line or spot there may ever be. This spot is the result of a most basic process of reduction as the Hubble constant is a most basic process of doubling up during a matter of time. By reducing the line constantly the only value that will eventually remain without dispute from any party arguing about the facts is Π. By only having Π and a radius as one square (the radius effectively becomes one holding any and all sides on one point) of any significant measure as the radius it will be an evenly spaced dot. From the smallest ever possible dot will grow a line in every imaginable direction relating to a prospect of Π not favouring one direction that puts all directions at equilibrium meaning that any form of what ever might develop from such a spot will have the end and the start being in the same position, which will also have to be a sphere as the flow outward will be equal in all directions. Please think clearly, is that not precisely the commitment we find in gravity, where gravity is flowing from singularity outwards but never favouring any side? This reasoning prompted me to look for singularity in such a spot because if the prime spot from which all came was a spot holding all, then the spot must hold the shortest line but more prominent it will hold the smallest form including the smallest circle or for that matter the smallest sphere. With gravity always being in the centre of a sphere where the space is least available in the entire structure (there is not even space left to fill) one finds a flow of gravity from that centre spot outwards in all possible direction even-handedly. The fact that the original gravity will begin as a circle or will be a circle is the direction it will take when being the first spot created. All progress will be evenly in all direction because no direction will stand out or be in favour above any other direction at first.

The spot forms a full circle, but the line running through the circle is forever present because that is the future radius of the circle that will one day develop the circle, which is equal to the present diameter. The fact of the presence of such a possible line in such a possible circle dividing the possible circle into two parts makes the centre line equal to the half circle. The line forms the half circle but not only that the line presents the half circle as much as the line is the half circle. The line then is 180^0 and the half circle is 180^0 because in singularity the two factors are the same. The same value is of course $\Pi^0 = 1$.

In this half circle of the future, which is no half circle as yet because of a lack of space there are three future points indicating the space less ness that will go on to become space filled with something. On top of such a circle to form must be a marker indicating an awaiting boundary or future border and at the bottom of the future circle there also must be a similar marker that is no marker as yet. Between the two possible points that are not there yet is a future line running that is not there yet. Then indicating the possibility of a position to come that will bring about the half circle being a future distance apart from the future line indicating a diameter that will one day be there a third such a marker must be established for the future. That forms a triangle with two more sides being connected by either a line being one or half pi being one. From singularity comes about that the line is the same as the half circle is the same as the triangle and all has one value being 180^0. From this come the most basic principles in as much as forming the ground rules of the law of Pythagoras.

When drawing a line such a line then starts off with a dot serving the spot that holds all sides equal. That means the line serving as the future radius will be equal to the half circle which is then Π. The only aspect of the point that stands in for the end of the single line forming the radius of the circle is that we then mathematically reach the single dimension. We decreased the line to where a circle being Π formed on the single dimension. This dimension also hold the circle dividing line because from there the radius must once again generate a value and by such a gesture that the extending would form the circle that forms the sphere that eventually leads to the formation of particles. This leaves a problem to investigate.

With no line possible there had to be another dot that formed since the Universe has many dots that formed lines. But let us not to get confused and lost in the range of possible diversions but let us stick to two dots. One dot was next to the dot next to the dot, but as I said we stick to one dot next to the second dot. M X M / r^2 is the first step gravity began with. That leaves us with a huge problem in as much as when r = 0 then r^0 = 0 and 0 dividing any value will leave 0 as the answer. If the particles were inseparable at the start it must bring about that gravity would not be forming since the distance will not permit any dividing. By allowing the distance separating the particles to be zero, the particles melt into a unit.

Again this is Mathematics and not my incoherency as some Academics dismissed my work. Let me run through the argument one more time because I have been insulted by Academics in the past telling me I am bending mathematic rules with my applying double values to try and produce some argument. The two particles formed by an inseparable unit separated by a sharing of a spot. We know that at least two spots formed because there are many more than just two that remained to become part of the visual Universe. Let us name the spots because that is what humans do best if they do not know what to do with what they have to do. Let us call the one em and the other one spot next to em we then call emtoo. Between em and emtoo there were nothing because em and emtoo were inseparable. By they're being inseparable we would naturally be inclined to think that the separation value should be nothing or at least zero. But putting zero in that place is a mathematical excluding procedure leaving future mathematics excluded. With m multiplying m_2 and then dividing ÷ r with zero (r=0) such a procedure will leave the lot at zero and with that nothing is going nowhere. That means although we think the space between the two parts are nothing the non-existing space has to be at least one to be a future factor.

Every part of the argument is sound but was never yet used. I repeat once more if my argument reflects on inconsistencies those inconsistencies are not about my work. In order to disprove my argument replace Mass one and Mass two with any number possible, then divide such a number with the square being zero. If there was no space then the value of the particles had to be one. If there was no space between the particles the particles then had to form a unit. But if there is a mathematical possibility of reducing a line to the single dimension then there had to be a factor representing r as a factor of one. Take $(M_1 \times m_2) / r^2$ and substitute any of the factors with zero and the result coming about has to be zero. The factors in the equation have to have any and all the elements at a value of at least one. Only if r was a factor of one can gravity bring about any mathematical equation developing from this argument.

That means the mass on both sides must have a factor of one being a limit, which does not allow such further reduction of r and any further reducing of r beyond the limit will not be tolerated. Only if r = 1 then r^2 can be 1 and mass can be apart. Like it or not but believing in the Big Bang must also bring about the accepting that the cosmos moved apart somewhat. The fact that r brought increase in the space separating the mass produces a problem that was solved already. About a century and a half ago Roche found just such a limit. Once again I were confronted by zero becoming growth. There is a huge hole that needs filling when bringing into a relation any forming of an alliance between a cosmos coming from nothing and filling with nothing and a cosmos growing spontaneously through balance shifting prominence. Mathematically the fact of applying nothing serving as a factor applying in the cosmos is not a strong and convincing argument. The minute one brings in zero as a multiplying factor forming a definite value working into the calculations of the cosmos, growth disappears. If growth was not a factor, the zero factors could be involved with some form of maintaining stability and where then further growth will accept the responsibility of zero

The region surrounding each star in a binary system, within which any material is gravitationally bound to that particular star. The boundary of the Roche lobes is an equipotential surface, and the lobes touch at the inner Lagrangian point, L_1, through which mass transfer may occur if one of the components expands to fill its lobe. It names after the French mathematician Edouard Albert Roche (1820-83).

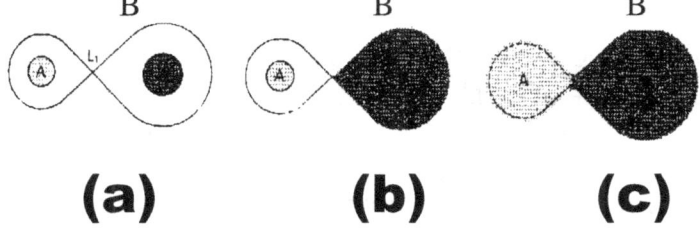

(a)   **(b)**   **(c)**

THE ROCHE LOBE: In a binary system, the Roche lobes of components A and B meet at the L_1 Lagrangian point. (a) In a detached system, neither star fills its Roche lobe. (b) In a semidetached system, one massive component, B, fills its Roche lobe. (c) In a contact binary, both components overfill their Roche lobes and share a common envelope. As with the graph I can see the two sides forming a connection therefore relevancy has to apply, all contradicting Newtonian claims of no connection but through

mass attractions. The mass does not attract but one interferes with the other total influencing the space surroundings.

The closest encounter worth noting we ever had with this law in the modern age of news and Television was the Shoemaker-Levy 9 incident during the previous decade. At the time and even in the present no one drew any similarities but after completing this book the reader should find why I could draw such a similarity, which there is between this incident and the Roche limit. Even the phenomenon called the Sound Barrier became clear when applying the Roche factor with the laws governing the influence of singularity.

The gravity Newton suggested and the gravity Mainstream science is in search of is not there. In fact there is no gravity if we consider what science would wish to locate as gravity. Mass does no instigate, initiate or create gravity by any means or measure. Gravity is the independence of matter in relation to its depending on matter in order to realise its independence. The motion that produces gravity is what inspires a new Universe into being

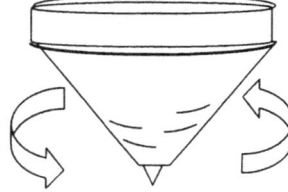
Every time a top or anything else starts to spin the spin create a newly establishe4d Universe which fills that Universe with the surrounding time in that Universe with a new centre and the centre is the gravity that rule and dominate that Universe.

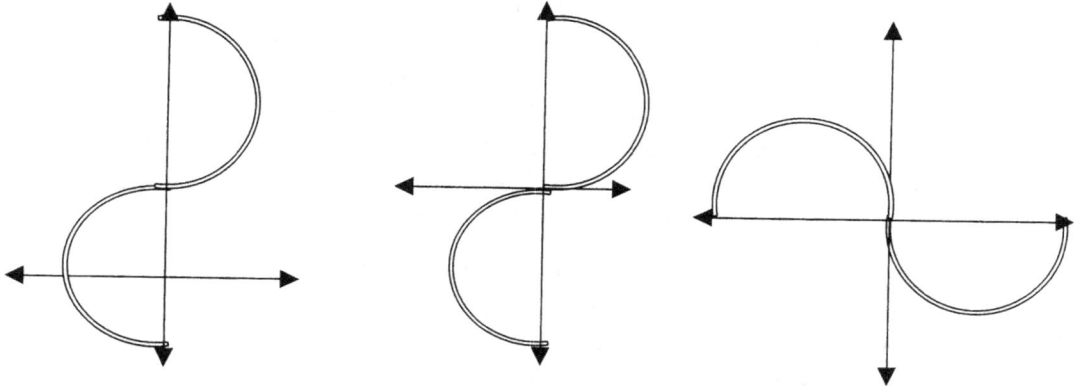

The graph is the result of motion where singularity form a centre but the centre formed is not zero. The centre formed is Π^0 and from such a centre space-time forms by expanding **space a^3** to the value of **Time T^2k. In the centre of the graph is Π^0.** This point, which I now am referring to, is the point where Π a fully appreciated value while the diameter D still remains a dimensional factor of one. His is the dawn of the second dimension where space was there but space was sparsely shared in some cases. It is when Π^0 shifted to become Π for the very fist time. The point without movement, the point holding singularity must have a value of Π being the eternal dot but since the dot has no dimension in having form the Π that indicates the dot must be Π^0. From such a point there has to be to the side of the centre point be a point where space do start. That point will then receive a diameter but that point will have form only in being a circle. In that point there is a shift from in relevance from Π to the centre Π^0 and for the first time it brought about two separate values for Π.

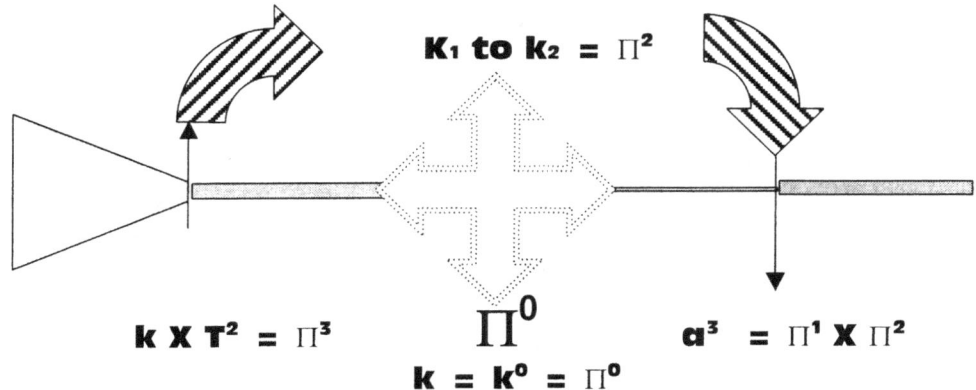

$$K_1 \text{ to } k_2 = \Pi^2$$

$$k \text{ X } T^2 = \Pi^3 \qquad \Pi^0 \qquad a^3 = \Pi^1 \text{ X } \Pi^2$$

$$k = k^0 = \Pi^0$$

Because the three points existed on equal terms in singularity sharing a same spot the coming out of singularity will enforce that equal value comes to all. That means the circle gets to become Π, the diameter becomes Π and the distance setting the structures apart will also become Π. This is what the coming from one point brings along. Only when being part of the second dimension can there start being separate values.

While the form was still being in the single dimension from the one side of the form the dots had to establish identities apart but not separated yet. The one circle had a factor of $\Pi^0 = 1$ and the centre had to have a value of (Π^0 / 2) extending past the very next object but also cutting such an object into a square double half value that was going to come about as soon as the other dimensions came into form.

The only definite place one will locate zero is in between the starting point of the lines going in opposing directions in the position the lines hold before there was the least of directions applied, but that is only because there is no such a position, not because any line is coming from there. The two lines are still one holding the opportunity of parting as an option but have not yet parted and therefore are on the very precise same spot. The line coming from there is already there because it already has the choice of going in any and all opposing directions and when it starts running it will place filled space in that location because the space was already filled with a line starting and not with a line not there at all. When reversing a line we might find a better idea of what is in place and where it is in place. Gravity is officially a force without limits going past and through borders and has an unlimited reach. It seems to remain even and this is conflicting with the flow of perceptions about mathematics.

The formula $F = G (M_1 \cdot m_2) / r^2$ is unable to explain the principle discovered by Titius and later by Bode and in contrary to all statements to that effect made by Accepted Science policy makers the Titius Bode principle is not coincidental. In fact it is one of the four most adhered and important cosmic pillars holding the cosmos structural in place. From the two above comes gravity. In a few pages I prove how one can arrive at the facts that prove how the Titius Bode Principle leads us in the direction the origins of the solar system. But before we can accept the influence of the Titius Bode Principle we have to deal with "Nothing" and as such dismiss nothing from science. "Nothing" in the Universe is coincidental; "nothing" in the Universe does not apply. Where mathematics connects to lines nothing disappears. Should any principle not match an accepted theory or change the accepted theory, the theory does not apply.

The content of my work holds a new view about Cosmology, which I have been working on for the past twenty-seven years and exclusively for the past six years. I always had a problem with the idea that space constituted of nothing, while I came to realise that lines mathematically couldn't start at zero because there is no evidence of zero as a factor in mathematics. Should you disagree with my statement the question in need of answering is this: What will the length of the shortest hypothetical line imaginable be and moreover, what would the total overall length be in that case? The shortest possible line (hypothetically) must be so short it must have an initial and ultimate point sharing the same spot. The two points must be one and only then can further reducing of any line not occur. If it used zero as a start, the zero part would not count, because the line will only start at a point past zero where the line then will start forming an infinitely small dot. I press this point in urging the understanding because there is such a point, but in an attempt to recognise the point, I have to convince the reader to abolish four or five thousand years of accepted and practised mathematical culture and that is no easy feat. Taking the line back as far as possible brings a dot because of the equilibrium that will stem from such a position. The dot is in infinity, however small, it is not zero. Zero ultimately means not existing and then that point, as a start does not exist. The smallest line has a beginning and an end at the very same spot located in infinity, and infinity may be beyond human scope, though infinity is still not zero. Infinity may constitute of something we do not yet understand, but we may not define our human misunderstanding not present in our minds and therefore as nothing. In this aspect lies the difference there is between arithmetic and mathematical science where arithmetic can have position such as zero since arithmetic excludes the cosmos calculating numbers only. Cosmology is not about numbers because no one can calculate the number of stars. Cosmology is all about lines and angles positioning objects, and in those there features no zero. No line can be zero long and forming a position of zero degrees in relation to another object.

A man may have that many oxen or so many sheep and even this amount of wives, (in Africa) or not have any therefore having then a total of nothing, but there cannot be nothing between the Sun and its orbiting structures. The having and have-nots are part of arithmetic. Light will indicate a line flowing between the Sun and whatever planet, following dot after dot thereby proving the existing of the possibility of something going about by a straight line, and any straight line in relation to other straight lines will be under the law of Pythagoras in as much as obeying the rules of trigonometry. There is no possibility of a straight line not forming in space. If there is space, there can be a straight line. The mere fact of two spots having different positions in space gives the two dots different values. If the line has the length of zero it is not present. If the triangle has one angle at a value of zero it is no triangle because the zero would dismiss all the other angles. Mathematics converts the values of integrating lines according to Pythagoras and arithmetic is about numbers to be added or subtracted. By mathematically excluding zero from cosmology a new Universe opens to the human mind. With the distance between the Sun and Pluto being roughly one hundred times more than the distance between Mercury and the Sun , the distance must hold something more than pure vacuum filled with nothing except one atom here and there occupying the vacuum between them and the Sun . If space supposedly comprises of nothing how can nothing then become plural forming more or be multiplied by a number as to indicate a growth in something not even existing. As the one becomes one hundred the one cannot substitute a value of nothing but then must be part of something. If the one substituted the nothing, all laws of mathematics will go in disarray because when one multiplies any number by zero it becomes zero placing both planets in the Sun . If Pluto was one hundred times closer than it is at present was it one hundred times nothing closer? $100 \times 0 + 0 = 0$. That is mathematics!

By allowing the three hundred a value the nothing must form one making that which is between Pluto and the Sun not to be nothing but there has to be something. This argument follows mathematics to the letter and in precise detail. With Pluto and the Sun being apart that being apart has to have one of something in place forming the being apart from each other's cosmic position one time multiplied by the many ones we find in that space standing relative to other space regarding whatever the space becomes what is between the Sun and Pluto. That factor cannot stand in for not one, which is the same as nothing as that is because one cannot take the place in the position that zero secures. By excluding nothing from the equation space becomes something bringing in a value lying inside the realms of the infinite that must form singularity. As the zero becomes a dot, something else becomes clear about the dot. Looking at the night sky we find darkness overwhelming the space in relation to the stars bringing across light.

My approach to cosmology shall prove to be somewhat unconventional but through the abandoning of the accepted, it enabled me in locating the precise location of singularity that forms the connecting basis of the Universe (and this I say with some degree of confidence). There **are two locations** but I shall **first concentrate** my explaining effort on **the prime singularity**. Singularity did not vanish into the unknown after the completion of the Big Bang development but is in a place science incorrectly valued and classified incorrectly and in that, there is something hiding the truth. If singularity was or is where the beginning is we have to go back and see just where such a beginning was. I cannot accept that the Universe started at zero and neither does anything else in the Universe start at zero. My excluding the possibility of zero includes that the Universe is not filled to the top with nothing and neither is nothing part of outer space. The Universe is about lines allowing light to flow from one point to another point and in following that line it has to continue in the line as the line has to represent something. The Universe is all in relation about lines indicating distances between cosmic structures. The cosmos is in short about lines connecting points in space being apart. It is about a line starting and continuing from such a start. But science advocates their opinion that such a start of a line flowing between any and all objects can hold zero because according to them the Universe are full of nothing. If the Universe in as much as outer space is a container filled with nothing at the present moment, and there is no place anything that was part of outer space previously could release to and there was no emptying of what ever filled it before, then it could not get rid of what was in the outer space when it started with what it started off with. We must then accept from what is not in the Universe was not in the Universe at the time during the start that at is present at present according to science because it then still must contain the same nothing and must have that same filling from the start to the present. If it was nothing it still must be nothing and that same substance being nothing is what it also used to grow using it as it grew because it filled outer space with nothing growing from and growing to nothing. Is that true? The filling of the Universe could not go anywhere so one has to presume it started off from nothing and from there it kept filling with nothing since what ever was in the Universe at the start had no place to escape to or no place through which to escape. That is only applying if it is nothing filling the Universe at large. Can nothing grow as much as a line is growing from a start of nothing? The answer is that such lines not only indicate a distance but since the Universe came from such a small space as science propagates with the theory of the Big Bang then all particles in the Big Bang Universe were rather cramped for space when the Universe started from that small line between particles and is now the same line but is now so big. In the past it seemed being so small and showing the space between particles to be awfully short at one time. It was short but how short was it? Did it start off as nothing? Is the line starting at nothing as science wishes us to believe? If it does then all lines must start from nothing so we better investigate this trend with the start of a line. In this following I show my argument with which I hope to prove the counter part of what science believes. I am about to prove that which science sees as nothing in space and in material is the very location of singularity.

Lines mathematically cannot start at zero because there is no evidence of zero as a factor in mathematics. Should you disagree with my statement the question in need of answering is this: **What will the length of the shortest hypothetical line imaginable be and moreover, what would the total overall length be in that case?**

Locating zero

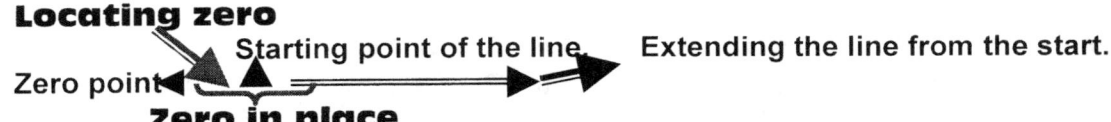

Zero point · Starting point of the line. · Extending the line from the start.

Zero in place

Let us duly test my statement by taking the line back as far as possible. The shortest possible line (hypothetically) must be so short it must have **an initial and ultimate point sharing the same spot.** The line that **cannot reduce** any **further** must be **so short** that **directions flowing away** from each other **are located** in the **same position.** Any theoretical line being the shortest possible line cannot have the line holding the initial starting point at point zero and advance from there. Mathematics simply will not allow it. If the point had zero as all it had to offer, such a point is not present. The zero means there is no such a position. If it used zero as a start, the zero part would not count, because the line will only start at a point past zero where the line then will start. Zero ultimately means not existing and then that point, as a start does not exist and where the line then stars is a point in existing. When the line **has a beginning and an end at the very same spot** and it wishes to extend the position as to further the possibility it has, which direction should it favour. Extending the line in any one direction will favour one direction without any clear reason not extending in other directions. The fact of direction being present only proves and is proved by another point established, which is placed in relevance to such a second pointing a position already established by the relevancy of two point located in a direction to one another. But if one point starts one line there is no favour of direction since there is no established direction yet. The only mathematically sensible option about extending any line starting at a pre-designated point without any other point to establish a pre determined direction will be non-bias progress in all directions equally in order to give a meaningful flow of mathematical equilibrium. Not one direction stands superior to other directions and all directions are equal with no bias anywhere. Of this statement the Pythagoras mathematical principle is proof of and that I explain later.

Let us dissect nothing, as we find nothing in the presence of the cosmos. The distance between the Sun and Pluto is roughly one hundred times more and if the distance between Mercury and the Sun , but both has nothing between them and the Sun . The space filling the distance from the Sun to Mercury has nothing more than the space between Pluto and the Sun . That means the distance between the Sun and Pluto is as equal in relevancy than the distance from the Sun and Pluto since both is the measure of nothing. If the one substituted the nothing, all laws of mathematics will go in disarray because when one multiplies any number by zero it becomes zero placing both planets in the Sun . The distance between the Sun and Pluto **is Pluto is 5900 X 10^6** kilometres of space, but in that statement we take it that the one of a kilometre is present in such a multiplication. The one constitutes the presence of fact being a statement of a value. By saying the distance constitutes of nothing we have to substitute the one factor with a factor of zero. Then the calculation must read **Pluto is 5900 X 10^6 X 0 = 0.** Including nothing as to state the presence of that part contained by the calculation delivers the total of zero. By excluding nothing from the equation space becomes something bringing in a value lying inside the realms of the infinite that must form singularity. Applying this logic to the Lagrangian system and interpreting that information to the law of Pythagoras a clear pattern comes about.

The reaction responding from my argument is that it is silly, but should that be your personal opinion too then test where the silly part applies. Bring the zero into the calculation, the zero that science so eagerly places in outer space and see the mathematical result. By applying the distance one accepts automatically that the figure become calculated with one as it represents one in being a calculating part of the cosmos. The calculation as all calculations normally are is in order to calculate something and the something will at least stand in as one in relation to the rest being part of the calculation. But saying that the factor of one in fact represents nothing since nothing is so much the part in the calculation being calculated, then the zero has to replace the one as the fact of being calculated.

The claim becomes obvious when observing the connection between the half circle, the straight line and the triangle, which could also promote all the qualities lurking behind the pyramid. Consider the connection between 180^0 sharing three different forms all part of mathematics where each is different in form, but equal in value and then one may realise in considering the very basic in mathematics being the Law of Pythagoras on which all mathematics are focused. The triangle stands in for one factor represented by one at a value of 180^0. So does the straight line become a factor of one and the half circle also becomes one where the factor of one equals all 180^0. All three are most seriously part of shapes in the cosmos. Revalue any one form to zero and the rest too must follow and share the same value. The Law of Pythagoras is about angles in relation to lines and not one angle can represent zero because that will reduce all the lines also to zero. The measure of angles between stars at a distance uses parsec as the indicator, but the parsec between the stars indicating an angle has to represent an angle whereby one may measure distance and such a distance cannot be zero because then the parsec will be equal to zero. Again it is multiplying the factor with the measure but if the measure is about a factor of zero, then the factor too becomes zero. That is as basic mathematics as I can present.

If the argument seems ridiculous it is not my mentioning such a fact that is ridiculous but the mere fact of the reasoning also becoming a recognising of an argument accepted by science making it as such ridiculous. If space is nothing then it has a number to use indicating just that value being zero or the capitol O indicating zero. Try and indicate what is measured and calculated in space, but not by simply not thinking about the fact and therefore simply ignoring that what is measured forming the sole value of space, but put the value of nothing as part of the distance in calculation because that is what is measured. When stating the distance between the Earth and the Sun place on paper what will allow the kilometres measured to represent the factor that is being measured. If represented by one being the total of one by hundred and forty nine million kilometres of nothing put that language in the International language of mathematics that spans all dialects spoken on Earth. Put it in mathematical terminology by saying there are 149 000 000 X 1 (multiplied by the kilometres) multiplied by what it is being measured which is 0 and what will the total come too… a full zero.

149 000 000 X 1 (km) x 0 (indicating what the km are made of) = 0

 Mathematics says it. If there is something to be measured then the least value the measurement can have in relation to what is used in the measuring has to be one. It cannot be zero and be measured…and we do measure outer space! It sounds as if something here is at fault. It is not with my mentioning the inconsistency one should find fault but the fault is with the fact that it is there and no one noticed! I am not to blame just because I am mentioning it, but the blame must go where it belongs.

I think it is by now little understood although I imagine not nearly accepted that by adding a million of nothing to one nothing there will remain one nothing and that is still nothing. Nothing cannot accumulate therefore I cannot accept anything holding the vastness of space being able to constitute nothing as the major component.

When reducing the circle in size one have to reduce the radius or the diameter because the pi is the indicator of the form as being a circle. Divide the r until there can be no dividing any further and that cannot in the end indicate zero because no matter how small, in that will forever be a value in place.

There will forever be subatomic particles building atomic particles because the reducing of space goes down as far and to infinity. There will always be some material forming a part of more material that builds into something which ultimately becomes the atom as a unit. In every centre of every subatomic particle running down into infinity there will be a centre composing singularity and the group will establish a centre governing singularity.

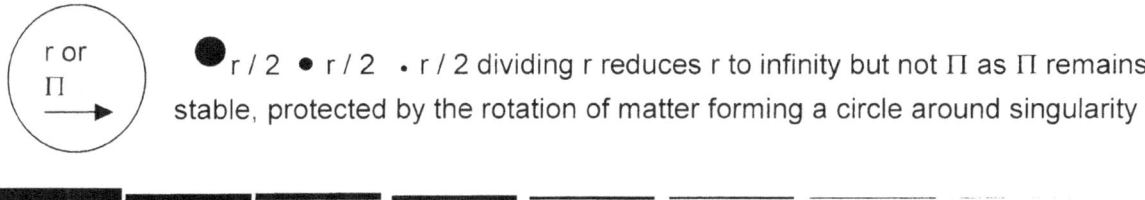

● r / 2 ● r / 2 ● r / 2 dividing r reduces r to infinity but not Π as Π remains stable, protected by the rotation of matter forming a circle around singularity

Taking that into account it is important to recognise that notwithstanding the size of a line, there is another line (or dot) eternally bigger as well as eternally smaller than the line in question. We can never grasp the size of a line that forms the utmost or the least of possibilities and therefore size belongs to the human mind forming conceptions of big and small, but it has no place in the cosmos at large. This concept not only applies to size, but to all limits and divides we wish to create forming borders we can appreciate. When looking at the circle in the conventional manner, we persist with errors brought about in culture and not by applying some significant modern logic.

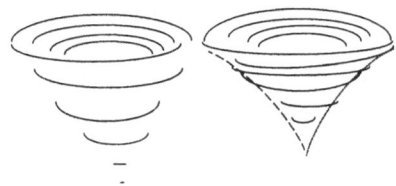

By reducing r indefinitely to the tune of half each time, r would become infinitely small, beyond human calculating means, however as mentioned in the case of the smallest dot holding one spot, r would become insignificant beyond human comprehension even, but never reaching zero and still Π would remain intact and dictating form. I believe one can begin to see where my suspicions are heading because the flaw comes about in the manner mathematics are practised for thousands of years. But before coming to the mathematics I would first like to bring your attention to the practical side. I am promoting a theory in which I am able to prove there is as much contraction going on in the cosmic Universe as there is expansion and the contraction is as much part of the expansion. The Universe rides on a balance and we have to locate such a balance. To prove my theory I firstly had to locate the centre of the Universe. Even admitting to such a notion sounds like madness, but please give me a chance to explain in more detail. If I wish to achieve success that would depend on my ability to convince all that outer space comprises of material and as such we can locate such material even if we are unable to see such material.

To find the invisible I had to locate singularity. I realised that my effort to locate the point holding singularity enabled me to backtrack the exploding Universe to its origins.

By applying some basic effort I have located the position from where all movement came and the direction it took moving forward in time

The reversing of the circle radius is not alien to nature at all. An observation coming instinctively to mind one may recognise is that the form reminds rather explicitly of natural phenomena such as hurricanes, water whirls and even the shape most commonly favoured to express the cosmic object referred too as a Black Hole. The similarity may be more than coincidental. Let us consider the statement in the reverse. In our calculating of a circle we apply two formula methods. The one use an r to indicate the radius and the other use a D to indicate the diameter, which is double the radius and therefore needs to be divided by a four to eliminate the Newtonian inverse square law amounting to the difference there will be between the two. The one using the radius is Πr^2 and the other formula is using the diameter is $\Pi D^2 / 4$.

In any circle or sphere the size only depends on the fluctuation of r in the square as a component to the circle or sphere but that does not affect the form by indication of Π in any way there may be. The conclusion from this is that no line can start at zero because that will be a mathematical impossibility. A line or spot starting at zero would therefore be shorter than the shortest line possible.

For obvious reasons can no line, or any line grow or extend from zero because such a line must then quit zero and become something, thus abandon its original value. That would mean the start of the line has a different value to the end and a line holds conformity through out. When any line is starting from point zero it can never leave zero because of the influence of being zero disqualifies any possibility of growth. If the line then had to grow in all directions at the same pace the line must therefore be a circle or being three-dimensional, a sphere. Flowing from this fact is that in the Universe there can be no zero point or unfilled space. In the case of the growing sphere the value of the circle is Π, and that is where creation started. That gave me the clue where to start looking for singularity.

One would find singularity in the value Π and the value Π will be in all things rotating in a circle. You might wonder how does that apply to the cosmos and moreover to gravity? In my search I stumbled on two accepted but not intergraded laws and when I found and located singularity the two laws became very much plausible and factual. Take a circle and reduce such a circle constantly to where it no longer can reduce. Reduce it to a point where only form remains part of the circle because the radius has gone beyond human measure and becomes so small it is not noticeable with what ever tools man may use, then what remains is pi since pi does not indicate size but indicates form, and form is all that then will remain.

I believe one can begin to see where my suspicions are heading because the flaw comes about in the manner mathematics are practised for thousands of years. But before coming to the mathematics I would first like to bring your attention to the practical side. I am promoting a theory in which I am able to prove there is as much contraction going on in the cosmic Universe as there is expansion and the contraction is as much part of the expansion. The Universe rides on a balance and we have to locate such a balance. To prove my theory I firstly had to locate the centre of the Universe. Even admitting to such a notion sounds like madness, but please give me a chance to explain in more detail. I realised that my effort to locate the point holding singularity enabled me to backtrack the exploding Universe to its origins. By applying some basic effort I have located the position from where all movement came and the direction it took moving forward in time…and yes, even time as such.

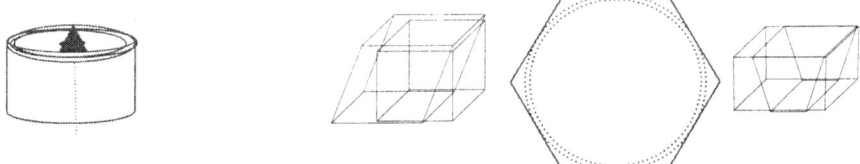

Anything occupying space in the cube will apply r and by r I mean just a distance not using Π because Π serves as a form indication while the collective product of r will determine form as well as accumulative dimension total. Notwithstanding the name used confirming the shape or r named as length width or height, it is all just a straight line bringing about the cube with all its other names that may find attachment to specific form but nevertheless still remains only a six-sided cube with connecting lines applying different angles changing in some cases. The normal perception is that any circle growing spontaneous would grow by the radius, which is r. In mathematics that may be true but it is not true in nature. In nature that cannot be the case because r is an indication of a straight line. By growing with the aid of a straight line from the centre to circle the influence that that would have on the circle would result in many circles following one another and not a continuous growth.

Gravity is the dimensional changing of space holding r as reference in the cube as to the sphere holding Π as the reference. In order to generate spin producing time in matter occupying space, therefore creating dimensional change, Π has to be a factor indicating the possibility of spin because implementing Π the circle sides will follow one another without establishing separation. The answer must be in finding Π, and thereby locating singularity. If singularity is in affect the original point of the cosmos birth, the reducing path we should follow will indicate the whereabouts such a point must be.

There are two standard formulas used to calculate a circle. The one uses an r to indicate the radius and the other uses a D to indicate the diameter, which is double the radius and therefore needs to be divided by a four to eliminate the Newtonian inverse square law amounting to the difference there will be between the two. The one using the radius is Πr^2 and the other formula using the diameter is $\Pi D^2 / 4$. However one looks at the mathematical expressions and Kepler's formulating of space-time, there is an exceptional difference between the two scientific uses. When investigating Kepler's formula one do find it appreciably differs from the normal Mathematical equation like $a^2 = r^2\Pi$ and $a^3 = 4/3\ \Pi r^3$. In the normally used mathematical expressions such equations tend to concentrate on the volumetric aspect. In the case of Kepler's expression it is something else that wants to surface. It is another idea that is coming to mind. In Kepler's formula a^3 stands to symbolise the third dimension and such a third dimension becomes equal to two other dimensions grouping and sharing value to equal a^3 efforts. It is not the circle of the rotation because with such a normal circle the radius is in the square and Π evaluates form. Here there is no mention of a factor Π, which one would suspect to be somewhere applying since the circle is Π and Π is the circle and the two are inseparable, but not in Kepler's a^3, where there is no mention of Π at all. The fact that there is a radius of some sorts used to indicate a position cannot hold the square as it normally does in the case of the normal equations. In the mathematical equation the factor indicating the position of the circle edge has the square value being called the radius or in some cases the radius doubles and which then is the diameter, and the circle indicator is Π. But in this event the formula value will bring about a square value to the answer one receives. It will bring a value to the surface of the circle. In Kepler's formula it specifically does not.

I realised before starting my quest that one possibility that the shortest line or smallest spot can never have is having a starting point on the zero mark. If the mark of zero holds the start it must also hold the end because the end and the beginning has the same position. If the position of zero then is the beginning, the end will also be zero leaving the line or spot without an end as well as without a beginning. Such a spot will constitute all of nothing. Any line starting from zero would inevitably start from a point where it ignores the zero mark because the fact of zero does not implicate a start or a size of value, but only the not being there of that position.

All lines would form a duplication of another line sharing value since there will always be a possibility of yet another line in the realms of singularity lying between the two lines in question reducing the size infinitely to either side of the divide we humans create. Boundaries therefore are human and as man made substances it does not belong to the cosmos outside the influence of man and must be discarded. No mathematics will ever measure the thickness, because as the line that is standing still it cannot have a width at all. The moment a width appears which one can measure or calculate, the line will become part of the factor forming the divided and not the divide. The instant when space connects, the spin direction will produce the partisanship of space and spin. Any form of space (even in the most minute) will expand as it favours a direction but changing the direction is by rotary motion.

The moment there is an area there is a measurable rotating brought about and no longer a non-interfering divide. Such a line holds space in a position that runs far beyond the boundaries and limits of the three-dimensional. Another factor of such a line would be that the radius (let us substitute the radius r with the using of Kepler's **k**), **k** would be immeasurably small. The factor **k** cannot be zero because infinitely close to that first **k** is the start of the third dimension where time plays the part as the fourth quarter. The presence of **k** is undeniable and recognisable yet it is not visible. The fact that **k** is there albeit stripped of any influence, disqualifies it from being zero and therefore not being there. With **k** already beyond any measurable space, leaving a^3 as a factor of one and not being able to pin any volume measure to that one **k** will have to be to the power of 0 being k^0. In Kepler's formula $a^3 = T^2 k$ the area a^3 would be one because of the dimensional non-existing of measured sides in any direction. If $k^0 = 1$ and $a^3 = 1$ the only alternative T^2 could possibly have is also one. The factor of T^2 identifies the time in the formula and when the formula indicates time as one, the time component must therefore be eternal. Only time in eternity does not change

The real formula applying when the calculation of the sphere volume is calculated is $a^3 = 4/3 \, \Pi r^3$ where it places one third of the dimensional (but lesser) factor in direct relation to another third dimensional relation held by the radius and all aspects about the factors being in relation is to acknowledge the form that is applying and serving as a sphere. However there is no criss-cross matching of dimensional accumulating. It places time in the square directly in relation to space in the cube in association where time shows two distinct qualities. The one factor is time in the circle rotating while the other is in the linear or the straight line implicating the position that the other would have. In all instances of measuring the distance the orbit travels around the Sun as the space displaces or space covered by travelling in the time it is covered and dividing such a ratio one find the distance of the orbiting object from the Sun the in relation to the other factors form one or very close to one. It is relevancies carried from the Sun and the Sun is the governing singularity representative for the entire solar system. This is about relevancies applying throughout the Universe. This balance is much, much more than what the figures say. It underlines and it explains gravity as a life form in the cosmos other than what we consider our life to be.

The German mathematician and astronomer KEPLER, JOHANNES (1571-1630) German mathematical and astronomer became Tycho Brahe's assistant in Prague in 1600 A. D. where he undertook to complete the tables of planetary motion Tycho had begun. Kepler first calculated the orbit of Mars. He spent much time trying to reconcile Tycho' s accurate observations of the planet with a circular orbit, but concluded (in Astronomia nova, published in 1609) that Mars moved instead in an elliptical orbit. Thus, he established the first of his laws of planetary motion. A theory that the Sun controlled the planets by a magnetic force led him to the second and third of his laws, which were published as part of his treatise on theoretical astronomy, Epitome astronomiae Coernicanae (1618-21). The Rudolphine Tables (named after Tycho's patron, the Holy Roman Emperor Rudolph II) of planetary motion appeared in 1627 and were still in use in the 18th century. Kepler also wrote De Stella nova, on the supernova of 1604 and Diptirce on optics and the theory of the telescope. The overall view followed in this book **Matter's Time in Space** places the true significance of his work in true contents. In KEPLER'S EQUATION is the equation that relates the eccentric anomaly of a body in an elliptical orbit to its mean anomaly.

The equation is $E - e \sin E = M.$, where E is the eccentric anomaly, M the mean anomaly, and e the eccentricity of the orbit. It is important as one of the mathematical relations enabling the position of a planet about the Sun , or a satellite about is planet, to be calculated from the orbital elements for any time. However this only relates to the solar system, and KEPLER'S LAWS only apply in the contents of the solar system. The three laws governing the orbital motions of the planets, discovered by J. Kepler is as follows: The first law states that the orbit of a planet is an ellipse with the Sun at one focus of the ellipse. The second law states that the radius vector joining planet to Sun sweeps out equal areas in equal times which as it says refers to time and not the circle. The third law states that the square of the orbital period of each planet in years is proportional to the cube of the semi major axis of the planet's orbit. The first law gives the shape of the planet's orbit; the second describes how the planet must continuously vary its speed as it follows its orbit, moving fastest at perihelion and slowest at aphelion.

The third law gives the relationship between the planets' average distances from the Sun and their periods of revolution. Instead of placing the true value to Kepler's laws, I. Newton placed his own interpretation to Kepler's laws, and in doing this he wilfully destroyed the principle working of the Creation. Through Newton's tunnel vision, he applied his own miss interpretations to the correct presumptions of Kepler. Newton reduced the implication that Kepler's findings hold, by using Newton's variation of what Newton wished Kepler's work would provide Newton when Newton was introducing his interpretation the law of gravitation. He then went about and changed it to three laws of motion. I. Newton generalized Kepler's first law, verified the second law, and showed that the third law should be amended to the form; $4 \pi^2 a^3 / T^2 = G (m + m_p)$. In this, the value of T and a are the period of revolution and semi major axis of the orbit of a planet of mass m_p about the Sun of mass m, and G is the gravitational constant. The major aim of this book is to correct these misgivings of Newton. I shall return to the statement about $4 \pi^2 a^3 / T^2 = G (m + m_p)$

What Kepler observed and formulated was more of a dimensional coming together by the cosmos in nature. He saw more of the principles guiding the cosmos than what one will find in the cosmos. It was practical mathematical forms of symbols finding value by form instead of true figures, which Newton wished to see. Kepler saw

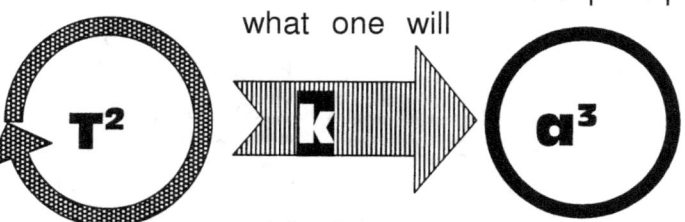

the first and second dimension forming the third dimension by allowing the third dimension space to flow and room to duplicate under the control of singularity.

In the argument Kepler made he had hidden so much more facts into one formula than what I think even he realised. Well, it is much more than what the Accepted Policy Protectors Of Science ever came to realise. He officially formulated space-time, he officially coined not the name but the origins of the Universe being the Big Bang and he was the first to put the speed of light in relation to cosmic development…and all of that with his rather simple formula. He said the space a^3 not the circle (a) or the circumference a^2 but in the circle a^3… where such a circle represents a factor in the third dimension.

The formula he compiled was not rather but very specific about the area being a third dimension area and to prove it beyond doubt he placed it in the relevancy of the formula in a ratio of presenting the third dimension in space. He said a^3 is equal to $T^2 k$. Newton and Newtonians came afterwards and played with mathematical toys as to challenge their mental capabilities. Newton introduced a $4\Pi^2$ to indicate the presumed circle on the one hand and on the other hand he brought this lot equal to $\{G (m + m_p)\}$ which he then presumed to be the general Universal gravity constant (G) and the sum total of the two structure mass. Newton saw a ring circling around a centre having $4\Pi^2$ to indicate such a ring outside a centre and he positioned $\{G (m + m_p)\}$ where the two mass factors combine the gravity effort in the general grand gravity constant in space. I have had so much resistance in the past from all Academics but that is not what I see what Kepler saw. I shall trace this back to the centre of creation.

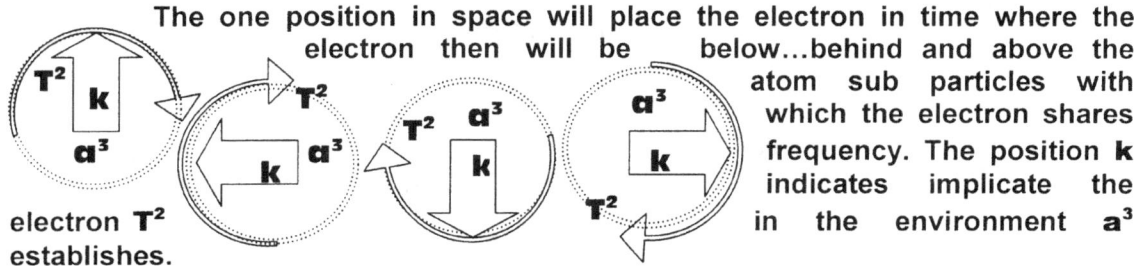

The one position in space will place the electron in time where the electron then will be below...behind and above the atom sub particles with which the electron shares frequency. The position **k** indicates implicate the in the environment **a³** electron **T²** establishes.

In their eagerness to calculate they calculated a formula to measure the circumference **a²** of a circle being Πr^2. I have seen an Astro physics examination where they use $4\Pi r^3 / 3$ as the formula to calculate the Sun and other stars volumetric space! They formulated the measuring procedure of the circle being in the third dimension that will show how big the volumetric space is of a sphere at **a³** being measured with the procedure being $4\Pi r^3 / 3$. This too was a fanciful devise allowing mathematicians to be much superior to the rest of the commoners and to dictate to the lowlife how and what they should think when they think and if they indeed can think of anything to think of. Then some Mathematician and an Englishman of Substance came onto the idea of gravity. Being a mathematician the Englishman placed the Universe at the feet of mathematicians. He saw circles where Kepler saw three dimensions. He saw three dimensions where Kepler saw nothing. He knew time had to be somewhere as something and then covered it by denouncing the circle as nothing.

What then is it that Kepler saw as he formulated $a^3 = T^2 k$. At the normal flow of time it takes the electron a certain time to spin around the atom. The atom uses space **a³** and the atom is a certain length **k** that forces the distance the electron has to travel in one cycle period **T²**. The atom **a³** connects the electrons travel **k** to gravity **T²**. The relevance **k** produces to support **a³** is to point **T²** to two positions the electron will be in the duration of one specific time. The electron travel will be cyclic and periodic in relation to the space the atom holds. The space stands related to the gravity with which the Earth confines the space of the atom to the space and speed with which the liquid heat confines the atom space.

$$a^3 = k T^2$$

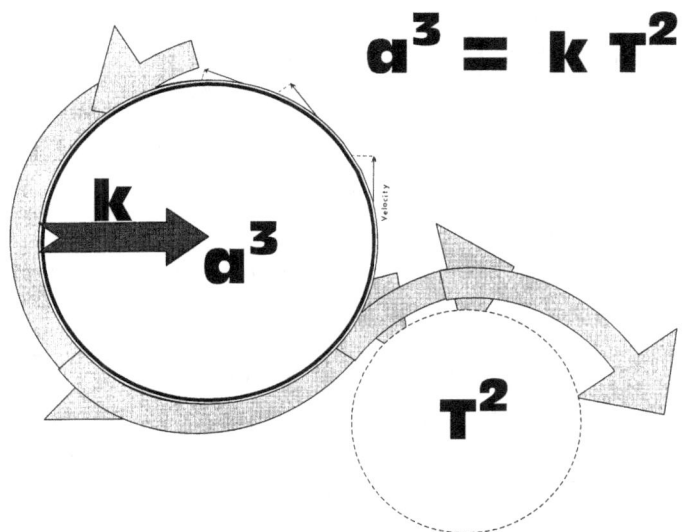

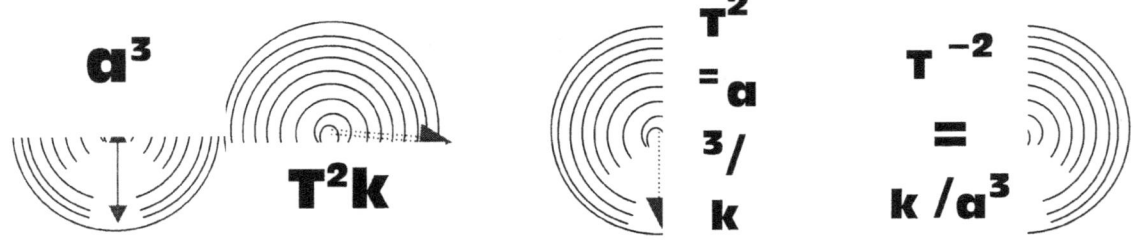

The Universe divides between space that was and will be where time is and time where it just released space is where it is accepting space. The one side is while the flanks are in motion of releasing or accepting the position space has.

Planet	Period T years	T^2	Distance k	Space a^3	Ratio
Mercury	0.241	0.058	0.39	0.059	0.983
Venus	0.615	0.378	0.728	0.381	0.992
Earth	1.000	1.000	1.000	1.000	1.000
Mars	1.881	3.54	1.524	3.54	1.000
Jupiter	11.86	140.66	5.20	140.6	1.000
Saturn	29.46	867.9	9.54	868.25	0.999
Uranus	84.008	7069	19.19	7067	1.000
Neptune	164.8	27159	30.07	27189	0.999
Pluto	248.4	61703	39.46	61443	1.004

At the first glance Kepler's formula seems to be numbers and positions applying between the sun and specific but different planets in the solar system.

The time frozen on paper in a single t is effective in remembering the viewer of an event but that is not the event in the present any longer. That was how the event occurred during the time from where the camera shutter opened T_1 to where the camera shutter closed T_2 and the time frame T^2 was then during the open period of the camera shutter. But afterward it represented t when looking at the picture and the looking of the picture became an event during a specific T^2 that went from where one is taking the first look to where one is looking away from the paper carrying the first dimensional image of an event gone by and that is at that stage a representation of t in another milieu of $a^3 = T^2 k$. The t in the single is when mathematically presented as only t indicating a mathematical single flat dimensional view of time and is then correctly applied because it represents a reminder of a four dimensional event $a^3 = T^2 k$ that went single dimensional because the moment in the fourth dimension was then frozen in a single dimension on paper while the fourth dimension $a^3 = T^2 k$ soldiered on and time will always be representing T^2 as Kepler stated in the square allocated to space having a cube $a^3 = T^2 k$ at a time even before gravity got a name.

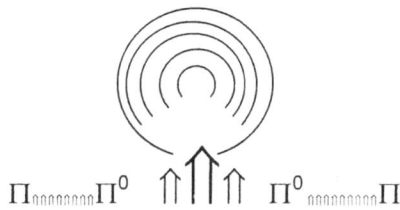

With singularity placed in infinity within the centre of every rotating object every atom and its relation to its surroundings including other atoms form space-time diverting from the point holding singularity as far as rotation goes because every object holds three relative positions in as far as where it was, where it is and where it will be in relation to singularity providing time. I elaborate on this else where.

Newton said a sphere is $a^3 = 4/3 \, \Pi \, r^3$

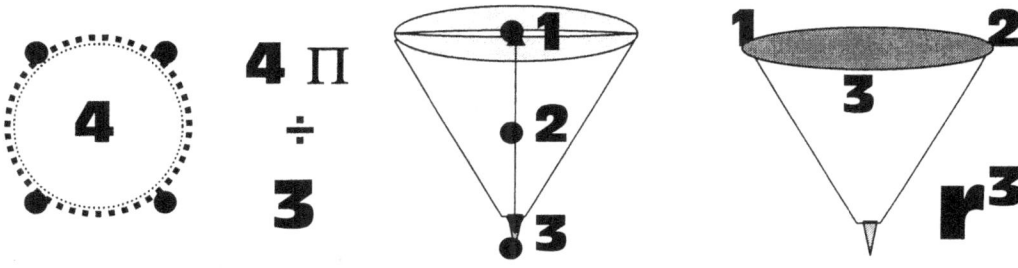

In $4 \pi^2 a^3 / T^2 = G (m + m_p)$

$a^3 = T^2 k$

$a^3 / k = T^2$ but at the same margin is

$k / a^3 = 1 / T^2$

$k = a^3 / T^2$ singularity

$a^3 / T^2 = G (m + m_p) / 4 \pi^2$

and $a^3 / T^2 = k$

then $k = G (m + m_p) / 4 \pi^2$

But I showed that $k = a^3 / T^2$ and Newton's claim is that $a^3 / T^2 = G (m + m_p) / 4 \pi^2$

The only definite place one will locate zero is in between the starting point of the lines going in opposing directions in the position the lines hold before there was the least of directions applied, but that is only because there is no such a position, not because any line is coming from there. The two lines are still one holding the opportunity of parting as an option but have not yet parted and therefore are on the very precise same spot. The line coming from there is already there because it already has the choice of going in any and all opposing directions and when it starts running it will place filled space in that location because the space was already filled with a line starting and not with a line not there at all. When reversing a line we might find a better idea of what is in place and where it is in place.

In the action of the inseparable drawing closer and moving closer gravity finds the dual value of linear and circular gravity. There is no separation of the two factors acting as one but both have different applications and values in the unit. This is the result of singularity having three parts acting as one but giving three distinctions in application. Gravity is as much part of dismissing space as it is about making contact with space in time.

But since the connection comes about as a circle, the connecting points will relate to Π as the value.

Due to the spinning nature of such a point with all surrounding the point will be alternating direction favouring change every second and in that the value of such a point can only be Π because of its constant changing. Using r would specifically oppose another r from every angle because the use of r will bring about a static relation to the previous and following instant and therefore it will cancel the constant spin flow.

The new direction pointing to a new location in relation to the previous point will oppose the previous point it had in relation to direction considering the centre point.

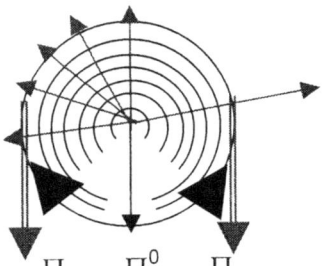

Pinpoint positioning of singularity Π^0 with Π positioning space to either side forming the border set by singularity
The new direction pointing to a new location in relation to the previous point will oppose the previous point it had in relation to direction considering the centre point.

Π Π^0 Π

The motion of a liquid confirms a centre and the confirming of the centre provides a flow of space-time in either direction, which produces the gravity. Without establishing and activating such a centre the centre is not active. It is there but it is inactive in being present.

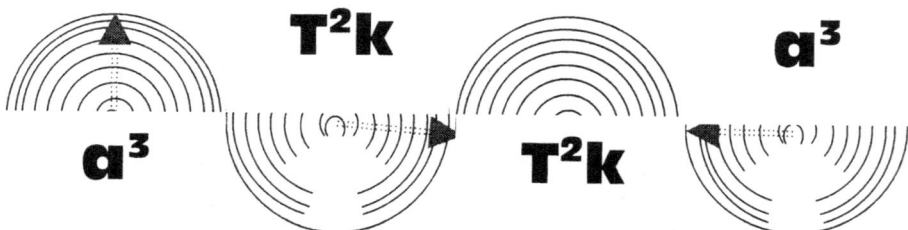

If the one side is a^3 then the other side is T^2k because the one side opposes the other side by providing control a^3 to the motion T^2k on the other side

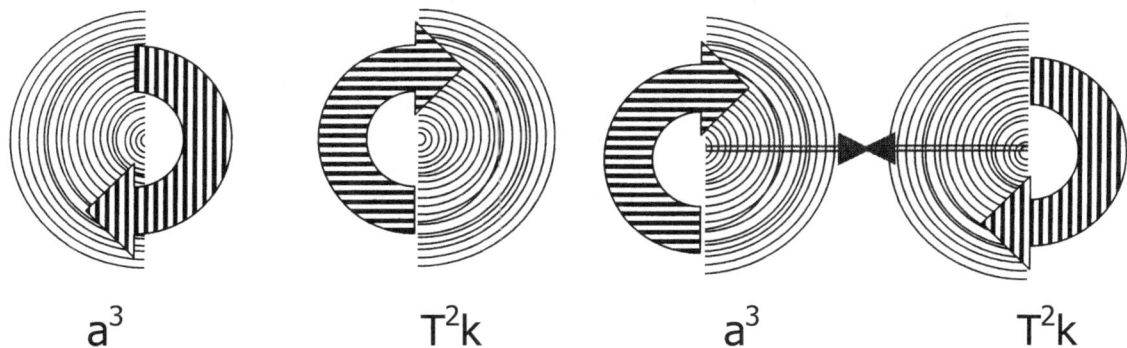

$$a^3 \qquad T^2k \qquad a^3 \qquad T^2k$$

The same motion is contradicting the motion it was or will be and because of that the contracting part is going to be expanding while the expanding part is going to contract. The one will accept while the other will release but the cycle can never be broken. Still in the end it is the same motion and the motion never interrupts the flow of time.

By taking **k** into a negative the space will reduce the time because the space cannot sustain the demand of space growth.

$k^0 = a^3 / T^2k$ $\qquad \Pi^0 = \Pi^3 / \Pi^2 \Pi$

$1 / k^0 = T^2k / a^3$ $\qquad 1 / \Pi^0 = \Pi^2 \Pi / \Pi^3$

$a^3 / k = T^2$ $\qquad \Pi^3 / \Pi = \Pi^2$

In all my other work, I make exclusively use of the value of singularity Π since it makes a lot more sense, but when I use the value of singularity, which is Π then no one seems to have a remote idea to which I am referring.

$k^0 = a^3 / T^2k$ forms

$k^0 / k = T^2 / a^3$ that becomes

$k^{0-1} / a^3 = a^3 / T^2 / a^3$

$k / a^3 = 1 / T^2$

The replacing of the symbols Kepler used with the value of singularity the mathematic equation comes into practise.

$\Pi^0 = \Pi^3 / \Pi^2 \Pi$

$\Pi^0 / \Pi = \Pi^2 / \Pi^3$

$\Pi / \Pi^3 = 1 / \Pi^2$

However, keeping Π as one ($\Pi^0 = 1$) we keep the Universe in the first dimension.

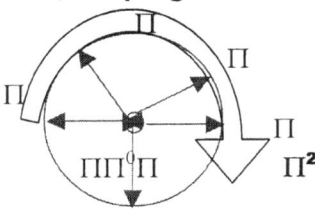

This point, which I now am referring to, is the point where Π is a fully appreciated value while the diameter D still remains a dimensional factor of one. This is the dawn of the second dimension where space was there but space was sparsely shared in some cases. It was when Π^0 shifted to become Π for the very fist time.

In the sketch below the circle to the right would come about from a straight line r growing influencing the appreciation of Π, but to influence Π would lead to a breakdown in r as Π and r are different entities. The circles to the left shows a continuous growth by extending Π every time and since Π is the same part as the previous Π, only extending that billionth of a millimetre or many times smaller each time, the circle will be truly continuous without any signs of a break. In the context of dimensions one finds coming from the centre Π^0 an established eternal flanking of Π to six positions since Π^0 forms the centre to the six sides and all six sides not having a diameter yet must apply Π to indicate specific value. In the very centre, which I am referring to, rotation must end or start depending from what vantage point the relevance is placed.

However, the equation looks far more sensibility when using the value of singularity

$k^0 = a^3 / T^2 k$ forms

$1/ k^0 = T^2 k / a^3$

$1/ (k^0 k) = T^2 k / (a^3 k)$

$1/ k = T^2 / a^3$

Expressing the equation by using the value singularity has instead of the symbols Kepler designated to the formula he introduced it makes far better sense expressed mathematically

$\Pi^0 = \Pi^3 / \Pi^2 \Pi$

$1/ \Pi^0 = \Pi^2 \Pi/ \Pi^3$

$1/(\Pi^0 \Pi)= \Pi^2 \Pi/(\Pi^3 \Pi)$

$1/ \Pi = \Pi^2/ \Pi^3$

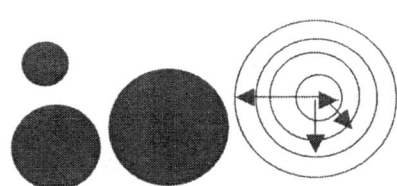

One should not try to focus on an image of such a spot or dot because there is no image. The line dividing the cosmos and that run through every particle, no matter how large or small is beyond our vision. Such a small line, so small it is not even noticeable is large enough to part the cosmos into sectors. It splits the biggest there is into particles and we are not even able to notice the precise location of such a split. In truth there is no top or bottom that we living in 3D can see. We shall have to use a general conception brought about by intelligence. Your intellect tells you about such a spot, but that is all because that spot is on the other side of the Universe (quite literally). From the centre of the dot there is a top and a bottom spot. From those points there is connection with four quarters. That produces six connecting points that are all aligning to the centre.

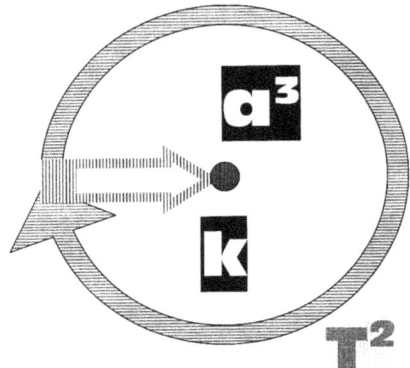

a^3 forms the space the atom claims while travelling in the Earth spinning all the way and travelling with the Earth around the Sun

k positions the electrons travel in the space relating to the space the atom holds while travelling with the Earth in the Earth around the Sun in relation with a specific position k will indicate in relation to the Sun .

T^2 is the time it takes the electron to be relevant to the position the Earth places on T^2 while the Earth captures the space of the atom by providing the space for the atom to be within while the Earth travels from one point T_1 to another point holding T_2 in frequency to the atoms T^2 relating to the

Earths T^2 in perfect harmony with the Sun having another T^2 relevant to all the other factors we call cosmic particles. Big or small it is only about cosmic particles holding space in time in relevancy.

Looking at the effect of gravity it shows the precise quality of no distinctive point, as gravity never seems to end at a point but flows all over affecting all that holds a position in its sphere of influence. The gravity coming from China meets the gravity coming from America at no particular spot but intermingles without distinction.

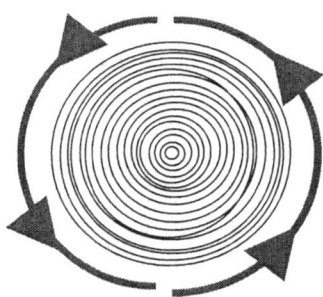

The very centre form an eternal divide that will not allow what is on the one side to present an influence on the other side. It divides spin. It divides direction of spin. It divides all rotation from the outside that one may detect and such divide is there because at one point spin will run to the left coming from the right and just immediately next to that point must run a direction from left to right. It cuts without contributing or participating in movement. It divides without any favour.

By expressing a wish to accomplish time travel such a person wishes to accomplish that material must collide with itself a mean feat if ever there was one. Such a person wishes to have one side collide with the other side as he stops and reverse time while the rest of time is motoring on.

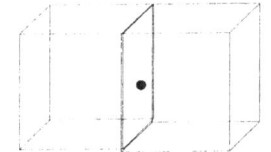

Taking the outlook from the point the sphere is holding from that centre out into space there are ten points connecting to the centre. In that are the dimensions of singularity connecting to space where five connects to space in the second dimension of singularity, and five connect in the third dimension of singularity

That is singularity not having a dimension of space and not having a dimension of time, or a radius connecting the rotating distance to Π. Every rotating object holds a centre from where the rest of the rotating direction will differ at any and all given points. Not one point is exactly the same, but in the very middle, the centre no one can draw, measure or see is a point not in motion.

The first condition for gravity is even-handedness through out the sphere holding the applying gravity and the second is to have most or the strongest gravity located where the space is least. That gravity then has a position in the very centre of the sphere and from that centre the gravity produces all the edges or borders that the sphere consist of. In the case of the sphere this factor makes the sphere much more dominating than any other form does. From the centre point controlling all sides is gravity and with gravity applying control the sphere has seven sides to the square in any other possible form having at least six sides.

The cube can come in whatever form there may be but the sphere adheres to precise measure and behind this principle is all that forms the Universe. The cube has six sides connected loosely and can change form just by changing the relevancy between one side (or more) in relation to the distance brought about by the other sides. The sphere being a complex circle stands related where the sides has to apply precise measure in equality. This becomes a law because in the precise middle one will find the strongest gravity as that gravity holds the object in form and true to form. If there is even gravity spread in all directions the form must be a sphere and the sphere insist on seven points relating to sides or borders.

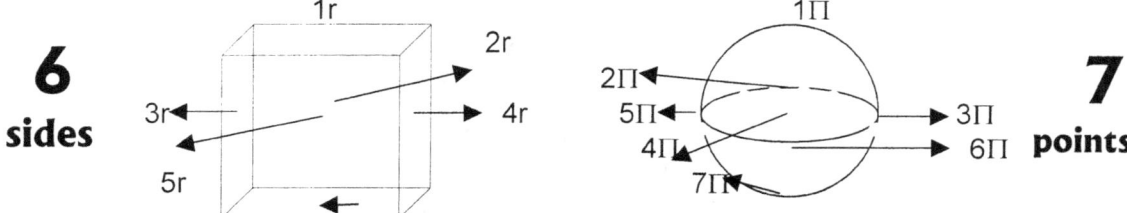

In the sphere there are no radius but only the extending of Π from the centre Π in six opposing directions relating to one another by the square but remaining Π because of the unity the matter holds in relating to space. It is not possible to draw a precise line that would form a precise ring and not cut some atoms in parts. Because there will always be an atom disallowing the precise positioning of the circle the circle continues on a solid basis holding Π as a positional reference and not r. In every sphere there then are the seven Π relating in precise dimensional and positional equality forming equilibrium to the centre Π as well as to one another by 90^0 and 180^0 implicating the dimensional positioning. Therefore the sphere holds 7^{Π} and the cube holds $6 \times r^2$. Where space comes into contact with the sphere the cube loses one of the six dimensions it has to the more dominating seven dimension of the sphere whereby the seven dimension in equilibrium will dominate the six dimensions loosely connected by r bringing about that the cube then has 5 sides to the seven of the cube. Because the space surrounding the sphere takes on the shape of the sphere and not the other way round where the sphere resolves in accepting the form of the cube, one may presume the form of the sphere is the most dominant of the two choices.

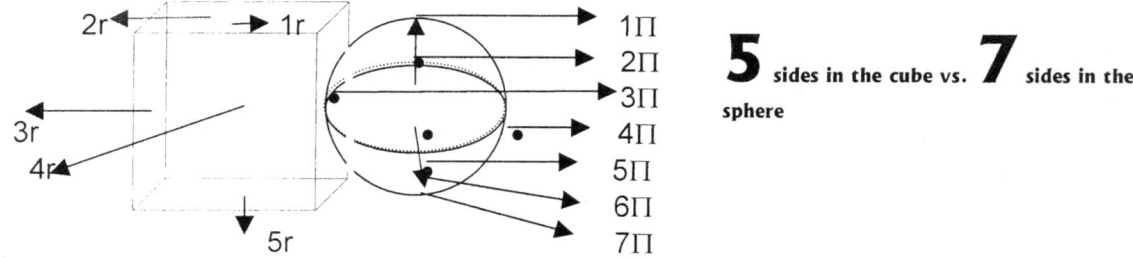

The sphere is a multitude of innumerable circles that forms one unit with all the innumerable circles that compile a sphere all put together. The circle is a constellation of Π where every Π flow from one into another and such flowing varies the number arrangement of r. In order to measure the surface of a sphere the radius carries the torch by going square. It is the radius, which is just another line that come from the outside and run all the way inside to bring the value of the line into the circle.

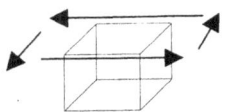

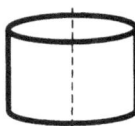

In a circle, there is a radius that initiates the circle. The calculation of such a circle is $\Pi \times r^2$.

Only the circle has a point on the outer edge of the precise distance from the centre and every point in measures a precise and equal distance from the circle, there is a radius that initiates the circle. The such a circle is $\Pi \times r^2$.

form that is at a the form outer rim. In a calculation of

The radius r runs from the circle outwards, from a circle centre point towards Π, the value of the circle. In the centre of the circle, there is a point where the radius starts. It runs outwards from that point in all directions towards the circle Π. Technically, there then has to be a point where r is infinite and not zero, an absolute infinite. However, the circle therefore remains Π. The circle does not disappear; it remains there for all to see. It is only the radius that almost disappears into the infinite, but it does never become zero! $\dfrac{\Pi r^2}{r^2} = \Pi$. If one removes the radius from the circle, the circle remains, only holding the value of Π. By removing the value of r, Π becomes singularity with no place to be. Singularity is the place where there is no space to be in place. However, Π remains because once r receives the slightest of space Π will find space. Then the circle will grow to Πr^2 and r would determine the space. Without space, there is no r but there is a circle with the value of Π. Singularity is in every single rotating object, be it the proton or the Universe. This situation is part of any and all circles and is therefore part of any and all spheres. The line will end at a point where the line starts going in the opposing direction and that value is indicated by the use of $r^0 = 1^0 = 1^1$

To that end the shortest possible line (hypothetically) must be so short it must have **an initial and ultimate point sharing the same spot.** Any theoretical line being the shortest possible line cannot have the line holding the initial starting point at point zero and advance from there. If it used zero as a start, the zero part would not count, because the line will only start at a point past zero where the line then will start. Zero ultimately means not existing and then that point, as a start does not exist. At one point the reducing attempt of the line would start making the use of mathematics seem silly. The reducing would seem tedious and leading nowhere. But as sturdy as mathematicians can be they would carry on (or so I am made to believe...). Then when the man doing the calculations gets carried off in a straight jacket, while the man is making funny noises, when he is totally cracked mentally, the calculations can still go on and on and on and...and that is where sanity prevail and someone says "drop the affair". It is at that point I would have loved to see Einstein carry on counting stars in so many galactica in his attempt to determine the critical density joke. That is not where we get into infinity. That is where man's brain gets blistered but infinity is still far off. As any one can see the Universe is far beyond some insignificant and senseless formulae invented to impress Academics while others are kept busy and free from boredom, but in the real Universe the attempt is not worth the thought it takes to disregard the attempt.

When the line **has a beginning and an end at the very same spot** and it wishes to extend the position as to further the possibility it has, which direction should it favour. Extending the line in any one direction will favour one direction without any clear reason not extending in other directions. The only mathematically sensible option about extending will be in all directions equally in order to give a meaningful non-bias flow of mathematical equilibrium. That is where one would have to go look for the beginning of the Universe. The Universe is a about lines connecting but where does that which does the connecting end. The Universe used form to this point in development, but then at some point the line came and established the presence it still has. The first form was moving from $\Pi^0 \Rightarrow \Pi$. Again I wish to press the issue, at that stage form was in use and not mathematics. The Universe was just simply too big to measure. If radius did apply, one could use r and r2 but since only Π was in use there was no radius to be used. There is forever one circle leading to the next circle, which is followed by the following circles. Where the light does not reflect the image that is there, we will still find a concentration of circles leading another and another and another. The end is eventually endless.

When working with concrete and heavy metalled solid objects r would show as a crack distinctly parting solid structures, while Π indicate a continuous flow of solidness giving the material an overall and continuous structural strength, yet engineering never recognised this difference. By confirming Π the circle employs singularity in all components and therefore proves to be a much stronger support as building choice than other shapes.

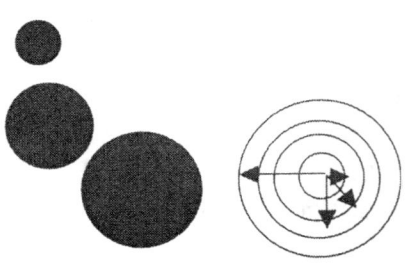

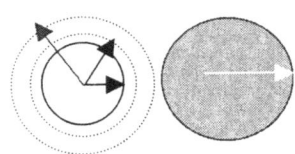

In order to understand the difference there are between Π forming g gravity and a radius, think of the difference there is in working performance between air pressure and oil pressure. Pneumatics use r as a pressure indicator and hydraulics use Π therefore air can compress and liquids cannot but act as the toughest solid found specifically because of it uses a relevance in the applying of Π and not r bringing conformity evenly.

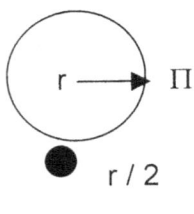

r $\longrightarrow$ Π

● r / 2

• r / 2

. r / 2

The circle to the left would come about from a straight line r growing influencing the appreciation of Π, but to influence Π would lead to a breakdown in r as Π and r are different entities. The circles to the right shows a continuous growth by extending Π every time and since Π is the same part as the previous Π, only extending that billionth of a millimetre each time, the circle will be truly continuous without any signs of a break. When **the circle reduces**, the **value** located to **r** will become implicated because **r determines specific size. Not so** in the **case of Π**, because Π in the true sense only **indicate that the circle is a square without corners** and therefore Π **dictates form and not size.** By **reducing size** only r **comes into contest** and will point to such reduction. By **reducing** the circle

radius r by half continuously will lead to an **infinite small circle** but Π **will remain because the circle as a form remains** even being infinitely small

The spot forms a full circle, but the line running through the circle is forever present because that is the future radius of the circle that will one day develop the circle, which is equal to the present diameter. The fact of the presence of such a possible line in such a possible circle dividing the possible circle into two parts makes the centre line equal to the half circle. The line forms the half circle but not only that the line presents the half circle as much as the line is the half circle. The line then is 180^0 and the half circle is 180^0 because in singularity the two factors are the same. The same value is of course $\Pi^0 = 1$.

In this half circle of the future, which is no half circle as yet because of a lack of space there are three future points indicating the space less ness that will go on to become space filled with something. On top of such a circle to form must be a marker indicating an awaiting boundary or future border and at the bottom of the future circle there also must be a similar marker that is no marker as yet. Between the two possible points that are not there yet is a future line running that is not there yet. Then indicating the possibility of a position to come that will bring about the half circle being a future distance apart from the future line indicating a diameter that will one day be there. A third such a marker must be established for the future. That forms a triangle with two more sides being connected by either a line being one or half pi being one. From singularity comes about that the line is the same as the half circle is the same as the triangle and all has one value being 180^0. From this come the most basic principles in as much as forming the ground rules of the law of Pythagoras.

 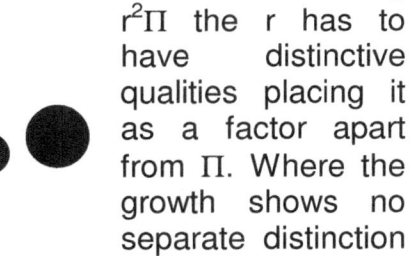

In the circle using $r^2\Pi$ the r has to have distinctive qualities placing it as a factor apart from Π. Where the growth shows no separate distinction but a continuous flow from the precise centre to the precise edge the flow would become in relation with Π depicting the circle and Π replacing r as reference to any point on the circle. By using r distinction in the circle is possible but by using Π there is no distinction possible.

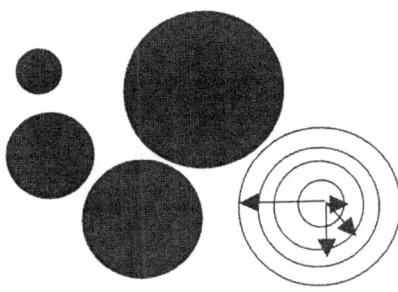

When working with concrete and heavy metalled solid objects r would show as a crack distinctly parting solid structures, while Π indicate a continuous flow of solidness giving the material an overall and continuous structural strength, yet engineering never recognised this difference. By confirming Π the circle employs singularity in all components and therefore proves to be a much stronger support as building choice than other shapes.

Everything at the time that was outside singularity and was in form at that time was equal. Think how big they were. They filled larger parts of the Universe than our brains can cover by thought. They formed the holding tanks that still hold us in the massive Universe. They were at the time too small to have size, but since then they grew into structures that are too big to have size. Those dots still are bigger than mathematics can apply because there still is no measure quantifiable mathematics can reach. Just because they compare with what we seem to preserve as small in cosmic relation they are too big and too large for us to comprehend. Even if they were immeasurably many they filled an immeasurable Universe in the same way they still fill the immeasurable Universe and we are so small we and our surroundings are measurable and quantifiable. They were so enormous there were no relevancies applying to compensate for distinguishing. Distinguishing only followed later when size started to matter. That meant the Coanda effect was in place without the Roche or the Titius Bode law. It was the start of the relevancy principle from which the atom later came and which is the result of the Kepler expression.

It eventually gets so small we humans can fit into it... remember we are seeing the reverse of the truth. It was never so big that it contained nothing because all that and we came afterwards being smaller than the dot that filled the dot.

With no line possible there had to be another dot that formed since the Universe has many dots that formed lines. But let us not to get confused and lost in the range of possible diversions but let us stick to two dots. One dot was next to the dot next to the dot, but as I said, we stick to one dot next to the second dot Π is the first step where gravity began with. That leaves us with a huge problem in as much as when r = 0 then r^0 = 0 and 0 dividing any value will leave 0 as the answer. If the particles were inseparable at the start it must bring about that gravity would not be forming since the distance will not permit any dividing. By allowing the distance separating the particles to be zero, the particles melt into a unit. Again this is Mathematics and not my incoherency as some Academics dismissed my work. Let me run through the argument one more time because I have been insulted by Academics in the past telling me I am bending mathematic rules with my applying double values to try and produce some argument. The two particles formed by an inseparable unit separated by a sharing of a spot. We know that at least two spots formed because there are many more than just two that remained to become part of the visual Universe. Let us name the spots because that is what humans do best if they do not know what to do with what they have to do. Let us call the on dot and the other one spot next to dot we then call dot two. Between dot and dot two there were nothing because dot and dot two were inseparable. By they're being inseparable we would naturally be inclined to think that the separation value should be nothing or at least zero. But putting zero in that place is a mathematical excluding procedure leaving future mathematics excluded. With m multiplying m_2 and then dividing ÷ r with zero (r=0) such a procedure will leave the lot at zero and with that nothing is going nowhere. That means although we think the space between the two parts are nothing the non-existing space has to be at least one to be a future factor.

At fist Π^0 =Π. Then after a while $\Pi^0 = \Pi^3 / \Pi^2 \Pi$ and gravity comes about forming space-time by motion of form. Being the sphere that formed the 7 holding relevance in the form of the sphere took shape. But the Universe is layer in dimension forming the next layer in dimension forming the following layer in dimension. The Universe was $\Pi = \Pi^3 / \Pi^2$, which is taken from Kepler's formula he received from the cosmos as $k = a^3 / T^2$. Where there is a sphere involved there is a natural tendency to grow by developing the sphere.

Every part of the argument is sound but was never yet used. I repeat once more if my argument reflects on inconsistencies those inconsistencies are not about my work. In order to disprove my argument replace Mass one and Mass two with any number possible, then divide such a number with to the square being zero. If there was no space then the value of the particles had to be one. If there was no space between the particles the particles then had to form a unit. But if there is a mathematical possibility of reducing a line to the single dimension then there had to be a factor representing r as a factor of one. Take $(M_1 \times m_2) / r^2$ and substitute any of the factors with zero and the result coming about has to be zero. The factors in the equation have to have any and all the elements at a value of at least one. Only if r was a factor of one can gravity bring about any mathematical equation developing from this argument. That means the mass on both sides must have a factor of one being a limit, which does not allow such further reduction of r and any further reducing of r beyond the limit will not be tolerated. Only if r = 1 then r^2 can be 1 and mass can be apart. Like it or not but believing in the Big Bang must also bring about the accepting that the cosmos moved apart somewhat. The fact that r brought increase in the space separating the mass produces a problem that was solved already. About a century and a half ago Roche found just such a limit. Once again I were confronted by zero becoming growth. There is a huge hole that needs filling when bringing into a relation any forming of an alliance between a cosmos coming from nothing and filling with nothing and a cosmos growing spontaneously through balance shifting prominence. Mathematically the fact of applying nothing as a vale applying in the cosmos is not a strong and convincing argument. The minute one brings in zero as a multiplying factor forming a definite value working into the calculations of the cosmos, growth disappear. If growth was not a factor, the zero factors could be involved with some form of maintaining stability and where then further growth will accept the responsibility of zero.

Any point will be opposing itself within the **rotating of 180°** where it **then change every aspect** of its **previous flowing** characteristics it had or **will once again have** in 360⁰ from there. While in rotation from the view point of a bystander it all may seem static and never changing but to the object in spin every next instant in time will be diverting from every aspect it had every second passing, and the direction it held in relation to the direction it held the previous mille, mille second will totally be incompatible with the direction it holds the very next mille, mille second of rotation. This is why we can use degrees measuring the circle by (6^2) (forming the square relating to matter through singularity) X 10 (square if space) = 360⁰ however it is always in motion. That proves no point can be static or constant, though it may seem that way to outsiders. Although matter is matter, matter can also be anti-matter and moreover form its own anti-matter at the same time. This degeneration of structure is very likely to occur with overheating. Revaluing Π to Π^2 will bring about a new contact point where Π meets **r** forming another relation in Π^2 **Time is** the **changes in relation** where Π

contacts a different r not withstanding the many r points there may form because **every r constitutes a different value** to the universe through other ratios and relevancies brought about **by heat and light. Time is the duration it takes Π to rotate between any two given points of r** and therefore must always amount to **a square (T^2)** moving from point to point through the **cube of space (a^3)** in that **duration of time (k)**. With that it proves **Kepler's a^3 (space) $=T^2$ k (time in the instant of motion)** but motion must continue through a specific value in space where the space-time is maintaining relevant equilibriums throughout singularity connecting.

In the circle using $r^2\Pi$ the r has to have distinctive qualities placing it as a factor apart from Π. Where the growth shows no separate distinction but a continuous flow from the precise centre to the precise edge the flow would become in relation with Π depicting the circle and Π replacing r as reference to any point on the circle. By using r distinction in the circle is possible but by using Π there is no distinction possible.

The sphere is 7 X Π = but from the other view the sphere is singularity relating to ten. That starts the Titius Bode principal which in relation to the Roche limit at the limit of Π use $\Pi^2 / 4$ to form gravity Π^2. However this is a little more involved, as it seems because

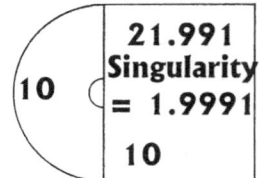

the one sprouted because of the consequence of the other bringing the one into the Universe as a relative that will sprout to bring about gravity.

However if the attachment was $7 + 10 + \Pi^2 / 2 = 21.93 / 7 = \Pi$ is the circle that serves as an attachment meant being a sphere and holding as well as sharing singularity. This is what Newton saw, but this is not gravity in the cosmic sense.

7 A sphere being formed by the six ends crossing as it incorporates the centre singularity
+ 10 anther sphere with an identifiable motion keeping the independent singularity apart but within the relevancy of the unit formed.
+ $\Pi^2 / 2$ Singularity by gravity shared by two in one unit.
= 21.93 / 7 still holding the unit to form where the overall containing form will be a cosmic sphere.

The sphere holds many dimensions relating to seven but also by the square of space which is 5+5 = 10. This brings about gravity generated by means of the Titius Bode law and is principle proof of this statement because it indicates an infinite number of numerical positions influenced by quarterly divided sectors around a point holding singularity.

Boys playing games will never realize scientific breakthrough explaining and grown ups do not play with toys. In this little toy played everywhere everyday by almost every one is the answer most brilliant of human Brainpower seek answers about all the cosmic riddles no one seem to understand.

Newton said the rotation delivers no work and therefore the effort of the rotation results in a zero. Firstly it bring us back to the zero idea where with all the reasoning in the world and all the leniency I allow I cannot find zero as a value being part of mathematics. Let's move back to the circle to try and find the zero Newton saw in the rotation.

$\Pi \times r^2 = CIRCLE$

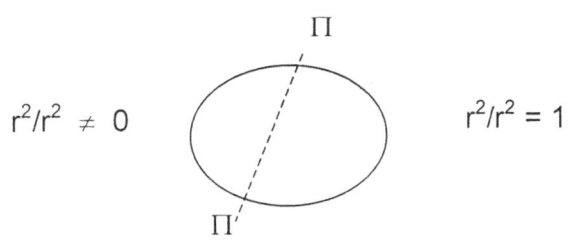

$r^2/r^2 \neq 0$ $r^2/r^2 = 1$

If you remove r it then is $\Pi \times r^2 / r^2 = CIRCLE$.

You cannot then say $r^2/r^2 = 0$ and therefore $\Pi \times 0 = 0$. That is nonsense. $\Pi r^2/r^2$ will always be $\Pi \times 1$, and that is the eternal circle. When looking at any rotating object, there has to be a point of no rotation and no rotation means "no rotation", not no existence. No rotation means a factor of 1, not zero.

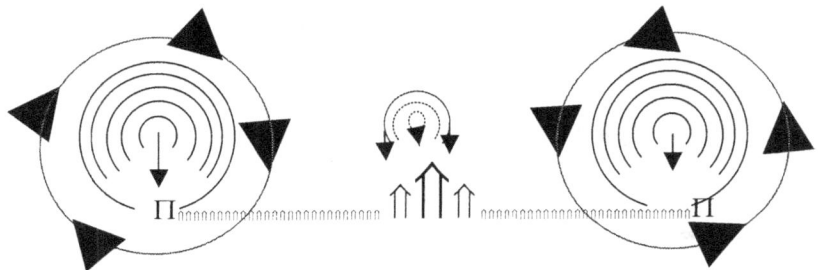

Not only does atomic individual singularity maintain self preservation, but in doing that it also sustain a governing singularity holding structural composition and form within a cluster of matter for example a star. As there is between stars so there are in the same manner a mutual or bonding singularity between atoms in stars, which we see as fusion.

Any object in rotation will have a middle point, a very specific centre point that does not spin. That point once again hypothetical but none the less must be standing still because every line running from that pint in opposing directions are also in opposing directional spin to each other

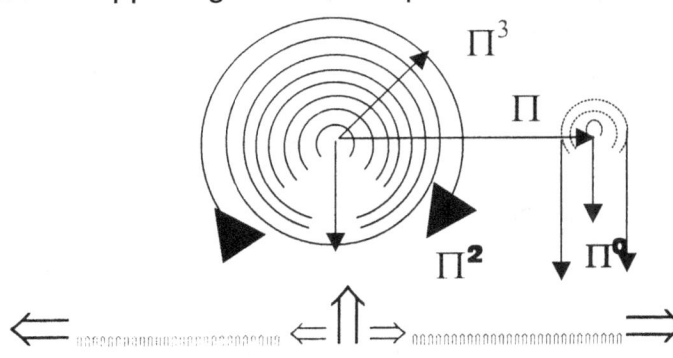

As the stop starts to spin the motion establish the centre line, which activates singularity, which activates space-time that activates gravity at a specific relevancy. Where we locate singularity there is not nothing because gravity cannot come from nothing because only nothing comes from nothing.

After all it is gravity that keeps the top as it is spinning in an upright position while it is spinning because it is gravity that stabilises the cosmos. Moreover, what is actually in progress from the top spinning is the Coanda principle activating gravity and that happens in accordance with Kepler's formula

This means that in the cube at the point of contact between the cube and the sphere the cube experience such a contact point as if the "bottom falls out" of the cube and without a "bottom" to support objects they fall to the sphere as objects does fall to the Earth. Remember that a body "floats" in space, but at one specific point it starts to "fall" to the Earth. That is gravity and it is a dimension change much more than any force. I shall explain this last remark later on. That too is the Lagrangian system with five cosmic structures holding relevancy to the centre structure where the centre structure stands in for seven positions diverting from singularity and the orbiting structures standing in for five positions in space.

In the centre runs an axis line that forms the division of rotation. No one human will ever be able to indicate the precise line, but such a line must exist because of our logic telling us about such a line. In the centre one will always find one more line smaller than the outside but forever also always bigger as it is towards the inside.

From such a point every other point will be opposing any other point not pointing in the direction to which the first point is pointing, whereby it extends the direction it holds. No matter what the point is or where the point leads, such a point holding a specific direction will be unique in the direction it is rotating because at that or any other specific point wherever, it will be directing not in the direction it spins but in the direction flowing from the centre point outwards.

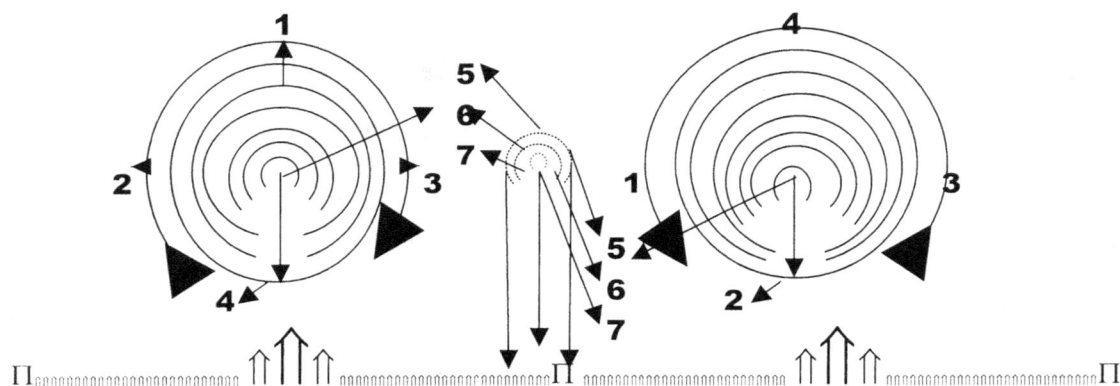

Any point will be it opposing itself within the rotating of 180° changing every aspect of its previous flowing characteristics it previously had or will once again have in 180° from there. While in rotation from the point of an outside observer all may seem static and never changing but to the object in spin every next second will be a diverting from every aspect it was in very second passing, and the direction it held in relation to the direction it held the previous mille, mille second will totally be incompatible with the direction it holds the very next mille, mille second of rotation. That proves no point can be static or constant, all though it may seem that way to outsiders.

In the very centre of the sphere the form of the sphere dictates that the shape will relinquish space as the line run from the outside towards the very centre. With this natural state of affairs the sphere are naturally inclined to dismiss all space that it can form in the form as the sphere holds space inside and the form will finally be without dimension. All that I attribute to the line shrinking by reducing actually takes pace in every sphere as the diameter reduces to the centre. In the centre where the radius line goes single the form relinquish the three dimensional form it has inside. Being without dimension in the very centre means that at a point in the extreme centre of all spheres there are a point that holds singularity because this point with no space has a mathematical position although it is invisible since there is no sides to such a point to give that point any dimensions. The shape of the sphere is calculated by using the formula $4\Pi (r^3) / 3$. By reducing r to a point where r is r^0 singularity steps in because only the form remains as Π. Going even further we find that there then comes a point where Π goes singular Π^0. At that point absolute singularity is present but so is absolute gravity present at that point. When holding the strength of the shape of the sphere in mind as well as taking into account that all cosmos objects of importance is in the form of planets or stars and they are all in the form of a sphere, we therefore may contemplate that it is where gravity originate. We now only have to find the reason why gravity will hold a base in a space less ness as Einstein predicted. It is clear to be seen that gravity is in the centre of the sphere controlling from the centre everything that is outside the space less centre. We can reason with confidence that gravity is the strongest where space is the least. We can further reason that it is gravity that is holding the sphere in true form and since the sphere allow gravity the best working opportunity, gravity can form the sphere in as strong a shape and form as the sphere seems to have. From every point on the surface of the sphere is where that point connects with the other side of the surface of the sphere by a line that runs through the space less ness of such a centre of the sphere. Such a line also connect by an angle of 180^0 as well as 90^0 to six other lines running from top to bottom, right to left, and back to front, where all join and cross in the centre of the sphere. There are therefore six lines crossing and connecting by a centre from any given point on the surface of the sphere. Such points connects in total six surface points on each side of the sphere while they all support one another through the space less centre. In that absolute space less ness in the centre holding singularity we find gravity supporting and controlling all space within the sphere as well as space connected to the sphere. That is where gravity control and guide the space, which falls in the parameters as well as under the influence of the form of the sphere. In the gravity centre space goes singular meaning space becomes space less or flat.

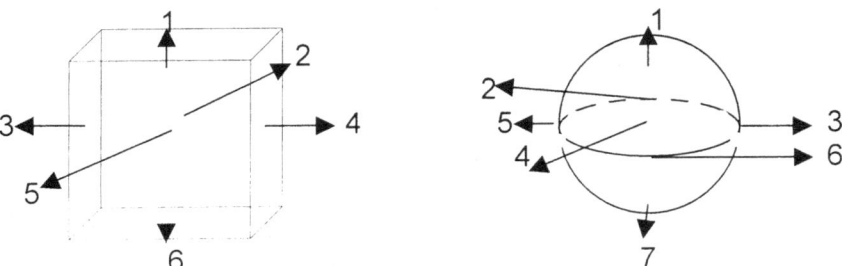

Also it is true that the entire form that is the sphere is controlled from a centre within the sphere. That centre holds the sphere in form and shape. Therefore the strong form is dictated from that space fewer centres where there is no space and no form left. The natural inclining is in the form of the sphere. It is part of the roundness that the overall shape of the sphere represents and this structural strength is carrying down to the very centre. Because the circle is forever reducing that reducing which is inherently part of the form of the sphere becomes a tool in distorting of space in the sphere and is eventually removing all forms of space from within the centre of the sphere. The very centre ends up as having no space because of the reducing that continuous down to become the space less inner centre. The all roundness is the ingredient that forms the backbone of the absolute strength that the sphere has and that is the component that the sphere is so famous for. The form the sphere has allows the sphere to have a control that is coming from the centre deep inside the sphere where the space vanishes and being without space seems to keep the entire structure rigged. The strength of the sphere comes from the centre of the sphere, which is inherent of the shape. That is why the sphere has such and the fact that all connecting sides refer to a centre brings the strength that the shape has. How does it work in its most basic analyses?

It is from the layout that the sphere uses as a natural form that we are able to locate singularity. In the case of the sphere the material naturally reduces by measure of the radius becoming smaller to a point where the radius is r^0. At that point the line that will form the radius has gone single dimensional r^0 and that is equal to 1^0, which is singularity.

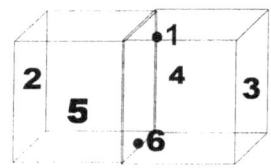
There is one more point in the sphere in the centre forming an addition in the sphere. That point holds gravity secure.
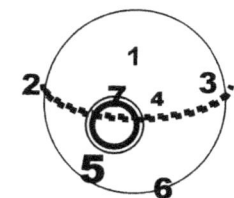

The cube has sides and the sides form a rather weak and flat surface that connects four corners. The flat surface produces a rather indifferent contact point with no special features on the surface. The corners connect to other sets of corners and those corners form a weak structure without any direct support coming from the other five sides. Without material to fill the body of the cube the cube has no direct connecting between any of the sides other than corners connecting at the edges of the sides.

Taking the vantage from the point the sphere is holding from the centre out into space there are ten points connecting to the centre. In that are the dimensions of singularity connecting to space where five connects to space in the second dimension of singularity, and five connects in the third dimension of singularity. On the other hand, the cube does show a very different characteristic, which involves only six sides (at least) connected.

In the very centre of the sphere the form dictates that the shape will relinquish all grounds in space that it can hold and the form will finally be without dimension. Being without dimension means that at a point in the extreme centre of all spheres there is a point that holds singularity because this point with no space has a mathematical position although it is invisible since there are no sides to such a point to give that point any dimensions. When holding the strength of the shape of the sphere in mind as well as taking into account that all cosmos objects of importance are in the form of planets or stars and they are all using the form of a sphere, we therefore may contemplate that it is where gravity originates. We now only have to find the reason why gravity will hold a base in a space less ness as Einstein predicted. It is clear to be seen that gravity is in the centre of the sphere controlling from the centre everything that is outside the space less centre. We can reason with confidence that gravity is the strongest where space is the least. We can further reason that it is gravity that is holding the sphere in true form and since the sphere allows gravity the best working opportunity, gravity can form the sphere in as strong a shape and form as the sphere seems to have. From every point on the surface of the sphere is where that point connects with the other side of the surface of the sphere. All other possible points connect by a line that runs through the space less ness of such a centre of the sphere. Such a line also connects by an angle of 180^0 as well as 90^0 to six other lines running from top to bottom, right to left, and back to front, where all join and cross in the centre of the sphere. There are therefore always no less than six lines crossing and connecting by a centre from any given point on the surface of the sphere. Such points connects in total six surface points on each side of the sphere while they all support one another through the space less centre. In that absolute space less ness in the centre holding singularity we find gravity supporting and controlling all space within the sphere as well as space connected to the sphere. That is where gravity control and guide the space, which falls in the parameters as well as under the influence of the form of the sphere. In the gravity centre space goes singular meaning space becomes space less or flat. That is where Einstein's Universe goes flat because that is where gravity is at its strongest. However my bringing up this statement brings me directly to the point where I get very confrontational about how the brilliant mathematicians treat those they suspect are less inclined to think.

By examining the form of the sphere, we find that there are 6 points on the surface of the sphere that is holding the form at a specific and equal distance from the centre. Lines run from the centre into space at $90°$ and $180°$ angles of each other from six opposing sides. There then are six lines at $90°$ and $180°$ connecting to the centre from six points on the outside edge of the sphere. As a result of the basic shape that a sphere has, there is a spot in the extreme inner centre of the sphere where the lines in $90°$ relevance cross each other and others connect by $180°$. There is also at that point a spot where all space relinquishes a position and only singularity 1^0 as form remains. At such a point we find the measure of the sphere being Πr^0 with $r^0 = 1^0$. That is where the line that represents the radius as a line disappears, as it becomes singularity r^0. After more reducing continue we get to such a point where we find only Π^0 left. At that extreme point is where space in all form disappears, as the circle providing the sphere the form the sphere has, removes all possible form by going into singularity $\Pi^0 = 1^0$.

Then in that area all form of any possible space disappeared leaving only the dimensions of singularity 1^0. I cannot delve deeper into the argument. However, from such a point there runs lines that connect to space on the outside where six points on the outside points connect to the space less point in the inside. In this book I take this argument much further but for now I leave the argument at that. Those lines carry the structural strength the sphere has. Contact with one point has support of six other points across the whole structure where the other six support every one of the six by singularity and the support runs through the entire sphere including the middle. Where there is no space, there must be singularity 1^0 just because the space filled with material removes zero and only material filled space is present. That means material fills the lot although in singularity 1^0. If zero was a factor where all space finally halted in zero as the value, then zero would be able to remove the space from the centre and such removing would continue to remove the space until all space was removed. It will finally abolish all space in the sphere and it would remove the sphere. Zero removes all possibilities of anything coming about. Since the sphere is there, a zero factor in the centre cannot be present. Only infinity can be a factor from where space may grow because infinity can extend and grow into and up to eternity.

The implication of this is that following the line down to the centre of the sphere we located the centre of the Universe. That is where gravity is. There is a lot more to that but be patient, we are getting there. In every centre we find a point, which is in truth not there but is the mainstay of all that is within the sphere. The mathematical value of such a point is $\Pi^0 r^0 = 1^0$ and 1^0 is singularity. That is the point where the Universe started and that is where the Universe will finally end. That is the Universe without space-time. That is $k^0 = a^3 / T^2 k$ which proves the Universe is without doubt a sphere...and we just located the centre of the Universe!

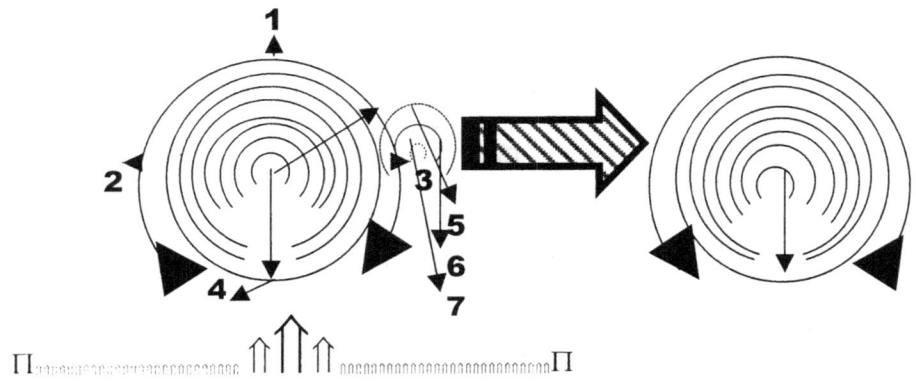

As one can see with the spinning top delivering the Coanda principle, every point overheating can spawn space-time by centralising singularity.

One can see from the top that singularity is established wherever spin occur. The motion generates a position of seven in relation to ten and singularity manifests as 1.9991 as is explained elsewhere. That means any point formed by the sphere spinning can and does start a centre in which no motion holds no space and of which motion surrounds such a point by forming space. Although everything at the time was in the form as a multiple circle, which results in a sphere, the sphere was not the only form present. This too has to do with singularity interpretations. We see a cube, as we know the cube but at first when form came about the cube were not yet a form.

While the one sphere forms on this spot where the dominating sphere secures an edge the dot may be reserved as an edge marker to the dominating sphere. To the forming sphere in progress of emerging heat gathers at that point because the rotation is a result of duplicating and duplicating is the tendency of naturally growing in space-time **$k = a^3 / T^2$.** In order to find duplicating coming about there has to be heat in order to duplicate what will form heat. The duplicating process is a process of one factor going softer or less solid and therefore more dynamic than the other. To have singularity is to have gravity but to have gravity there has to be a point of motion and a point of sturdiness. The point of sturdy may be in the centre of singularity, but then the solid must be motion. However even today it still apply: what moves forms liquid in the presence of a solid and at that point singularity presented the solid therefore what we might think of as solid was the liquid because it moved around the solid. Where the one factor is duplicating the other factor is compressing $k^{-1} = T^2 / a^3$

The points duplicating is four moving around a centre by the square of gravity.

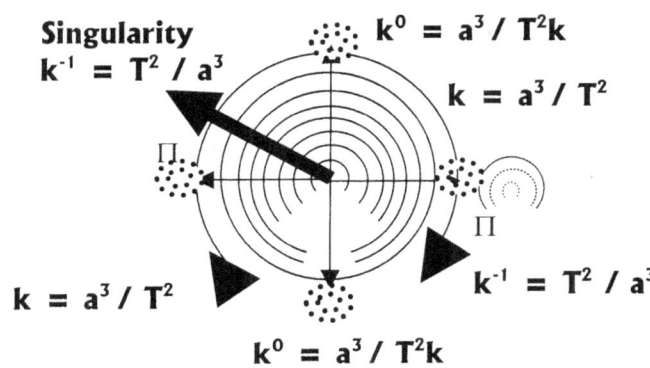

Singularity
$k^{-1} = T^2 / a^3$

$k^0 = a^3 / T^2k$

$k = a^3 / T^2$

Π

Π

$k = a^3 / T^2$

$k^{-1} = T^2 / a^3$

$k^0 = a^3 / T^2k$

The motion is the sources of heating because the heat is bringing about the movement. The heat growth therefore provides the action because the action is what energises the points to provide the motion. The motion is purely is space-time duplicating and the duplicating is feeding heat to the centre from the four points overheating thus the points that shows expanding.

But also the duplication leads to the spawning of one point of singularity that provides the installing of the next centre for the next sphere.

Because of the principal in which the Coanda works the motion will centralise a new sphere and by appointing six position around the centre three points will not move while four will move about the three points forming the centre line. The result is that the four points by duplication will reserve the point moving as the next point in singularity because of $k = a^3 / T^2$ singularity will be a natural result of the motion. Then that point will secure a position $k^{-1} = T^2 / a^3$ which will secure six points about such a centre. The centre will bring about four points spinning around three points holding a line singularity. The line in singularity will stand in relevance the contacting factor $k^{-1} = T^2 / a^3$ and the duplicating by expanding points will be four and serve the relevancy by contributing $k = a^3 / T^2$ as space-time only in form. From this the rest of the Universe burst into the next phase of Creation.

The gravity is in relation to the spin, which is in relation to the four points spinning which are $\Pi^2 / 2$ and that is the Roche limit. It is the dividing of singularity sharing space-time just as we on Earth share singularity by division between the Earth and us others that is not part of the Earth. The total that forms from the point that spawns is seven plus five plus pi square in division of four totalling twenty one that stands related to the first seven and once again another sphere formed. However this is an eternal relevancy that can never break.

Any object in rotation will have a middle point, a very specific centre point that does not spin. That point once again hypothetical but none the less must be standing still because every line running from that point in opposing directions are also in opposing directional spin to each other. Although the points had the same characteristics only seconds before, they oppose the characteristics it had just before and just after the very second in which they are and to which they relate by similar points also in rotation. Due to the spinning nature of such a point with all surrounding the point very varying second, the value of such a point can only be Π because of its constant changing.

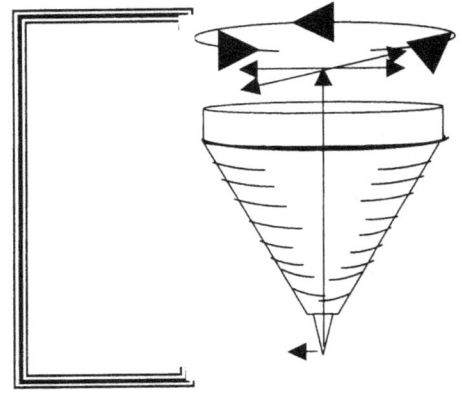

The sphere has seven points. The cube without truly being a cube but is just in consideration of having a cube in form holds five points to singularity. In the centre runs singularity to the value of Π^0, which means that which surround Π^0, holds a position of Π^2. The spinning sphere activates the seven points, which places gravity in relation to a centre. Outside the centre there are five sides by dimension. The sphere has seven points of which four is spinning. The four spinning stands related to the gravity of spin,

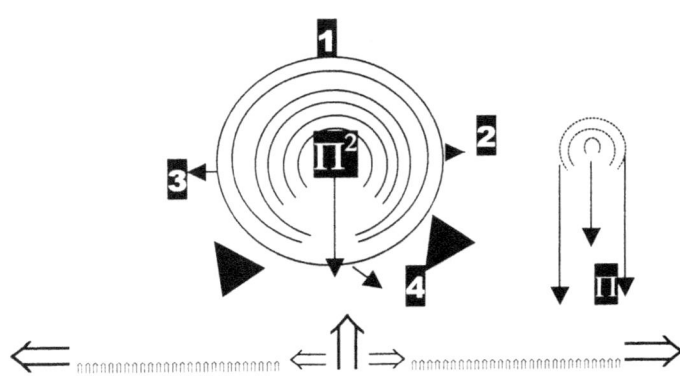

The extending of the motion also produces an ending point where the limit will either attach or dissolve the material next to the motion. Such a point liquefies what ever is in the location. That is the principle behind the Roche limit but also not only that; it is also the principle behind the Coanda effect. All that is closer than what the Roche limiting point at $\Pi^2/4$ will allow becomes dissolved as liquid and removed into the unit forming the sphere.

Using r would specifically oppose another r from every angle. From such a point every other point will be opposing any other point not pointing in the direction to which the first point is pointing, whereby it extends the direction it holds. No matter what the point is or where the point leads, such a point holding a specific direction will be unique in the direction it is rotating because at that or any other specific point wherever, it will be directing not in the direction it spins but in the direction flowing from the centre point outwards. Any point will be it opposing itself within the rotating of 180^0 changing every aspect of its previous flowing characteristics it previously had or will once again have in 180^0 from there. While in rotation from the point of an outside observer all may seem static and never changing but to the object in spin every next second will be a diverting from every aspect it was in very second passing, and the direction it held in relation to the direction it held the previous mille, mille second will totally be incompatible with the direction it holds the very next mille, mille second of rotation. That proves no point can be static or constant, all though it may seem that way to outsiders. Although matter is matter, matter can also be anti-matter at the same time.

At this stage time was still eternity being interrupted by infinity. To say the Universe is or was 13.5 X 10^9 years old is shear Newtonian thinking. Was it 13.5

$X \times 10^9$ years and how many days in the year of our Lord and what about all the years that passed since this date was revised? Time was flowing according to interruptions in eternity changing from what was to what is to what will be. Time is a norm that comes as things in the Universe change about things that are places around and scattered throughout the Universe. We may presume time at this point somewhere became a factor since sphere sprouted from points on sphere edges and differentiation in development came in place. Considering the role that the Roche limit played one can see how points in singularity grew from contraction and secured ever stronger centres by divulging hear points within the realm of singularity control. When a point in form developed at a position that was close than the original Π^0 to Π, the singularity in control took control.

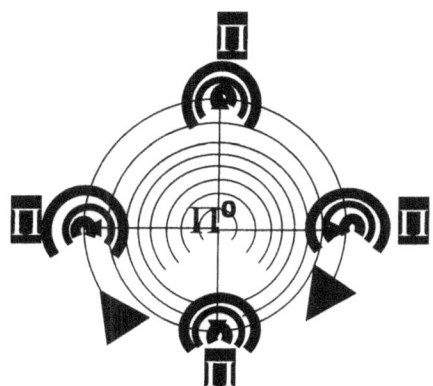

With every one of the four points taking position on the side of Π^0 running to the allocated position of singularity extending that forms the value of Π at a measure of $\Pi /2$ each brought about the Roche value of $\Pi^2 /4$ in relation to the developing centre. One has to remember that the star of today takes on the characteristics of the form of that era.

Consider what happens to a star that developed closed than the Roche limit of Π to $\Pi^2 / 4$ would allow, it is easy to see how the singularity centred grew by concentrating the heat the points in singularity brought about

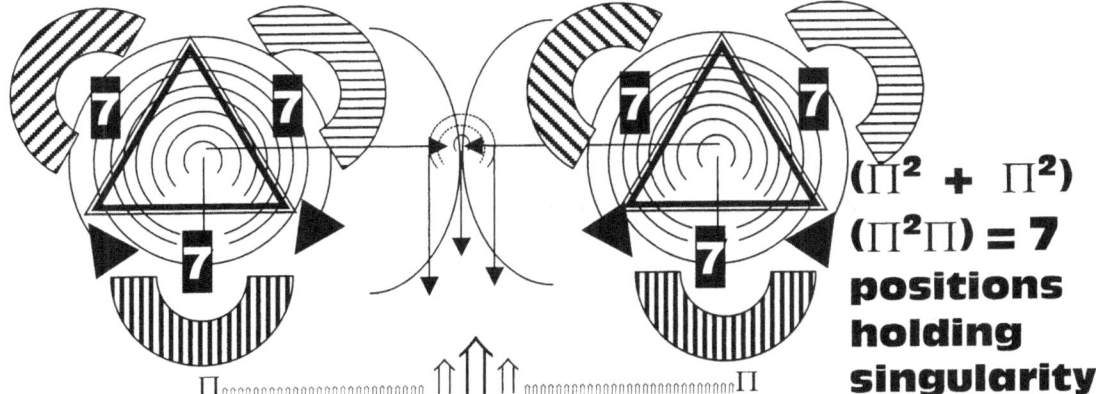

$(\Pi^2 + \Pi^2)$
$(\Pi^2 \Pi) = 7$
positions holding singularity

$(\Pi^2 + \Pi^2)$ $(\Pi^2 \Pi)$ 3. =1836, which after wards the atom was about and the Big Bang proceeded

I again wish to repeat the centre in the sphere and the centre of the sphere because in this is where the realisation comes from how the Cosmos started. If there were gravity at the very first instant then there was a sphere at the very first instant because gravity can only be in the sphere because of what the sphere represents. If that which extends from singularity does not meet the form of singularity by measure of $\Pi \Rightarrow \Pi^2$ in precise duplication it will tend to destruct and that we call imbalanced spin.

The truth about gravity in the cosmos and that gravity will contract the cosmos one day is the fact that the sphere has singularity as a natural substance. The entire form that is the sphere is controlled from a centre within the sphere. That centre holds the sphere in form and shape. Therefore the strong form is dictated from that space fewer centres where there is no space and no form left. The natural inclining is in the form of the sphere. It is part of the roundness that the overall shape of the sphere represents and this structural strength is carrying down to the very centre. Because the circle is forever reducing that reducing which is inherently part of the form of the sphere becomes a tool in distorting of space in the sphere and is eventually removing all forms of space from within the centre of the sphere. The very centre ends up as having no space because of the reducing that continuous down to become the space less inner centre. The all roundness is the ingredient that forms the backbone of the absolute strength that the sphere has and that is the component that the sphere is so famous for. The form the sphere has allows the sphere to have a control that is coming from the centre deep inside the sphere where the space vanishes and being without space seems to keep the entire structure rigged. From the centre the sphere shape shows strength that the shape as tough as it is. How does it work in its most basic analyses?

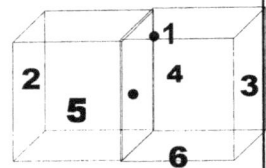

> **There is one more point in the sphere in the centre forming an addition in the sphere. That point holds gravity secure.**

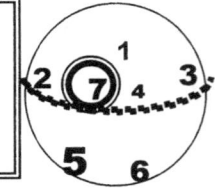

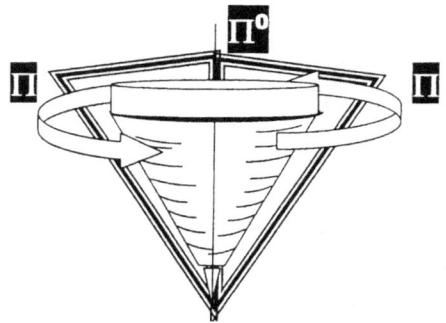

If the cosmos had any other shape the might be available contraction in the end as the final conclusion would then not be possible. Only in the sphere and more so in the circles all forming a sphere is singularity present. Singularity comes as part of the construction we find in the sphere. By singularity forming the base of the sphere the Coanda effect will forever be present and with the Coanda effect gravity applies.

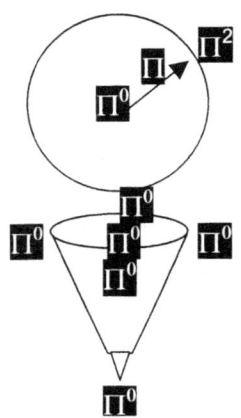

As is evident from the experiment of the Coanda effect there will always be an attraction by singularity controlling the space-time from the centre of the sphere. Such control has the name of gravity, but it is no force. It is a combining of liquid (10) and solid (7) in the presence of three in time that will produce such control by motion. But why is the Coanda effect arose by rotation and moreover by motion?

It is because one Universe arises with opposing qualities where the one attract as the other repulse. It is in the motion that the action if gravity is vested.

The spinning of Π^0 around the centre Π^0 establishes Π and Π is what produces the singularity. Singularity is always present in the form the sphere presents however; in the sphere singularity presents no influence except in the case where the sphere starts to spin. In the spinning a line comes about which caries the term if an axis. The axis has never been acknowledged as the most vital piece of any particle in the Universe manly because the role that the axis plays has been neglected and subdued by the incorrect attributing Newton placed on the motion and the axis.

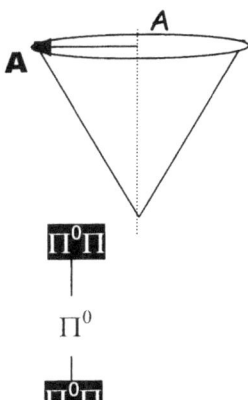

The moving of Π^0 to Π involved relegation and not motion as we consider motion. It was Π^0 getting a side and that is all. There was no true side but only a form that came into place. **Singularity (A)** received singularity (**A**) and no more of anything but the shift to comply with having a relevancy forming in relation to singularity. The dots had no sides, had no length or diameter. There was not measurable space or measurable time involved. The time could have been a micro, micro second as much a trillion millennium because time had no relevance. It was eternity interrupted by infinity, as it still is the case, however the line that eternity followed

was no line because there was no space to hold the line.

The line was interrupted by infinity, one there, there was The lines were not relations to sides being formed. momentarily however with no no one to notice. lines but

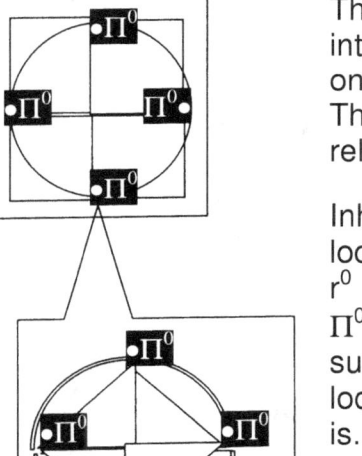

Inherent to the form the sphere offers, there is a specific location of singularity where the radius first goes single $r^0 = 1$ and then form goes into the realms of singularity $\Pi^0 r^0$. The cube also may have such a pint bur having such a point does not connect directly to six points located on the edges of the cube or any other form the is.

In relation to such a centre where $\Pi^0 r^0$ forms singularity there are always four cubes related to such a centre where the centre is part of seven points in total representing the sphere.

Every cube gas lost one side to a point of the sphere where the sphere takes control of form and removes one side of the cube. In relation to the time factor that is inherently part of singularity by the extending of singularity there are five sides connecting to four points standing related to singularity by the Π^0 factor and that gives 5 X 4 = 20. That is always directly in relation to seven points singularity offers.

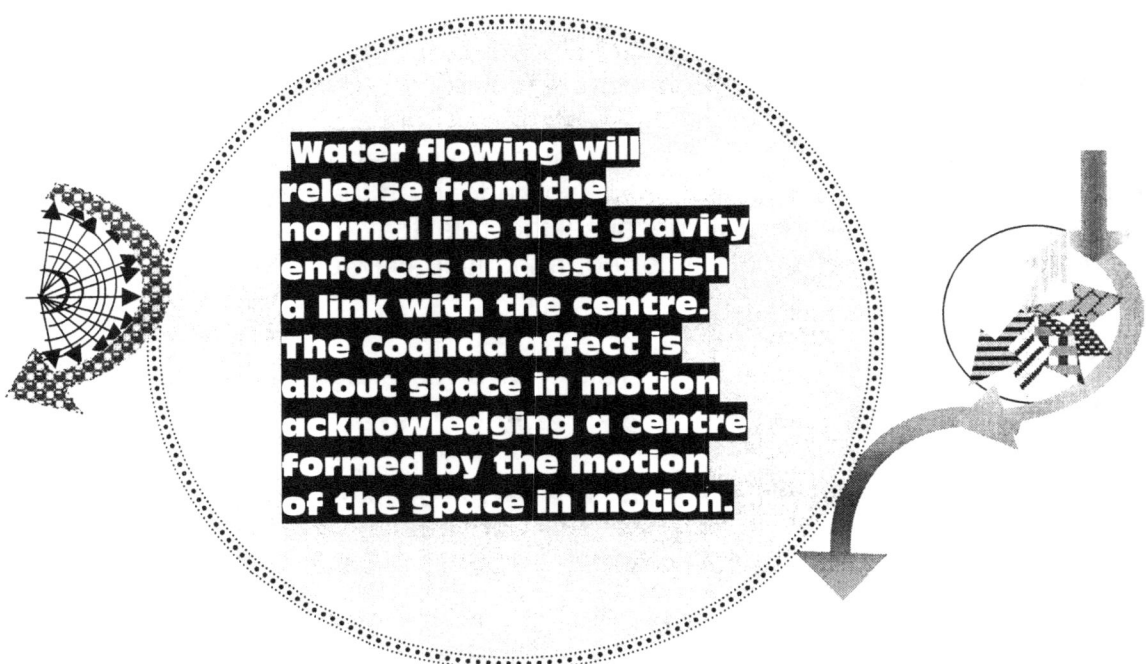

Water flowing will release from the normal line that gravity enforces and establish a link with the centre. The Coanda affect is about space in motion acknowledging a centre formed by the motion of the space in motion.

This should lead any person to investigate a centre that forms because evidently, there is a centre but that centre comes about by motion of rotating around a fixed point serving all points in motion. That leads us to the centre of everything in rotation because everything in the Universe is in rotation. As Kepler said the Universe is centred by $k^0 = a^3 / k T^2$ and we have to find $k^0 = 1$

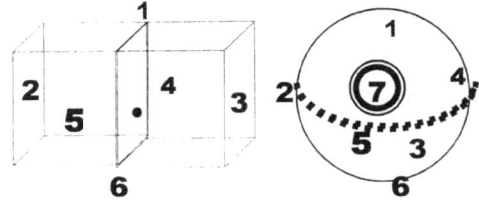

Kepler's formula also indicates that a sphere is within a cube that is holding a sphere at singularity $k^0 = a^3 / k T^2$ **with all the centres being all the same** $k^0 = 1$

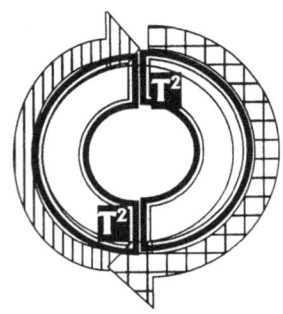

At this point Newton's second law come into affect. Motion by means of the Coanda effect introduced space as motion introduced time. For the first time ever time was interrupted when motion provided time the space to interrupt. From motion by the way of the Coanda principle gravity came about as a centre formed a point where motion surrounded space, By motion space-time was established in relation to singularity

If the universe did start from one single point and time matter and space flowed from that point, then that point must have a relative connecting base because such a point holding singularity must be eternal as space matter and time link eternal. There therefore must be one point linking the entire universe when regarding the fact of singularity. Then according to the theory off relativity there has to be one exact point holding time in a relevance notwithstanding the fact that time depart from that position and relate differently to all space-time away from such a point.

Every person I have discussed facts about creation recollects images in the trend depicted in a presentation as one may find to the above. That would be the most unlikely way Creation came in place. The recalling of pictures representing images about creation must have form, but to mathematics it had no form. From this thought the very opposite arise where Creation came from nothing but such an idea is mathematically simply not possible.

The thought of nothing is just what it is, a thought of nothing and although it is in the human mind common nature to present nothing as a value in the recalling of something, nothing is a presentation of the figment in the human mind. There can be no number such as nothing and that was (possibly) Newton's biggest error. Nothing represent non-existing and that is just what nothing is, it is non-existing.

In order to prove my point I wish to ask the reader to define the shortest line there can theoretically be. If he should answer anything but that the shortest line will be at a point where the beginning and is the very same spot he will be wrong. The shortest line that can ever be anywhere must have a start and finish holding the exact same spot. The line will be humanly impossible to create but we humans are capable of very little.

When the line has a beginning and an end at the very same spot and it wishes to extend the position as to further the possibility it has, which direction should it favour. Humans in the west would naturedly think of extending from left to right while in the east humans may want to go from right to left. Some persons will tend to go up or down, but all of the options are about human preference and not mathematical conclusions. Extending the line in any one direction will favour one direction without a conclusion about not extending in other directions. Such a conclusion has no sound mathematical foundation. The only option about extending will be in all directions equally in order to give a meaningful non-bias flow of mathematical equilibrium

The shortest line in the realm of possibilities must have a start and finish holding one spot and such a line will also be a dot or a circle. Not favouring one direction puts all directions at equilibrium meaning that any form what ever may be can develop from such a spot with the end and the start being the same. This reasoning prompted me to look for singularity in such a spot because if the prime spot from which all came was a spot, then the spot must hold the shortest line but more prominent it will hold the smallest form including the smallest circle.

One possibility that the shortest spot can never have is having a starting point on the zero mark. If the mark of zero holds the start it must also hold the end because the end and the beginning has the same position. If the position of zero then is the beginning, the end will also be zero leaving the line without an end as well as without a beginning.

While very line is circling bringing about time in space to the value of Π repeating Π to form Π^2 at the same time Π is extending in one specific centre to the value of Π^0 and only the spin value keeps Π not becoming r The spin keeps the immovability from becoming Π and maintaining Π^2 by performing duplication, but with any slightest reduction in spin reducing Π^2 to $\Pi^{2/4}$, Π will start extending and as one can see from the behaviour shown in the Roche limit, the heat will be concentrated at the centre and the singularity in the centre will grow four time in concentration. Only at points exceeding Π in diameter was time as Π^2 able to retain form and also grow. From that space slowly developed because at Π could Π^2 bring about a form which provided motion. In the centre there developed Π^2 and Π^2 kept all form at a safe distance of Π to bring about the needed solid immovable centre with the form $\Pi^2\Pi$ about the double Π could $\Pi^2 + \Pi^2$. That secured the makings of the atom by applying the Coanda principle of enticing gravity in the centre of motion, which then provides space-time by measure of $(\Pi^2 + \Pi^2)(\Pi^2\Pi)$. This totalled seven in dots and with three of those seven circling singularity at 1.9991 the atom came about. **I again wish to repeat the centre in the sphere and the centre of the sphere because in this is where the realisation comes from how the Cosmos started**. If there were gravity at the very first instant then there was a sphere at the very first instant because gravity can only be in the sphere because of what the sphere represents.

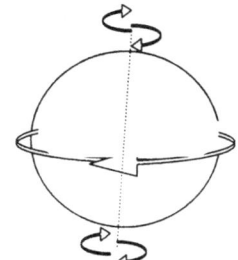

On the inside, there are the seven markers of which singularity is the focus point in the centre of the centre. The markers are representing one aspect of space, which for argument's sake let us call it cold. Then there are three more markers on either side being part of the space but not captured in the space. It is space in motion by the influence of the motion of the Earth.

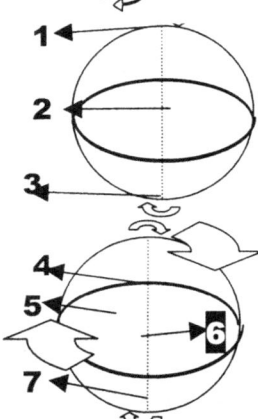

By not having motion the lines also have no space as the space extends to form space forms space and the line includes serving the three points to the outside. Where there is no motion, there is no space and where there is little motion, there is little space. The only space the line may relate to can be a point that is on the border of the sphere that is crossing singularity and connecting the two edges on either side of the sphere that is forming the sphere. That means the line from one point holding singularity to another point holding singularity that line will cross the centre line which gives the line in singularity valid space-time to control. Singularity does not have the ability of motion therefore singularity does not hold space. Singularity is also eternally indifferent to motion and motion can excite singularity but singularity cannot be shifted by motion.

Three points form the line.

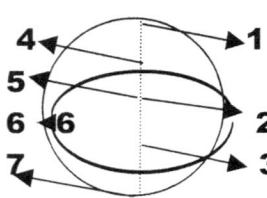

Every time motion takes place, the centre line holding positions 1, 2 and 3 stands still. There is little generating going on and reducing to a point where there is no generating going on. On the outer edge of the rim however there are four points that do shift. The points shift from one location to the next location by generating space

All this is happening while the crossing is all concerning singularity moving from one sector of singularity to the other sector of singularity which is $(\Pi/2) \times (\Pi/2)$.

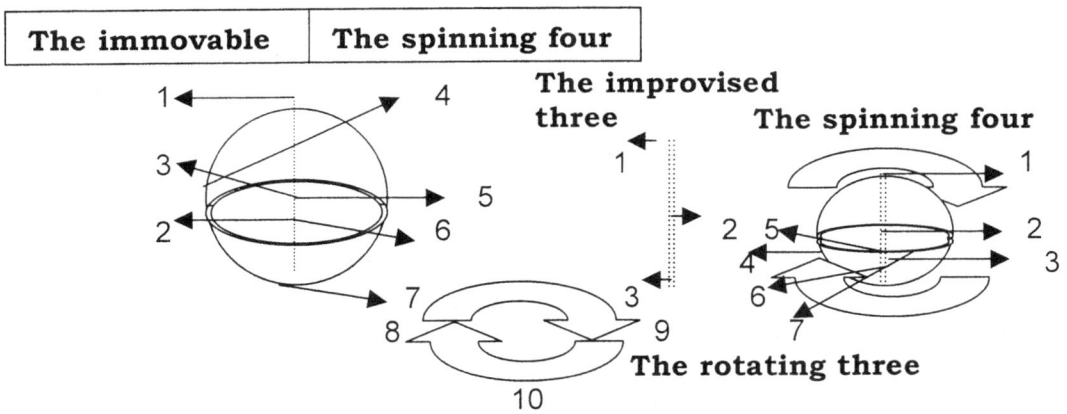

The relevancy forms part of the duplicating and dismissing displacement of space-time we call gravity. In that, we are looking at relevancies and no precise specifics. However, the Universe was built block by block in this manner. As it was but is no longer only form that applies in the Universe but concrete measurements also come into play therefore even the relevancies may apply in different relations as they switch over to compensate for other factors alternating as they are coming into prominence. The lesser developing sphere orbits the dominating sphere and between them, there are definitive relevancies. The centre circle singularity line of three is unaffected by spin which I shall call the immovable three. However, the immovable three holds such a stout position as far as the centre sphere is concerned. In relation to the orbiting circle the centre line is part of the building and destructing process that manifests as duplication as the centre singularity maintain domination and control over the orbiting structure. In addition, it has a major part in the motion building of the sphere in orbit and moreover building by generating the singularity line that generates the lesser and the orbiting sphere. In that relation the centre sphere reflects the centre line to serve the orbiting sphere by supplying the reference needed to establish motion in the orbiting sphere as one Unit. In that there is an undisputable reference of seven orbiting the centre and the centre providing three as a reflection of the seven which in all accounts for ten relating to the four which also is spinning as time and in total forms the seven taken in relation from the orbiting ten.

Points seen from the side form a line that never moves and is a line in singularity holding three points

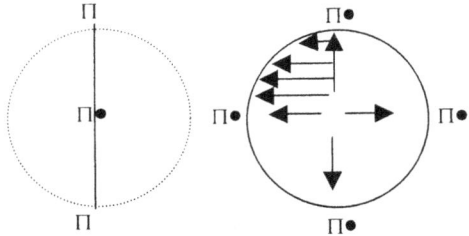

The seven can never totally separate from the ten, but by singularity being the same but being on the other side it is withdrawing space-time altogether. See it as seven (let us think of that as the cold basis of space) spinning or turning in the ten (which then will represent the hot part in the cold basis) and the ten is part of the seven but the seven is not part of the ten. The third factor is the axis around which hot as well as cold will turn. Therefore when reading the next page, please envisage a cold base turning in a hot and cold space. The purpose of this is not to define whether the argument is correct or not but it is to help the reader gain understanding of the process principles involved. But motion also converts space to relate to space by changing relevancies through motion matter is in relation (part of) to the total dimension of space but is not the total dimension of space.

Space-time is allocated in progress from the line singularity offer. That space-time

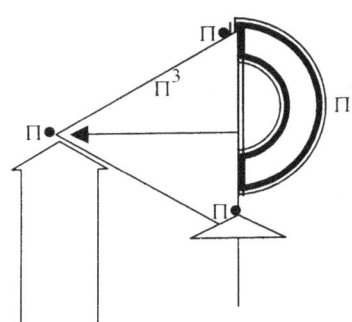

consists of heat and can therefore store more heat, which strands in contrast to the cold than singularity presents. By providing, more space-time at the equator there is more space-time to allocate to heat being there

In this one can clearly see that it is the motion that sets the top free and independent from the gravity of the Earth. By motion the top generates individual gravity that allows the top an individual gravity and that motion frees the top from the gravity by which the Earth restrains the top. The motion gives the top independence that the top immediately loses when the motion subsides. This fact is the utmost important issue of all physics in the Universe.

The top is one of (perhaps) the easiest and most common examples one can find to demonstrate the cosmic generating of gravity. By spinning a "force" which is no "force" keeps the top erect and it is only motion that accomplishes the act of gravity.

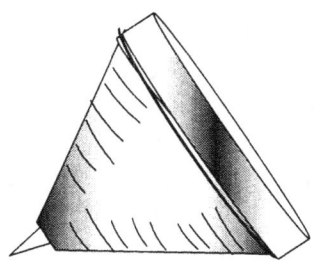

The top lying still holds the same singularity principle that the sphere holds because if the shape the top has. The roundness protects singularity at a seventh position deep inside. However the top is a dot Π or even going down to a spot Π^0 and is only by the form the material has which puts singularity in place.

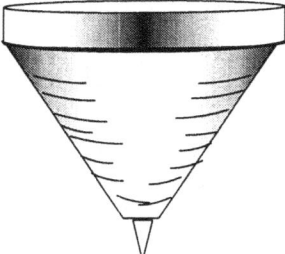

According to Newton it takes no effort $\dfrac{dJ}{dt} = 0$ to get the top from where the top was motionless to where the top is spinning. I say this on the work that Newton suggested comes about from the effort it takes the top to circle.

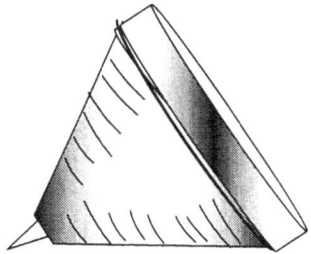 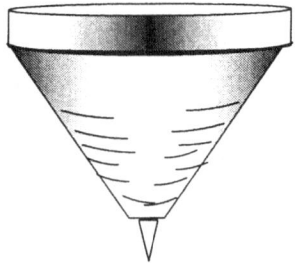

From a position where the top is lying down being collapsed the top generates a position by spinning and the motion puts the top erect. It takes motion and not nothing to establish such an independent and secure position. It puts the top in the centre a newly established and independent Universe as the top then finds courage to fight the gravity of the Earth up to the last "breath" is fought.

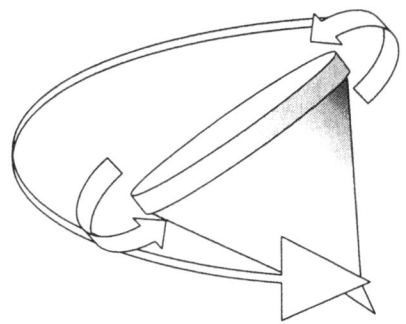 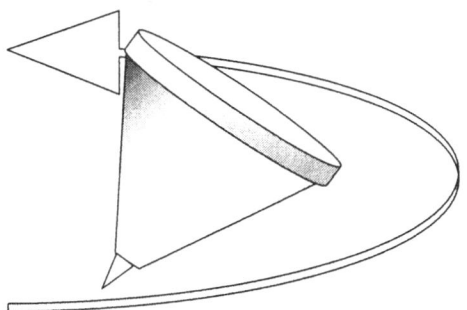

In the motion a line comes to life running in the centre of the top. This line is not just another line but can focus the top to spin upright and erect. The line was not there when the top was on its side. By motion the line can concentrate an effort that will unleash such dependence to the top that the top will come into a position where the top has the tenacity to take on the Earth gravity in a struggle for life and supremacy.

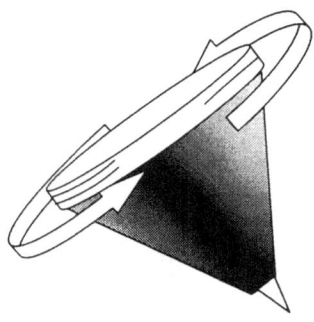 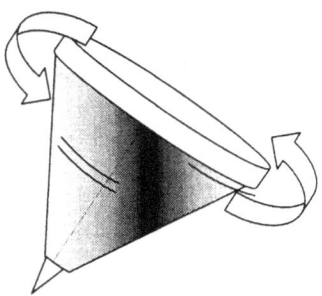

From whichever angle one looks at the top, the top seems possessed and I can even be slightly forgiving towards Newton for calling it a force because although not a force the stance the top takes when spinning upright leaves on with an impression of forcefulness being part of the situation. One must not see a force but one should see the manipulating qualities of life extending to the top and by life's ability to manipulate space-time and control motion in space-time with space-time, the throwing of the top is as little a cosmic event as the apple Newton saw falling from the tree. In every event, the top as well as the apple the drive was life controlling events and as far as there is proof there is no possibility of such an event taking place anywhere in the Universe by something as small as the top or the apple. In the case of stars such development does start the star on a course of independence and it sets the star development on the rode of becoming independent from the galactica. The drive generates gravity but the driving that allow such rotation is inspired by the accumulating spin of the entirety of all the atoms in motion within the young developing star. The heat blanket in which the star cradles has a lot of influence but it is the atoms becoming a driving factor that inspires the motion and such inspiration allows the top the initiative to form independent gravity as a star. In the case of the top the spin is completely cosmic unnatural. If you start to imagine about life in the Universe you may just as well start believing in ghosts fairies and all other fantasy creatures. Science must decide whether they wish to speculate about the life's abundance and being a dime a case, found all over and everywhere you look throughout the Universe, but in such an event distance their fantasies from science and reality, or stick to science in reality and believe only in facts as science presents facts.

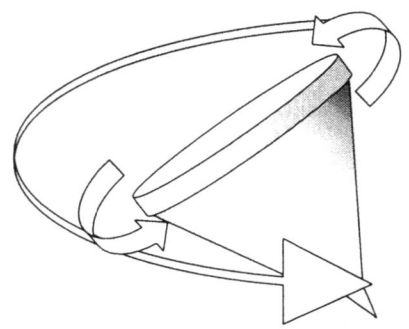 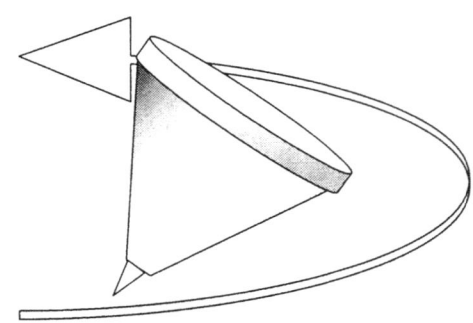

Let's consider what are facts with the top spinning as the top does. This is no fantasy or life coming from same imaginary source but it is a cosmic reality which life found a way to manipulate.

6 sides
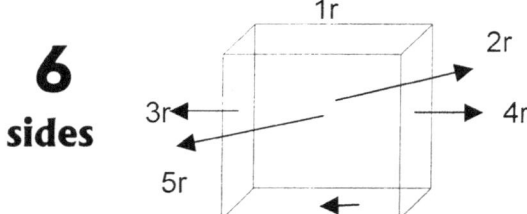

7 points
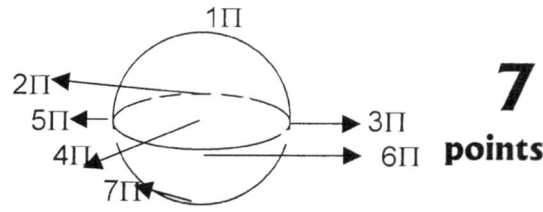

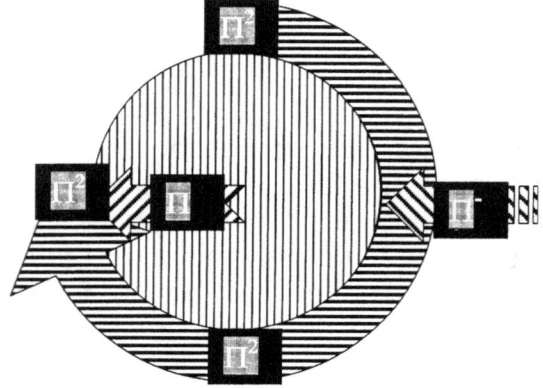

The effort it takes the top to spin gives the top a distinction of extreme significance. The top is promoted by the motion initiative to that of a star in motion because it charges singularity into existence where singularity then controls space-time. There is no difference between the top spinning and the Coanda effect and in both cases singularity is generated by motion that is charging a centre to activate a Universe.

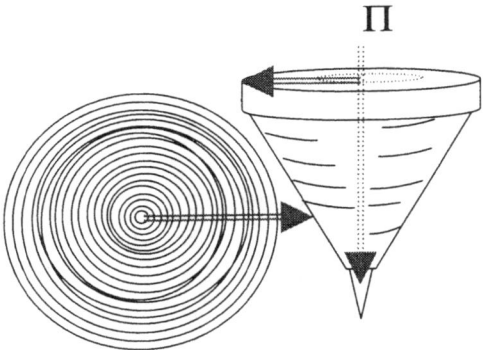

Singularity is a mathematical point hidden in every sphere. Singularity is very much inactive in every sphere but to keep a centre of structural bonding in every sphere. Something happens to the top in having a singularity just like any other sphere has to a point that takes charge with all the cosmic dynamics the sphere may show.

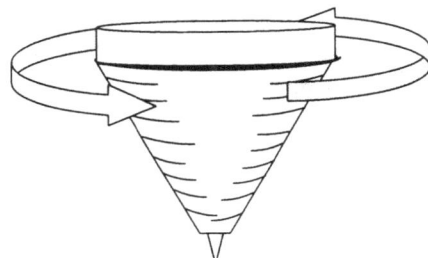

The top is charged with an energy, which not only takes charge of the top as well as the body of the top, but also the immediate space surrounding the body of the top. When a person with skill manages to put a high degree of spin into the motion of the top the top then spins in the surrounding space so vigorously the top stars to whistle vigorously

The linear remains linear because the linear redirects its intentional direction because of the rotational change that the linear motion always ends up doing. The line forms an eventual circle because the linear line must constantly entertain the centre.

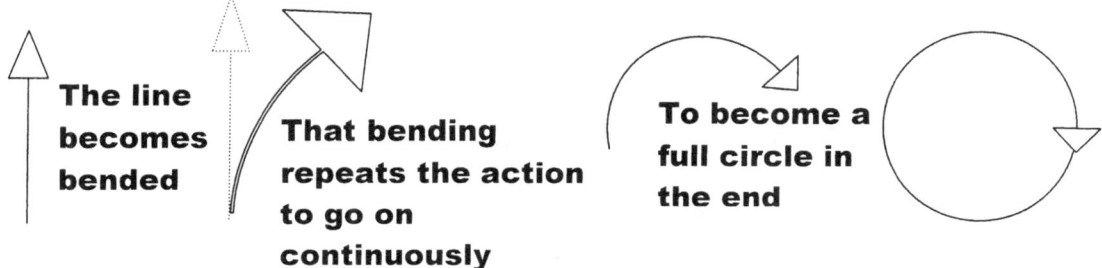

The line becomes bended

That bending repeats the action to go on continuously

To become a full circle in the end

Our gravitational falling to the Earth is a result of a circle going straight and forcing us straight down to an everlasting directional alternating circle, we have as we spin with the Earth as we spin around the Sun. As we fall straight down we, change direction while we are falling straight down because that point we are heading to what we are falling to is changing too. From the centre of the axis, everything seems neutral. The axis does not spin at all, because the axis brings about spinning motion changing eternally. That is in nature and not man-made motion.

As the top comes to motion the top finds the characteristics that the top shows very indicative of the characteristics that all moving objects in the cosmos show. The top spins in a straight line that bends by a 7^0 inclination.

Apparently an idea concerning the subject of gravity Einstein came to was about him falling off a multi story building and the gravity mass that he then would experience. Remember, this was long before flying and parachuting with free fall acrobats showing on TV how a man and a car with a man in the car can descend five or six kilometres while falling side by side. This happened while Einstein was still being a patent clerk in his younger days. Apparently Einstein was looking out a window of the multi story patent office, when Einstein suddenly realised that had he, Einstein fall out of the window from the roof to the ground of the patent office where he was working at the time, then he (Einstein) would feel as if he was weightless during the time of his fall. By falling with him, those articles would feel equally weightless should they accompany his fall down as being part of the falling process in his imagination. As the objects were travelling alongside Einstein down the building to the ground the lot would travel at the same speed from the top to the bottom of the building. Then I went one step further by supposing the Einstein group's falling was real and no imaginary thoughts were set in the fall, then what was the imaginary factor then? Let's pretend Einstein did fall with his pen, his chair and his desk and Einstein was not imagining his fall. Einstein as a human being can imagine but his falling companions can't. If Einstein was imagining his weightlessness, it might be psychological, but in the case of the other travelling companions it was not possible to imagine anything. There is an immense difference in size between the falling companions and that notwithstanding they travelled the same speed while descending. If they travelled the same speed as Galileo proved and they all hit the Earth the same time, which then indicated that their weight and mass, that which gravity used to drive and what propelled them downwards and that which was causing the drawing of what the mass was instigating to allow the motion of fall to commence, was equal. Kepler found space a^3 being equal to the motion thereof T^2 in relevancy to a centre point **k**. Kepler found space had to move.

When reading this that evening so many years ago, I came to realise that Einstein could only feel weightless if it was true that he (Einstein) was weightless. He could not feel as if as if was part of his imagination because he was truly falling, and in truly falling the falling was then without his imagination doing the pretending. Einstein had to feel his weightlessness as a cosmic fact in the true sense because if he was truly falling, then the part, which was the falling experience, was what he was experiencing in reality by three dimensions with one dimension in time. If Einstein was experiencing weightless ness, it would be because he was weightless while falling, then Einstein would not imagine the weightless ness because Einstein was truly falling, thus carrying out his cosmic state he was in. His body being in motion ($a^3 = T^2k$) was at that moment truly weightless while experiencing unrestricted gravitational motion. Einstein, the pen, and the chair had the same weight since they were all weighing the same in falling. If there were any mass differences there had to be speed differentiation for the force of the one would generate more motion than the force of the other onto the different mass components but since there is not mass discrepancy amongst the falling while falling, the lot is having the same state of weightless ness, they adopt the same speed in the fall. After all it supposedly is the mass that is doing the pulling and more mass does more pulling…except if the mass is not doing the pulling in the first place. All four items including Einstein, would be equally weightless during the falling…that was what Galileo found because objects of different size and different mass travel at an equal pace (distance over time or space moving divided by time flowing while the object changes position in relation to the Earth $(a^3 = T^2k)$) while descending. The bigger objects do not fall quicker than a smaller object and that can only be attributed to one fact; it can only be true if the four weighed the same while falling and no one weighed anything while falling. That means the gravity applied while time flow in relation to the space that was applying the motion, which was what gravity is $k = a^3 / T^2$ according to Kepler. The single line falling is represented by the factor k being the relevance of space a^3 that was relocating its cosmic position while all that was happening in relation to the motion of the Earth T^2, which was in relation to the Earth spinning around the Sun and that rotation gives us our time T^2. While in motion the four different objects weighed the same since they travelled at equal speed downwards. By standing still the objects had mass differences and when they were in motion they weighed the same. When the motion became frustrated by being blocked by another space that was also filled with material and that was holding the spot too where the motion was directed, they then had different weight. The pushing resulted from the bodies striving to remain independent. The two objects were in a fight to claim the position each desired, and that was to fill the centre of the Universe. Being $(a^3 = T^2k)$ was being in the centre of the Universe because the centre of the Universe was $k^0 = a^3/T^2k$. Then one may conclude that gravity is motion of space and mass is the restricting of the motion of space. Having mass does not bring about gravity but it does restrict gravity's motion, which is what brings about the mass and weight. Gravity produces mass but mass does not produce gravity or in fact mass produce weight but mass is not responsible for the intended motion. The intent on moving while being blocked by another object is frustrating the motion of gravity in both cases and the higher the frustration on motion is, the more mass there is coming the way of the bigger object who then has the greater desire to move. The reason why it has the desire to move and why space is equal to the moving in time of the space in relevance to the centre of the Universe (which at that point might be the Earth or be the Sun) is what the have the effort to explains. Mass is the restraining of motion and gravity is material moving about by committing gravity. Mass only comes into the application thereof when two objects filled with space moves into a position where both want to claim the very position in space the other occupies.

It is the motion and the independence they show to hold onto their individuality as independent cosmic structures that prevent them the sharing of space which in turn prevent further motion that causes mass. Gravity is in essence where mass is present, still in a tendency to commit motion but is then in the frustration of motion and gravity at such a point is the commitment to move once the blocking of space is relinquished. Because the one object that has more "mass" would put in a more assertive effort to move in relation to a smaller object and the effort to move will constitute to a greater resisting effort by the blocking objects in a fight not to relinquish its position on the space both object claim that the tendency to move and the tendency to block the movement will bring the effect of greater or smaller mass being present during the effort and in line of resisting the effort. However while any space is in motion, the gravity of motion is equal to all and puts everything on an equal basis. Therefore there are no big and small and the big Sun does not pull the small Earth closer. Mass is when the motion is prevented that a differentiation in motion effort becomes part of the picture.

Do not be fooled by the seemingly innocent explanation that space is the motion thereof which is what gravity produces because of all things the cosmos creates, motion of space through time is the utmost complex manoeuvre and without bringing a restraining of mathematics into science, it is so complex there is no viable explaining in physics about how the cosmos produces the act of motion of space in time. In order to get every atom to spin as every atom follows the lead of the atom in front. This gives direction to follow to the atom just behind while giving coherency to the structure. By following the one in front and being followed by the one behind the lot of atoms are holding as an individual unit times the units there are going around in the entire Universe. The measure of this complexity is beyond what the human mind can absorb. While the atom in front is vacating space to fill the space of the atom in front is vacating at that instant, the atom behind is filling the space that the atom in front has vacated in order to vacate and relinquish the previous position in favour of the following position to honour the direction gravity is insisting upon. Times that with every atom there is in the Universe and one may grasp the significance of the calculation. Removing material from space by filling material into a position of new space sounds simple because the complexity has never been realised. I am in the hope that in this matter I will be able to will reveal what the factors are in understanding the commitment of material to move through time. This was all a result of understanding the dynamics of Einstein's arguing about gravity and mass. Then I kept this information in mind and it helped me to further realise gravity is motion differentiation between objects.

It is the independent motion providing a different speed while sharing a common centre of attracting that allows a discrepancy to establish mass under specific conditions applying between the two in relevancy. While falling the gravity applies as moving of space that is putting time in relation to the distance travelled. That means there is a speed relevancy between particles in motion and synchronised motion, which would bring about equal orbit around a shared centre. That is the result of gravity functioning. While the object falls, the motion confirms gravity. When motion ends mass sets in and becomes the constraining of the object preventing further motion. The motion is still there but now it is reduced to a tendency to move thus establishing the object mass as the limiting of further motion. Preventing the motion by implementing mass is the resting of objects against each other by resisting the motion to continue, which then is where the mass takes the place of the motion. Where a confronting of objects restricts gravity, the action then implements an introducing of the mass as a substituting factor to motion that then replaces motion as substitute to the motion that would be and the mass is providing the tendency of gravity being the motion of space.

However mass then restricts motion and becomes motion in a tendency to apply motion. While falling, gravity applies and motion neutralizes size, mass or weight. Mass counters motion being when the Earth restrains further motion of the falling object and the moving object is stopped from further movement where mass is then preventing or hindering gravity. This is the result of objects claiming an individual and personal claim to space occupied in a dual or in fighting for their individuality and independence of each other while wanting to be in the centre of the Universe. While falling or moving there is no opposition to the body being independent. When the motion seizes, the falling object remains individual and still tends to move while Earth individuality resists further movement of the falling body's movement. Further movement is disallowed as other material fills space that falling body wants to lay claim to. The only manner to remain independent by the falling object will be to relinquish to motion in the securing of mass as a substitute to motion where it then finally comes to rest. Mass then sets in not causing the motion but substituting the motion and from that motion restriction becomes resistance that becomes mass. While falling the object is experiencing gravity because the object is in gravity but when on the soil the object experience mass which is the restricting of gravity or motion by other space filled with material.

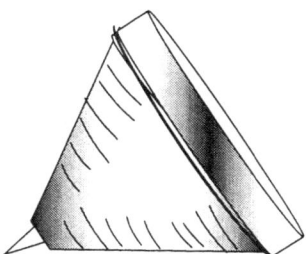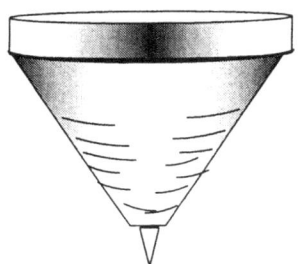

Looking at the top spinning and not spinning, brings a question to mind: why would the motion of the top beat the gravity of the top and that of the Earth hands down when the top is spinning. Surely the mass is in effect while the top is spinning just as much as the mass is in effect when it is not spinning and yet when it is spinning the pulling subsides to give way to free the top from the mass restriction and charge top with much excitement. The excitement is so much it seems to relieve the top of the pulling there is between the top and the Earth. That even strengthened my suspicions more about the fact that gravity is motion. By spinning the top finds additional motion on the side of the top that brings the side of the top in a more favourite position than was the case before the spinning commenced. The top finds additional gravity in the spinning and the gravity in addition has to bring reconsideration to the position the balance in gravity sets in margins. The top secures a better margin or stretches the parameters of location because the top inherits the Motion of the Earth and in addition to the gravity motion that the Earth provide, the two secures more gravity motion. With more gravity motion in addition to the Earth's gravity motion the top has to have more gravity motion than what the Earth has. That will allow the top the spin while facing such enormous disadvantage there is on the side of the top in mass difference when considering the size disproportions. Let's face it, if it was about mass pulling the top in relation to the top pulling the Earth, then the top had no chance ever to move by the very slimmest of chances there may ever be.

However if it is the gravity motion that the top inherits and with the aid of what motion life can add in addition to the motion already the which is the motion the Earth already contributes and considering the mass the Earth provides then it will not require that much a bigger effort to get the top going. Singularity charges motion by instigating motion without ever moving. Coming from a spot to a dot and then producing a line running from dot to dot though a spot signify the birth of the cosmos. A line holding time by the square and by the square of ninety degrees announce a birth in the Universe of a birth of the Universe. That which was not there suddenly is there by not being there. That which was undetectable suddenly is detectable by being undetectable. It is not my forte to write riddles but in this case there is a Universe within a Universe, which is not in the Universe and does control the Universe from a point no one may ever locate inside the Universe. The Universe is built up by innumerable dots and each dot is charged with being the Universe while being in a representative position since it is not in the Universe.

Space-time is a four dimensional position of the Universe where the position of an object is specified by three coordinates in space and one position in time. According to the theory of special relativity there is no absolute time, which can be measured independently from the observer, so events that are simultaneous as seen from one observer occur at different times when seen from a different place. Time must therefore be measured in a relative manner as are positions in three-dimensional Euclidean space, and this is achieved through the concept of space-time. The trajectory of an object in space-time is called world line. General relativity relates to curvature of space-time to the positions and motions of particles of matter.

In view of the definition of space-time I wish to elaborate on my view of singularity and my deriving of space-time from the likeliness that singularity may produce space-time. In the past singularity was mentioned in the manner one would speak of a ghost hiding in a haunted Black Hole. Let's put singularity in the clear. Singularity is within every sphere due to the natural shape or form the sphere is committed to.

While the toy top is, spinning one will find singularity by moving the rotating line or radius progressively to the middle by reducing the length the line has from the edge to the middle. At one point all further reducing must end but the ending cannot include zero or nothing because the rest of the line still attach the rest of the top.

Locating and finding the presence of singularity

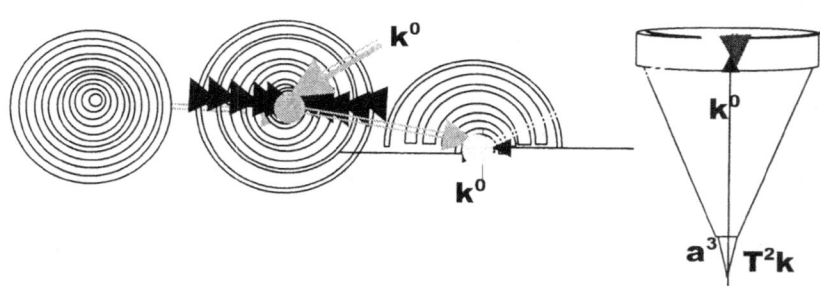

What is in the Universe is spinning. In the **precise middle** of all **objects in rotation** is a precise centre dividing the object in sectors that will **start the spinning initiation** from that centre point.

$k^0 = a^3 / T^2 k$ **states that whatever is, is also spinning in order to be present.**

Thus, the spinning object **will have a middle point,** a very specific **centre point that does not spin** and only holds Π as a specific value because no radius can apply. But also the one value such a line **cannot have is zero** because the line **is there and holds contact** to the rest of the material bringing about that **zero does not start any** line and therefore the **value of the line must be infinite**, just as described in accordance and by **the definition of singularity**

As I am introducing a very new idea, I wish to explain in better detail what I try to convey. **That point** albeit hypothetical, is also as much a reality none the less and is placed where that point **must be standing still** because every line **running from that point in opposing directions** is also **in opposing directional spin the other or opposing side.**

As the rotating direction moves inwards, the rings holding Π will become smaller and smaller. The reducing of the radius r will eventually end where the spin direction ends at $Π^0$. However that point where the directional spin ends is the point where the actual spin takes place. The spinning is on the precise location the point is not spinning.

The definition of space-time is as follows:

According to Einstein singularity is a mathematical reality within the Black Hole but much more so in every sphere. Einstein may be the first to name it and Galileo (unwittingly) may have been the first to define it as Kepler was the first to formulate singularity, but in mathematical terms singularity is the most basic principle. At this point I wish to establish a fact that seems lost in all other grandeurs of cosmology. When tracing the radius down into the sphere the radius stars where all lines start and a straight line cannot begin at zero or nil it can only start at infinity. Such a statement will hardly seem appropriate but the relevancy of this fact has no limits. If gravity is motion then motion starts with a line. Let us follow the line as motion abides by the rules of the line.

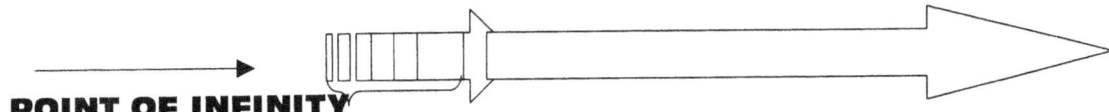

POINT OF INFINITY

If the line started at zero there was no line to start because zero multiplied by whatever results in zero as the answer. That must also be the cosmic starting point. Einstein introduced such a point and named that point singularity. When looking at the cosmos from whichever angle all indications lead to the fact that the whole cosmos is in motion in its entirety. It is forever spinning and it is going too

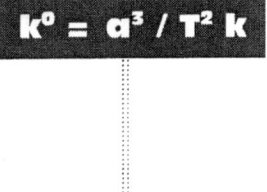

$$k^0 = a^3 / T^2 k$$

as much as it is coming from. Everything is on the move and always encircling something of greater importance. A top can spin but the parameters of its spin are limiting the motion it can apply. By not spinning the top is still spinning as the Earth is doing the spinning on its behalf.

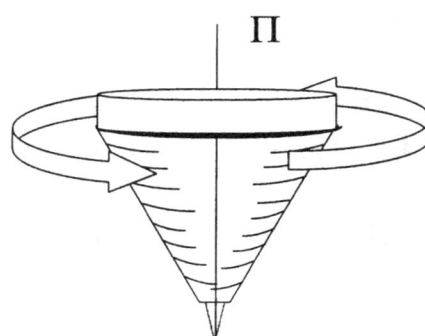

When the top starts by spinning to fast it is clear that the top is in a fight with something that is restricting its spin. This we can see by the re-aligning and the swaying the top manoeuvres with to try and circumvent the restriction. As the top spins something starts to tarnish and erode the spin. When the top starts to spin too slowly the top tries the same manoeuvres but in that case it then seems as if the top is in a struggle to keep the spin alive. These manoeuvres the top display triggers questions in need of answers. Why would the top stand upright when spinning? It can only be that the spin activated singularity into manifesting the gravity of motion.

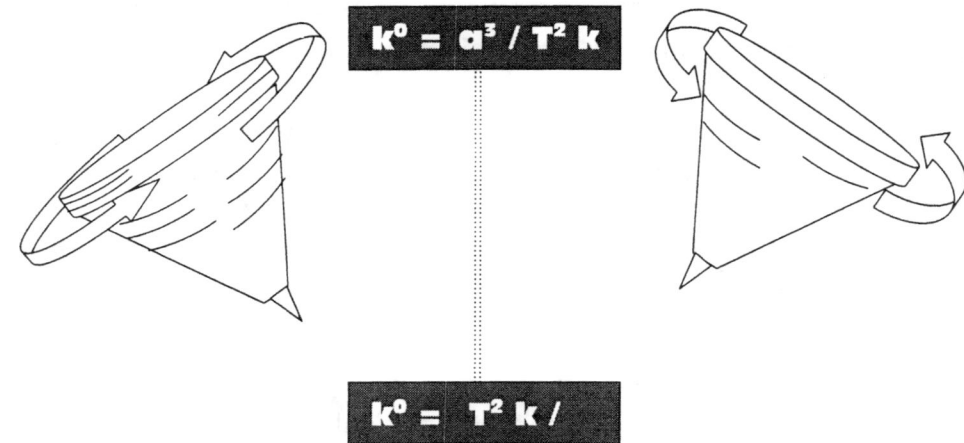

The spinning of the top is all the evidence one needs to come to a conclusion that the motion establishes a drive and the drive establishes independence of the surrounding space the top finds a control over. The top divides the Universe by establishing a generated Universe independent of the Universe we think of as the Universe.

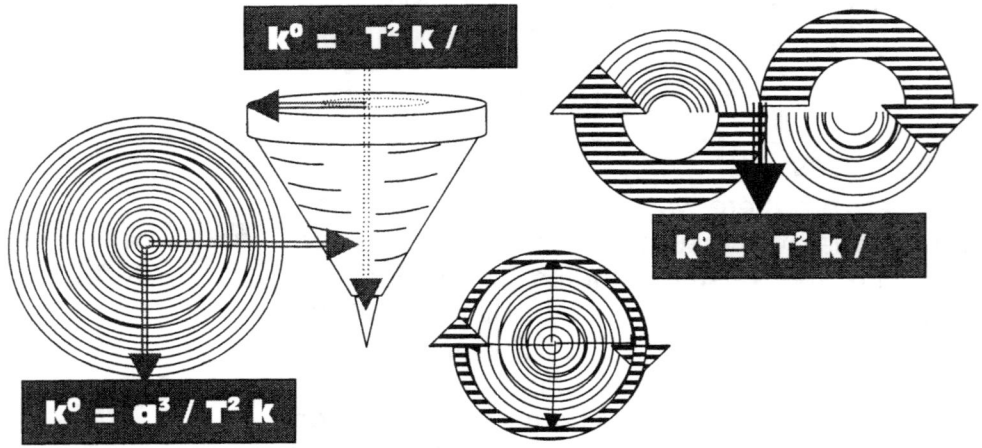

The centre is never there because the centre is eternally there. The centre is drawn into action by motion but the centre does not change in principle. That which activates the centre changes by directional contradicting of its nature.

The motion establishes singularity, which implicates the Coanda effect as much as the motion establishes the Coanda effect. The spin realises a space limit while the space limit attaches the motion in the form of time onto the space of the material, which then allows time inside connecting to time outside the space filled with material. Judging from the behaviour of the top while the top is spinning it seems very obvious that not only is there a singularity running inside the centre of the top but another point holding singularity forms next to the top. It is this singularity point that seems to carry the restraining the top wishes to fine release from.

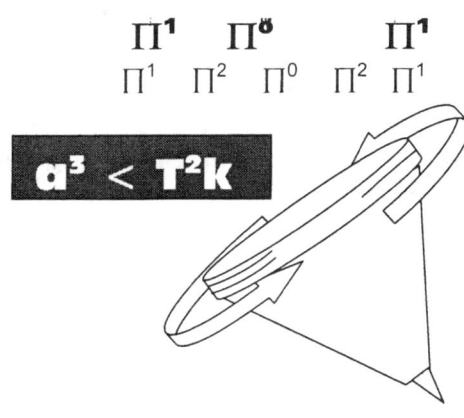

$$\Pi^1 \quad \Pi^6 \qquad \Pi^1$$
$$\Pi^1 \quad \Pi^2 \quad \Pi^0 \quad \Pi^2 \quad \Pi^1$$

$$a^3 < T^2 k$$

In dimensional terms, which I explain later on the value of $2k$ **relates to** T^2. **That relation extends to the next value where** T^2

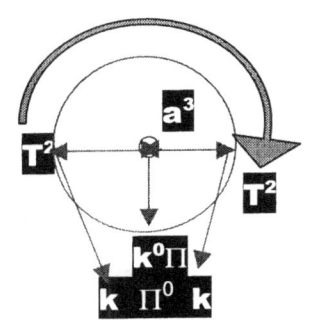

relates to k, **which relates to** T^2. **The first space in the circle will then be** $T^2 k$. **From the centre being in infinity, one can realise by applying mental power the single dimension factor not seen but present all the same. Extending that into the 3D comes six** k **and any one of the six will further extend to form a seventh point as** T^2 **All this is a multiplying of** k^0

$$= a^3 / (T^2 k) = 7$$

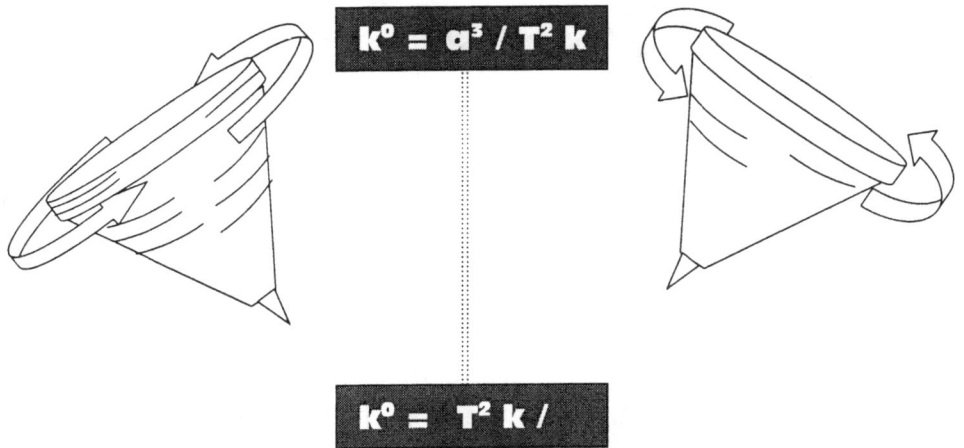

$$k^0 = a^3 / T^2 k$$

$$k^0 = T^2 k /$$

The Coanda principle indicate that the gravity described in the previous page is generated by motion of liquid in relation to a solid anywhere motion can produce gravity. There is no mention of mass because mass is a derogative of the gravity which the motion creates. A centre is formed where the surrounding space-time forming the one group is relating a position from the "centre point". That forms one inclusive relevancy between points within the gravity field. The gravity field is holding "back" and "front" running through "the centre" where the other line is relating from "side" to " side" running through the "centre point". The fact of the line in the centre is that "it is there", but we cannot see it. Try as you may, no one will be able to calculate the very position that forms the lines, but as they change all particle characteristics, the lines are a reality as the spin of the matter is real. Being to small to hold atoms, the space holding such a centre line is no space at all and with that knowledge we may presume then therefore what ever the line constitutes of must become part of singularity, where singularity is a spot in the centre with two lines crossing the spot at an angle of 90^0. That is the basis of singularity, and since all the positions still relate too a centre of a circle, forming a part of a spinning circle, Π must form the basic value. The second major reality that one has to recognise is that the only way singularity was broken was by motion. The only way motion can come about and break space less ness is by establishing heat which establishes expansion and the Universe became a possibility and later a reality by expansion. The heat swell into space and the space swelling is the motion that produces the gravity we find visible in the Coanda principle. The space at first was presumably filled with material because the expanding could only be material. The coanda principle alters time and establish with such alterations to space-time a new Universe with borders and all. By introducing motion it sets a new time standard by which the space created will apply a newly generated gravity.

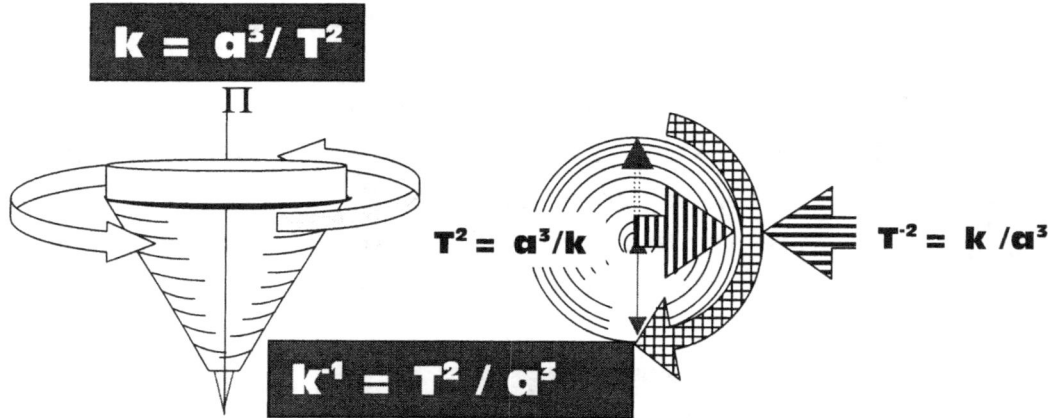

$$k = a^3 / T^2$$
$$\Pi$$
$$T^2 = a^3/k \qquad T^{-2} = k /a^3$$
$$k^{-1} = T^2 / a^3$$

The motion activates singularity but also establish singularity by creating limits and borders. Those limits and borders serve as the gravity applying a differentiation the spin direction brings about by normal flow of rotational spin. This gravity establishing factor we call the Coanda principle and the division also divide the liquid relation as a factor from the solid as a factor. Wee find that the centre secures a point of control and the borders form an expanding of the limits. When the spin exceeds the limits, the expanding of the borders tries to find a way of release or relieve from the controlling centre. When the limits of rotation can no longer sustain a motion, it is the borders that become unable to balance the control and the centre control diminish the spin. The liquid reduces as it wishes to contract while the space claims as it expands.

We have to be clear about what we think of when we think of the Universe. Most people think of a picture recalling the black night sky when thinking of the Universe and that thought is most incorrect. Einstein was most correct when he declared the Universe was going flat where gravity is at its utmost, but the concern we should have is not with the mathematics being valid or not but with the vision about the Universe being what we think of and where we place the Universe. The Universe is in the centre of what is spinning and the biggest single particle that is spinning in total independence of the rest of what forms a total Universe is the atom. The atom spins and by the motion the atom evokes the Universe forming what must be the group effort of all the atoms then spin by the motion the atom renders the rest of the larger Universe. The Universe is the part that allows the rest of what the Universe establishes to spin. What spin you may ask. Kepler said it without saying it: $k^0 = a^3 / T^2k$ and not even Einstein with his super human mathematical skills could say it better or more accurately.

The motion established by singularity results in the implicating of the Coanda effect as much as the motion establishes the Coanda effect. The spin realises the space limit while the space limit attaches the motion to the space in the time within the time.

With the top spinning the Coanda effect steps in and do justice to Kepler's formula.

Time is always a displacement of space in relation to the implication of singularity, and comes about between two points in space relating to the centre of singularity as positioned by k, either to the value of k or to k^0.

With the top spinning as it establishes the Coanda principle it brings justice to Kepler's formula.

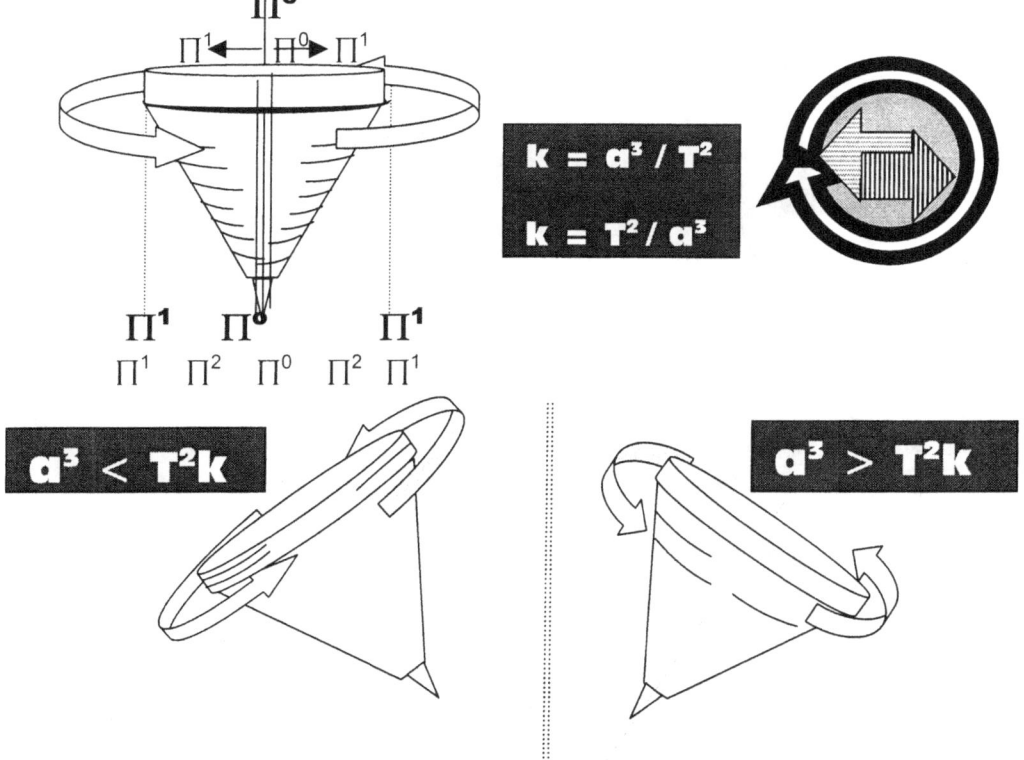

$$k = a^3 / T^2$$
$$k = T^2 / a^3$$

$$a^3 < T^2k$$

$$a^3 > T^2k$$

Time is always a displacement of space in relation to the implication of singularity, and comes about between two points where time forms time by being divided by space. Time in the centre is generated by parting from time in motion where that establishes space as k departs from k^0

We not only have to start thinking exclusively space-time but we have to stop not thinking space as a substance standing apart from time because it is the same thing. The one is not merely complimenting the other it is providing the other with substance of being something. Without space there is no time and without time there is no space. It is having a left side when there can only be a left side when there is a right side to match the left side and without the right side the left side disappear. We cannot accept space without accepting dimensions and by accepting dimensions we accept space because time is the duplicating of sides to form the space that time is matching. It is space doubling or it is time halving but it is the same thing. When space falls away inside a Black hole this comes about as time goes eternal inside the Black hole. The Neutrons leaving the Neutron star provides the Neutron star with time to dismiss the neutrons as it finds the time to provide the neutron to space to leave. It is not only modern and cool to think space-time (as I one day herd a lector tell his class) but it is of utmost understanding the concepts to think space-time because there is no other way of thinking than to think of space as time and time as space.

Singularity by Time

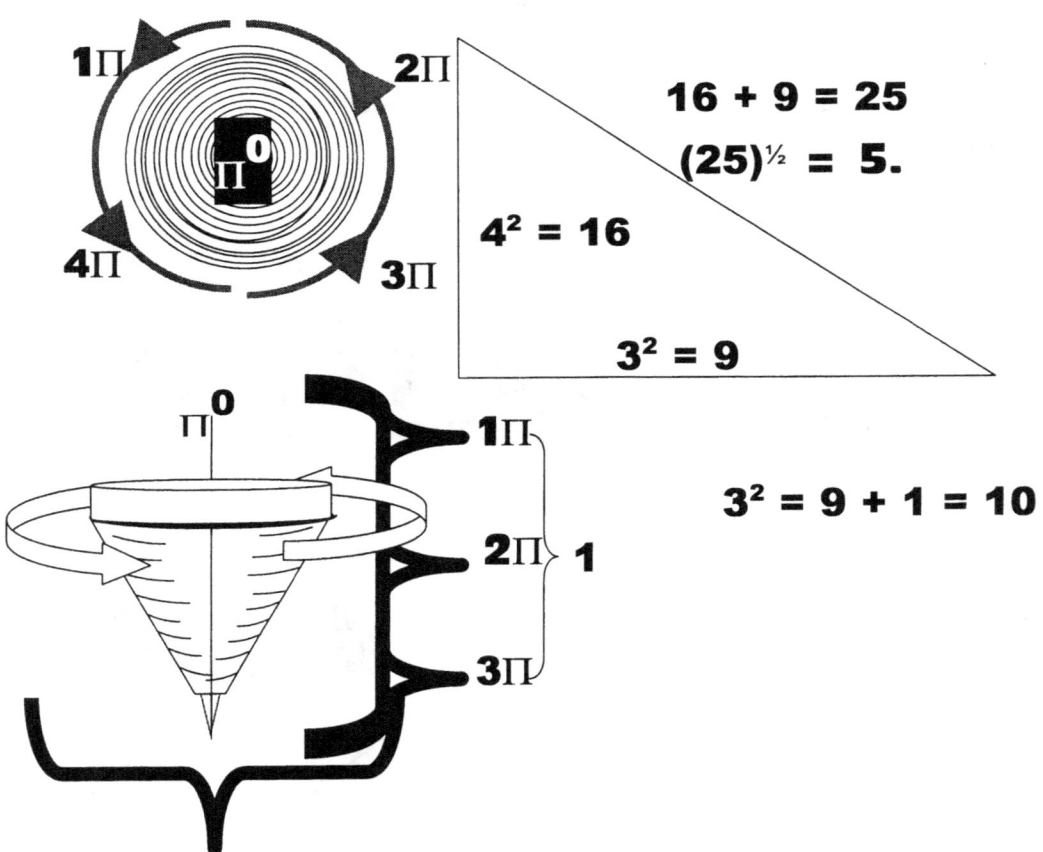

$$16 + 9 = 25$$
$$(25)^{1/2} = 5.$$
$$4^2 = 16$$
$$3^2 = 9$$

$$3^2 = 9 + 1 = 10$$

At the point where space began time began because time and gravity is the same thing. It is motion that is creating space. That is why the universe is so today is what it is. It is antigravity applying heat and will always expand. It is securing heat to cool the rest through applying gravity or space conservation. But material grows as space grows and as **k** extends so will T^2 and a^3 extend. When the dimension walls reduce space reduces but so does time increase because the gravity providing the density claiming the space increases. Space compressed is heat denser and in that is time because time is motion of heat in space. By accepting that there is some conductor (not the ether of old) between two cosmic structures can one accept that there are certain invisible undetectable influences on the edges outside the surrounding of material. One then can see how space conforms as it converts to liquid heat. In the same effort one can see how material confirms heat from space to material. By reducing and confining the heat drawn to the centre the space becomes more concentrated as the heat levels begins to rise. Take any bicycle pump and compress the plunger and the result will be that the heat created by such action will burn the finger you use to cover the valve hole. The heat comes about from concentrating the space, which holds the air. But the air does not concentrate because one does not bring in more air than there was before. The relevancy changes as the space reduce to change the space back to heat. This action is the very opposite of an explosion. But reducing pace the action brings about that the space turns to heat. This is most crucial in accepting because this is the precondition about the understanding as much as accepting a new concept, which I try to introduce, and at the same time I try to produce a concerned effort in dismissing myths from cosmology.

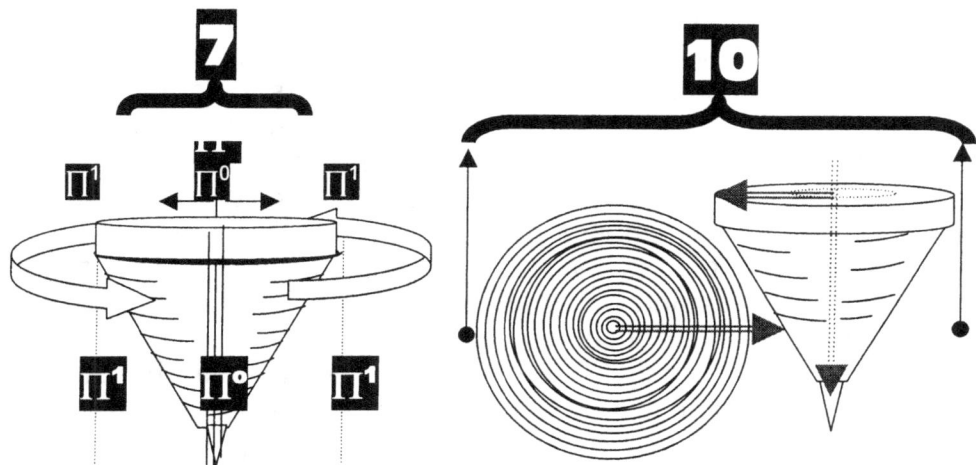

Gravity is about turning space into hotter denser space as it reduces space and that is the reason why there is no visible or measurable stronger gravity in the centre of objects. The centre does not indicate more gravity because the gravity that is the measure of the accumulation of heat that the gravity produce in that space $a^3 = kT^2$. The heat increase as the gravity becomes more intense because the more intense heat is the gravity increase. By the reducing of the space coming down towards the smaller area such coming down leads to space reducing, which brings about heat increases. The stronger gravity personifies in the denser heat produced by the reduced space. As space increases heat dissipates and that we find is what happens in an explosion. The bigger the heat release the more space becomes available as winds (shock waves to use the name hiding the truth) blowing across fields. The winds come about as space multiplies through the release of heat creating new space that was not there before the time. In the explosion will the heat decrease be the decrease of gravity that was before the explosion bounding the heat into condensed space and the explosion is the release or antigravity reducing the heat as it increases the space. In my search I stumbled on two accepted but not intergraded laws and when I found and located singularity the two laws became very much plausible and factual.

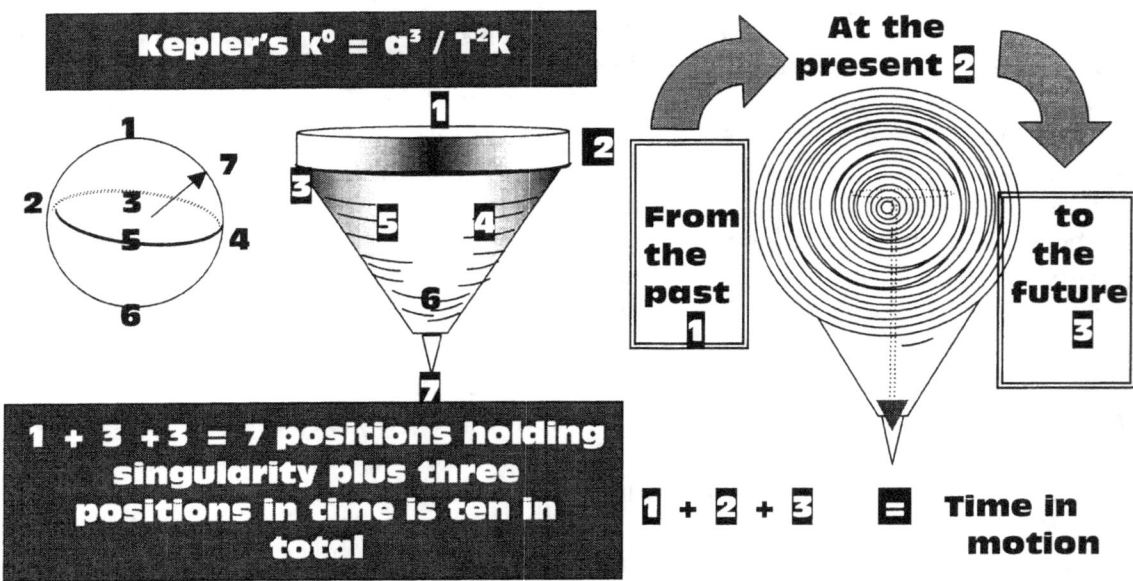

Take a circle and reduce such a circle constantly to where it no longer can reduce. Reduce it to a point where only form remains part of the circle because the radius has gone beyond human measure and becomes so small it is not noticeable with what ever tools man may use, then what remains is pi since pi does not indicate size but indicate form, and form is all that then will remain. I believe one can begin too see where my suspicions are heading because the flaw comes about in the manner mathematics are practised for thousands of years. Space is nothing because that means space is a standard fit all issued out before the time. Space cannot increase and winds are ghost blowing their breath. Winds are as much antigravity returning reduced space back to increased space. Before coming to the mathematics I would first like to bring your attention to the practical side. I am promoting a theory in which I am able to prove there is as much contraction going on in the cosmic universe as there is expansion and the contraction is as much part of the expansion. The universe rides on a balance and we have to locate such a balance. To prove my theory I firstly had to locate the centre of the universe. Even admitting to such a notion sounds like madness, but please give me a chance to explain in more detail. I realised that my effort to locate the point holding singularity enabled me to backtrack the exploding universe to its origins. By applying some basic effort I have located the position from where all movement came and the direction it took moving forward in time…and yes, even time as such. Gravity is the dimensional changing of space holding r as reference in the cube as to the sphere holding Π as the reference. In order to generate spin producing time in matter occupying space, therefore creating dimensional change, Π has to be a factor indicating the possibility of spin because implementing Π the circle sides will follow one another without establishing separation. The answer must be in finding Π, and thereby locating singularity. If singularity is in affect the original point of the cosmos birth, the reducing path we should follow will indicate the whereabouts such a point must be.

In the normal applied mathematics there are two standard formulas used to calculate a circle. The one use an r to indicate the radius and the other use a D to indicate the diameter, which is double the radius and therefore needs to be divided by a four to eliminate the Newtonian inverse square law amounting to the difference there will be between the two. The one using the radius is Πr^2 and the other formula using the diameter is $\Pi D^2 / 4$. However one looks at the mathematical expressions and Kepler's formulating of space-time there is an exceptional difference between the two scientific uses. When investigating Kepler's formula one do find it appreciably differs from the normal Mathematical equation like $a^2 = r^2\Pi$ and $a^3 = 4/3\ \Pi r^3$.

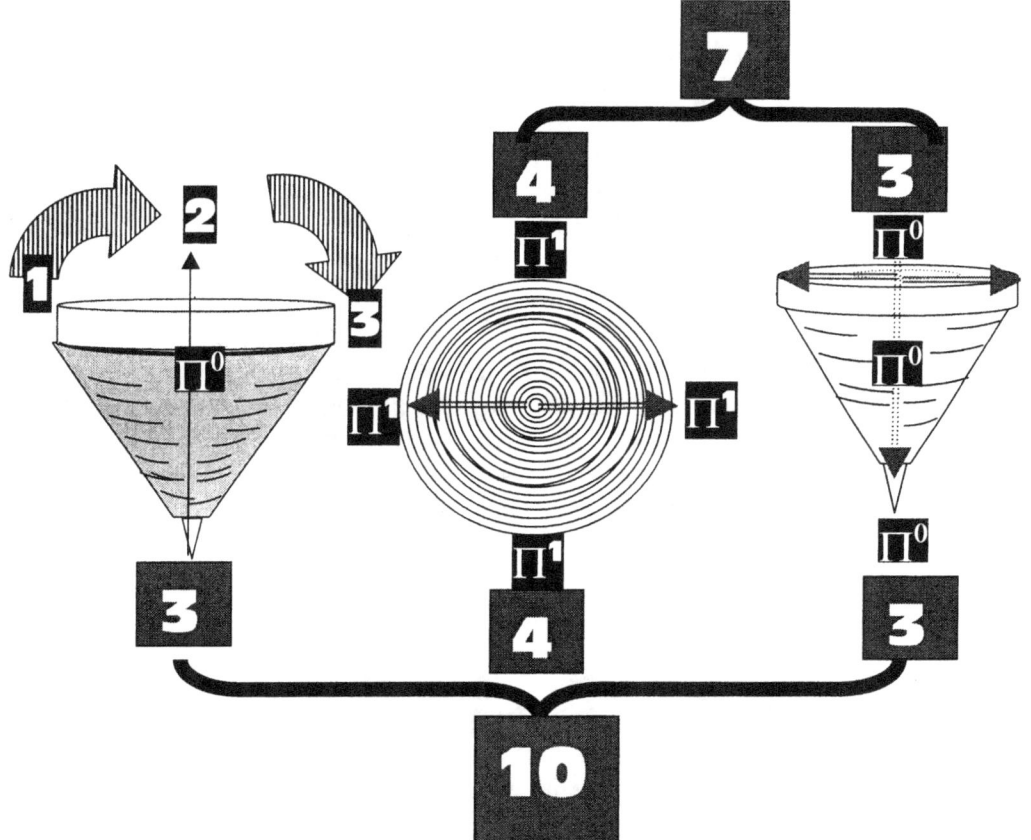

In the normally used mathematical expressions such equations tend to concentrate on the volumetric aspect. In the case pf Kepler's expression it is something else that wants to surface. It is another idea that is coming to mind. In Kepler's formula a^3 stands to symbolise the third dimension and such a third dimension becomes equal to two other dimensions grouping and sharing value to equal **a^3** efforts. It is not the circle of the rotation because with such a normal circle the radius is in the square and Π evaluates form. Here there is no mention must of a factor Π, which one would suspect to be somewhere applying since the circle is Π and Π is the circle and the two are inseparable. But not in Kepler's **a^3**, where there is no mention of Π at all. The fact that there is a radius of some sorts used to indicate a position cannot hold the square as it normally does in the case of the normal equations. In the mathematical equation the factor indicating the position of the circle edge has the square value being called the radius or in some cases the radius doubles and which then is the diameter, and the circle indicator is Π. But in this event the formula value will bring about a square value to the answer one receives. It will bring a value to the surface of the circle. In Kepler's formula it specifically does not. I am not the one that brought Newton into disrepute. Before me the cosmos did. The comets with they're not colliding did, and so did Roche and Lagrangian principles. Hubble was another one and it becomes apparent that every one that made a study about matters in the cosmos was in some disagreement about Newton. By Newton's effort to improvise on behalf of Kepler Newton made a statement that Kepler never made. In all honesty nature reacted strongly against the claims Newton made on behalf of Kepler and not about Kepler's work but about Newton's modifying of Kepler's work. In short: how can a comet sail past the Sun time after time without colliding and still apply a contraction in the manner which Newton suggested by the one claiming a freezing grip on the other? This strongly contradicts $F = G (M.m) / r^2$ How can five structures as the LAGRANGIAN POINT form around a centre structure while the centre structure keeps the five in position at equilibrium? By rejecting Newton's improvising this strongly contradicts $F = G (M.m) / r^2$

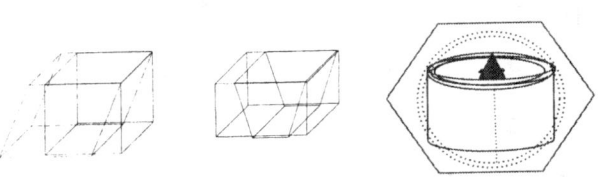

Anything occupying space in the cube will apply r and by r I mean just a distance not using Π because Π serves as a form indication while the collective product of r will determine form as well as accumulative dimension total. Notwithstanding the name used confirming the shape or r named as length width or height, it is all just a straight line bringing about the cube with all its other names that may find attachment to specific form but nevertheless still remains only a six-sided cube with connecting lines applying different angles changing in some cases.

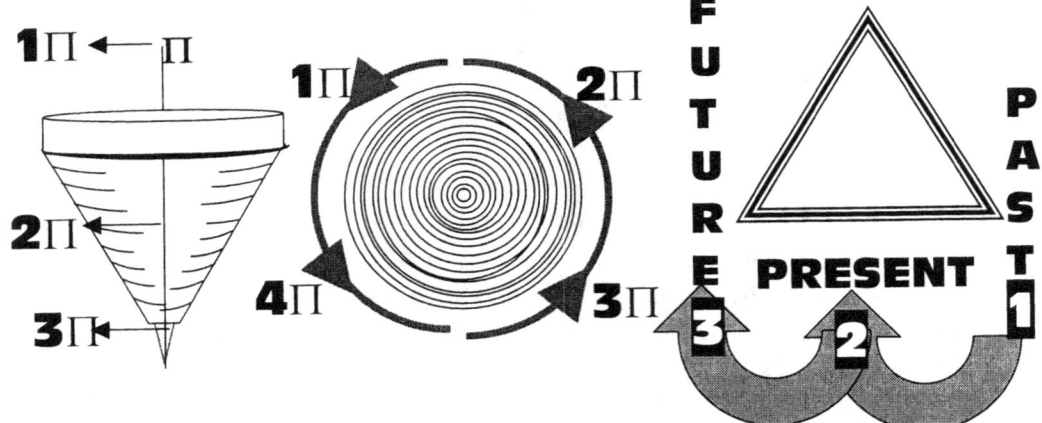

The normal perception is that any circle growing spontaneous would grow by the radius, which is r. In mathematics that may be true but it is not true in nature. In nature that cannot be the case because r is an indication of a straight line. By growing with the aid of a straight line from the centre to circle the influence that that would have on the circle would result in many circles following one another and not a continuous growth.

If we wish to believe in the Big Bang and we wish to accept the factor of singularity then we have to accept that there was a period where there was no space. We can backtrack the space to a point where the space is no longer space on the precondition that the space is coming about from motion. If more heat comes to such a centre the centre will produce more motion. The motion will produce more duplication of space and the duplication of space is the gravity we experience as a contracting direction of motion where as heating is the expanding of space through motion. But it had to have started with a space less motionless dimension-less Universe wrapped in singularity. The differentiation coming from motion is a dimensional barrier that changes many aspects in cosmology. The dimensions came about as the Universe came about and each had its individual introduction period. Space and time parted at $(\Pi^3)^2 = 961$, material formed identities at $\Pi \times \Pi^2 \times \Pi^3 / 5 = 192$ and $\Pi^2 \times \Pi^2 \times \Pi^2 / 5 = 192$ where space either had material or had heat without material and space separated from heat and matter at space holding $10/7\pi^2/2(\pi^2+\pi^2) = 139$ material $7(\pi^2+\pi^2) = 138$ and space having liquid within $7/10\ \pi^2/2(\pi^2+\pi^2) = 136$ This is suggesting that these are meaning this was the first time liquid became part of the cosmos while all were still part of the same unit as the Roche principle would suggest $(\pi^2/2)$ as well as the $(7/10)$ and the $(10/7)$.

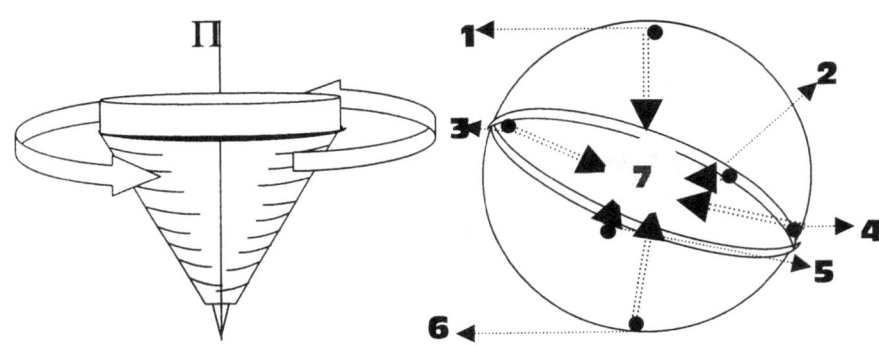

Material seems to get glued to the earth by some force, which holds the name of gravity. Moving such a gravitated particle needs some drag by motion. The secret of lessening the effort in applying motion to the object in need of shifting is reducing the drag that is not drag at all but motion not in motion and hiding behind the name of mass. Let us find this drag in nature and work from there to find a better natural understanding of being in a solid state on ground or a liquid flowing down into the ground or a gas floating above the ground. It is accepted by science that water can rub together and form static electricity. Can you believe respectable men say this shameless! The scientist that thought this one up had some or other big problem with his hair and thought that rubbing his hair with a plastic comb will be the same as water rubbing against water. I cannot believe that science can indorse such shit and shit it is! How can water form static electricity because lightning is the product of gas going into liquid by motion by turbulence of heat. Let's remove gravity and find mass.

It is well documented that where heat turns to space and space becomes more than the space which the matter is occupying before when the heating. By heating material there is an introducing of space because the space needed after heating becomes more than what the space would have required before the heating started. By heating the material with a sudden burst will bring about so much space available it brings along the destruction of matter in the position and form it holds. This destruction we know as an explosion. The advancing of the newly formed space reconstructs the position layout the matter holds. With the knowledge and countless demonstrations brought about by war and other destruction, the opposite must also apply. With the reduction of space in forming heat, the reconstruction of matter therefore also must come about from such a manner. Where matter removes heat to reconstruct its element worth and element position in value, the reconstruction is the direct opposition to the deconstruction of matter by explosion, therefore the relevancy changes and with the relevancy changing the result therefore must become reversal to the explosion.

In outer space an object floats. On the moon nothing solid will float but there is no liquid air either. Those not familiar with this statement must think what the difference is of space in outer space and space in the atmosphere and why objects entering the atmosphere suddenly acquire the ability to heat up and burn out. The reason why outer space is a gas is because the density applying between space and material is very little in comparison to what we are use to in the atmosphere. Outer space is not colder but hotter, much hotter. In space all objects are very loosely connected and move quite freely about. When this occur it reminds one of a gas because in a liquid there is much more density in the matter relation and when solid the matter is as close as can be found. Therefore the conditions in outer space form a gas and a gas is the hottest of the three conditions there are available to substance material. Comparing the likeness with anything we can compare too with our vision of what is on earth, we must move to something we all consider to be a natural in all three forms, one being solid ice (very cold), two being liquid water (less cold) and three being gaseous steam (very hot). Conditions in outer space come down to steam because there is much space between the particles bringing about more space in the density than there is material. By introducing heat to water, water changes from being a solid we call ice where there is much more material in the ratio between space not filled and material filling space whereas with through liquid such as we call water is more space unfilled than there is in solids but much less there is than in gas and gas we call steam.

By introducing heat to water we change water from a solid (cold) to a liquid (less cold and more hot) and with the introducing of much more heat we get the heat to become a gas such as it is in outer space. By introducing heat to water we get clouds forming. By introducing heat to air we get clouds moving, moving excessively, where the movement in fact displays a density increase. Because there is a density increase the wind can uproot large trees. To suggest that something as light as say oxygen and nitrogen can blow down a tree with the quantities present in such a density as one find in space proves how little science are able to think! With more increase of density in the wind we find spiral motion in lateral movement. With wind circling it has terrific density because in such a form it not only uproots trees but also takes on houses and much of what man can build. The wind in access blowing extensively produces the same qualities by producing destruction damage than water flooding in a river can match. When the density increase by adding motion much of the increase goes along with vapour, a form of air that is thick with water, (I distinguish the terminology because why not only use one word, either steam or vapour. After all it is the same thing!) In clouds we find lots of vapour but we find little water. The difference between water and vapour is that vapour has more unoccupied space and less space filled with water material in ratio. The thick density is there, but the air is so thick the vapour and the air combines to form a gaseous liquid we can see as a cloud. Remove the heat in the cloud (which the cloud needs to be being vapour) the water returns as a liquid and produce a form of heat we named lightning, The vapour liquefies too electricity as liquid heat and fall as a solid in the form of water we named rain. The liquid which was the water did not vanish but became liquid air which we call lightning. The question about density increase always comes from material being more prominent and more abundant in such a space. With the increase of density it always accompanies the increase of heat and the discharging of heat.

If one would think that it is vapour in the wind that increased to such extend that the wind can uproot trees, then why does hail with such a lot of solid water not uproot the trees. There is a world of difference between windstorms and hailstorms because of the abundance of electricity or more bluntly phrased heat in spinning motion. Windstorms having the ability too uproot the trees have a very sticky substance between the molecules and the more sticky evidence there are, the bigger the ability to cause damage. The air substance shows a bigger resistance to part or create space than the tree shows to remain secured by its roots. The substance can be sticky to the point where it breaks braches that will require an effort of many hundreds of Newton meter to break.

Even spraying the tree with water which man created artificially by pressurising the water would hardly break the branches and less hardly uproot the tree. If it is done, the water flow will be enormous but I have seen many times trees uprooted without one drop of water visible. The only logic remaining supplier of such a density increase must be the heat. In this there are changing relevancy dynamics, which I then introduced as equal to the substance found in atoms.

 When one takes Kepler's equation into consideration the whole process of motion starts to make sense. It is motion that keeps the top erect and that was accepted from the time of Newton. Now however, by close scrutiny as well as considering Kepler's equation where the statement emphatically reads that the space a^3 is equal= to the motion of the space in relation to a very specific centre. But this relation works both ways and not only from one side. It has nothing to do with pulling because if the two were pulling the top had no chance of any motion. The top uses the motion of the Earth to the advantage of the top and then on top of the Earth's motion, the top applies individual initiative and claims even more independence than the structural independence it had before. If it were merely gravity of mass and nothing more, then the gravity disappearance the Earth produces would be such an imbalance that the top would never stand a chance of committing individual motion. However should my view be correct about motion and mass being the frustration of motion hindering the motion of gravity, then yes, the top by motion free the distorting of the Earth mass, which is a frustration to the top gravity motion and that would enable the top to spin even with such slight energy applied. Merely taking into account that the top has to overcome the considerable mass f the Earth brings the Mass theory into dispute because how can the top with such slender energy find freedom from the enormity of the Earth mass.

Every round object has a point establishing a very centre, a middle dividing one side from the other. That division determines the space from one side away from the other side. At one point there must be a point that does not fall on either side of the divide. Such a point will still be a circle, because from that side the circle divides into two sectors.

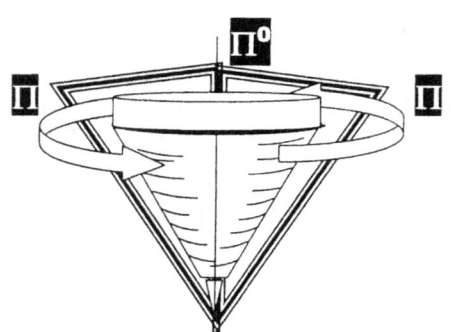

In every object there is a center but the center we find in the sphere is the only center that can taker complete charge of gravity because it is the only center that is controlling every point on the edge of the circle at any point the circle can offer.

In all units there is a singularity seeking independence in relation to singularity elected seeking dominance. From one side and any singularity there will be present in the unit one factor of singularity carrying the value of 1 but also there will be one divided by space square singularity absent because only one singularity can apply to the unit in dimension. Therefore, one tenth of the space is absent where singularity is one and holding .9991 valid as part of the other side of the Universe it is attached to but not connected to. With the unit being connected to motion the motion will stand related to seven from the one side as well as the full ten from the other side on both sides of the Universe. That relates as ten in space-time on both sides of the Universe (10 + 10 = 20) plus one factor of singularity present and one factor in the tenth not present (1+.9+.09+.009+.0009+.00009-.0001=1.99991)

Many dots spawned from the spot and such spawning was in flurries but equal.

Then the four cosmic pillars set laws and progress started happening as the cosmos applied and stuck to conditions set under these principles.

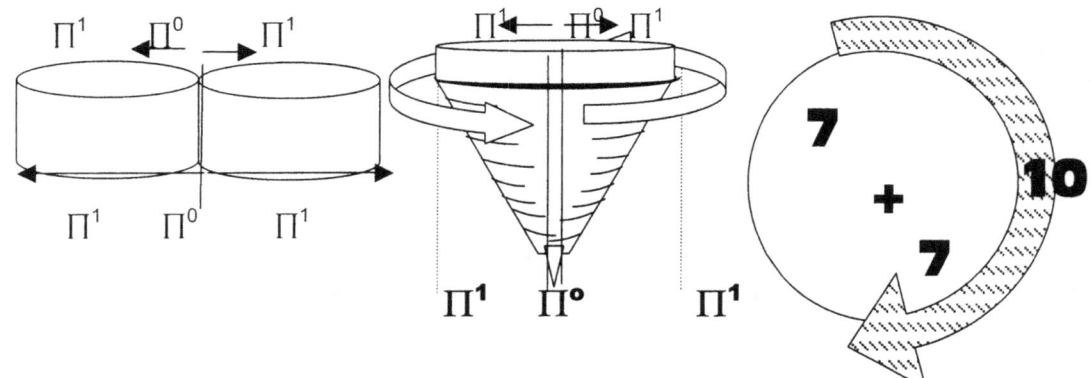

Some singularity formed a dominating role as they were in dominating some particles formed a subordinate role but space-time formed since there had to be motion and motion provide form which can be space. But I wish to place a distinction between form and space because by calling what was in progress space, the mediate human connection would be to adapt space by dimension to what was in place. If there were space it was Π, which is form. There were no mathematical equations because for mathematics to be in place there has to be space. There was no space of that particular type yet invented. Please think clearly as this is very important, is that not precisely the commitment we find in gravity, where gravity is flowing from singularity outwards but never favouring any side? This reasoning prompted me to look for singularity in such a spot because if the prime spot from which all came was a spot holding all, then the spot must hold the shortest line but more prominent it will hold the smallest form including the smallest circle or for that matter the smallest sphere. With gravity always being in the centre of a sphere where the space is least available in the entire structure (there is not even space left to fill) one finds a flow of gravity from that centre spot outwards in all possible direction even-handedly. The fact that the original gravity will begin as a circle or will be a circle is the direction it will take when being the first spot created. All progress will be evenly in all direction because no direction will stand out or be in favour above any other direction at first.

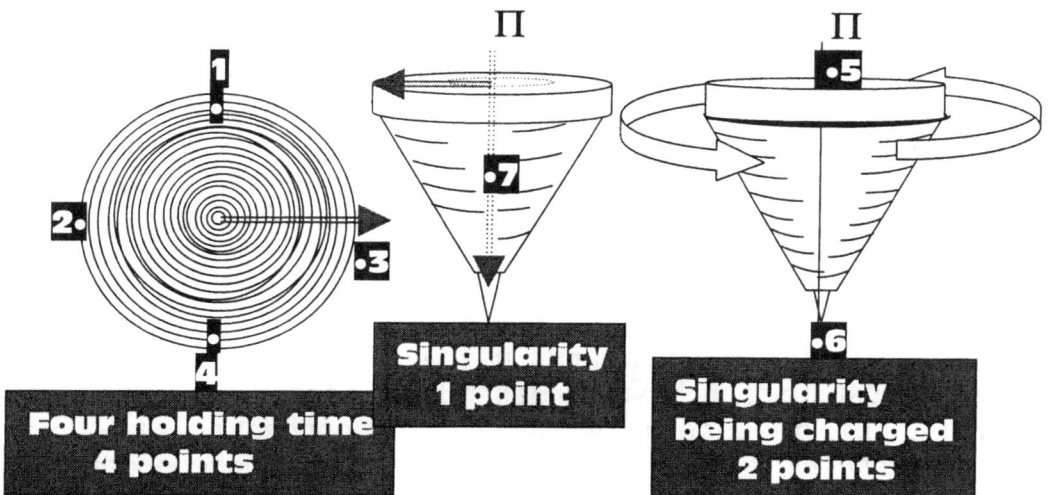

Four holding time 4 points

Singularity 1 point

Singularity being charged 2 points

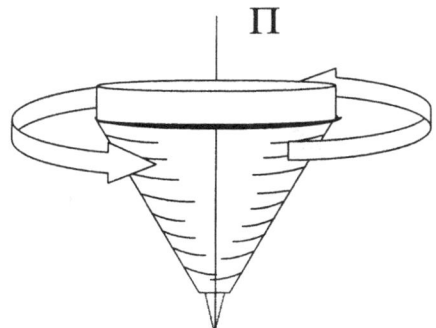

When the line **has a beginning and an end at the very same spot** and it wishes to extend the position as to further the possibility it has, which direction should it favour. Extending the line in any one direction will favour one direction without any clear reason not extending in other directions. The only mathematically sensible option about extending will be in all directions equally in order to give a meaningful non-bias flow of mathematical equilibrium.

That is where one would have to go look for the beginning of the Universe. The Universe is a about lines connecting but where does than connecting end. The Universe used form to this point in development, but then at some point the line came and established the presence it still has. The first form was moving from $\Pi^0 \Rightarrow \Pi$. Again I wish to press the issue, at that stage form was in use and not mathematics. The Universe was just simply too big to measure. If radius did apply, one could use r and r^2 but since only Π was in use there was no radius used.

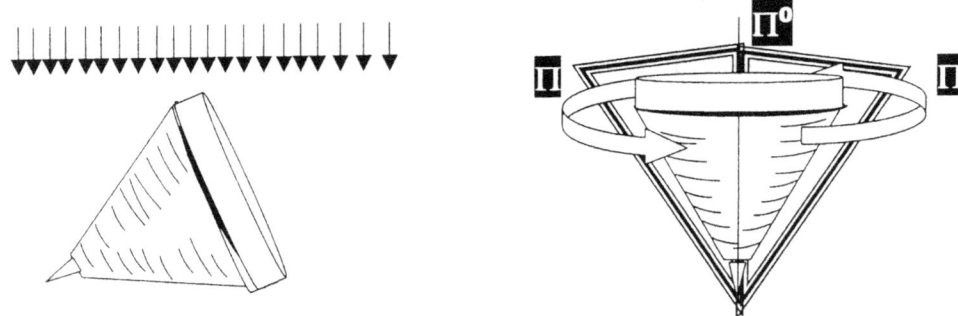

There is one Universe of difference between the top lying down showing no independent motion and the top spinning erect. A top on its side that is not spinning exists as part of another Universe. The top's entire Universe collapsed when the motion seizes to activate singularity in infinity. While being in a motionless state the top submits to the singularity lines running towards the center of the Earth. There are gravity lines invisible as they are still they are there running through the top at 180^0 to the Earth placing the lines at 900 angle in relation to the top. These lines suppress the structure of the top to confirm in mass the gravity the Earth applies. The top succumb to the flow of the lines running to the center of the Earth and it is the responsibility of these lines to eventually burry the top as part of the structure of the Earth. That is the purpose of the mass because by applying mass to the top the Earth will eventually have the top relinquishing its structural independence to the Earth and totally submit all individuality to become part of the Earth. But when motion is added to the structure the top holds another dimension of independence are accomplished and the top receive almost the same cosmic independence as the Earth has. It is the task of the top to uphold the motion and maintain independent singularity while the Earth singularity will fight to submit the chances of independence by restraining the top even to a point where the top is liquid.

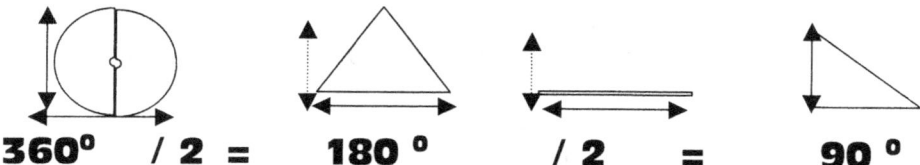

360⁰   / 2 =   180⁰   / 2   =   90⁰

The circle is a square holding a round shape, as the straight line is a square holding one side to infinity. Calculating a circle involves two aspects where the one is either the radius or the diameter that is double the radius. The other is the factor Π

Because gravity work both ways and not singularly in one direction as the Newtonian myth would have us believe, there is the interaction in the neutron position between the total of material in relation to time formed in space as space and time formed in space in relation to the total of material.

In contrast to Newtonian view about a spontaneous effort there are in the cosmos of joining and sharing, quite the contrary is true. There is a natural tendency to remain independent t and away from each other and where the tendency of staying apart is bridged, there is a tendency to destroy and conquer, to control and delete the lesser by an onslaught of the more superior. There is no mass fighting to join mass and to become one in all. That part is fiction as much as the part about a force is fiction. There is a struggle for superiority and there is a fight for freedom from dominance. The whole idea about masses joining and uniting runs very much against the basic fabric of cosmology and in particular the Big Bang theory, The sound barrier principle, the Coanda affect of motion bringing about space-time control and so many more.

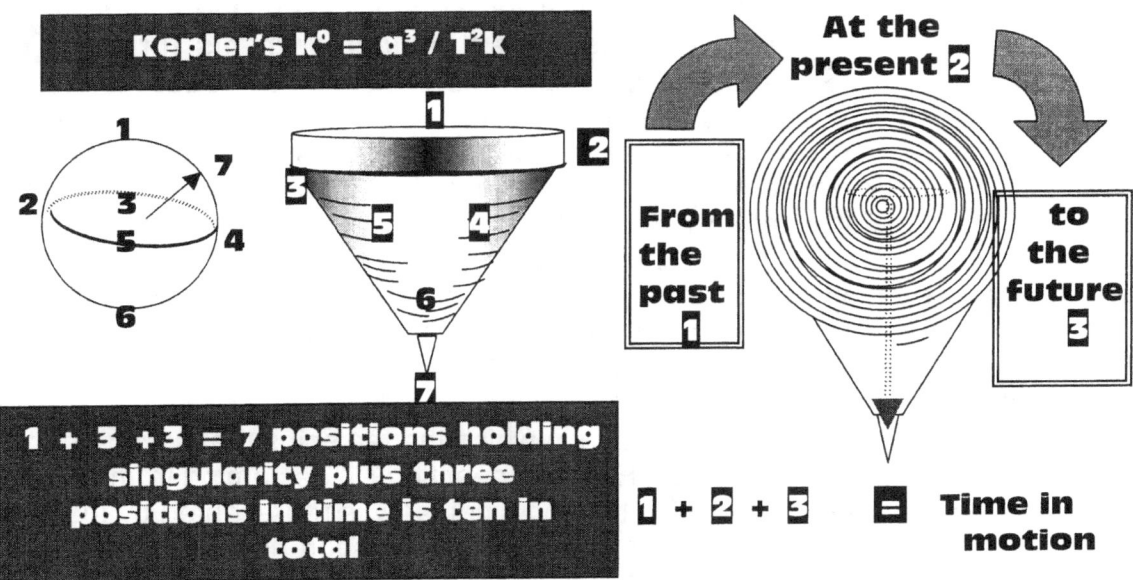

If the joining were with merit, we would by now not have known a moon orbiting apart and on its own coarse around the Earth around the Sun around the Milky Way. While there are those attachments they are only attachments and not obsessions of joining and uniting. The moon holds a separate identity, which it refuses to relinquish. This refusal we call a lunar cycle. The moon is on a running spree ever since it's Independence Day. The moon is taking a route that would progressively carry the moon further way from the Earth as the Earth is rerouting its orbit further way from the Sun. The question to ask is what makes the Sun, the Sun and the Earth, the Earth and the moon the moon. It is $a^3 = T^2 k$ or better put it is $\Pi^3 = \Pi^2\,\Pi$. It is the space-time as collected by singularity using the Coanda affect. If the joining were with merit, we would by now not have known a moon orbiting apart and on its own coarse around the Earth around the Sun around the Milky Way. While there are those attachments they are only attachments and not obsessions of joining and uniting. The moon holds a separate identity, which it refuses to relinquish. This refusal we call a lunar cycle. The moon is on a running spree ever since it's Independence Day. The moon is taking a route that would progressively carry the moon further way from the Earth as the Earth is rerouting its orbit further way from the Sun. The question to ask is what makes the Sun, the Sun and the Earth, the Earth and the moon the moon. It is $a^3 = T^2 k$ or better put it is $\Pi^3 = \Pi^2\,\Pi$. It is the space-time as collected by singularity using the Coanda affect

The spin was going on for eternity because the spin does not apply, it has a value of infinity and infinity at the time was combined with eternity.

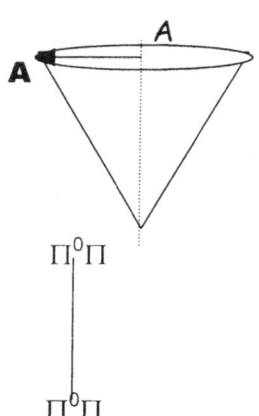

The moving of Π^0 to Π activate a line that was not there before. It was Π^0 moving to Π that evoked a line but the fact of the matter is that the line is still not there. The line shows a presence in *Singularity* (A) establishing (**A**) and that shift announced the rise of another Universe. The movement did not do nothing as Newton would indicate, but the motion evoked the birth of an entire Universe surrounding singularity. The motion brought the top into space just as Kepler announced with his formula $a^3 = T^2k$ where it says space is produced in equal measure of the motion of the space...and the top is the undeniable proof of Kepler's statement. The top by motion brings space into the Universe and without motion the space is denounced as a Universe by the Earth motion. Should the motion of the Earth end all space accountable to the Earth will seize. It is once more proof that time cannot stand still as Newton, Einstein and Mainstream science would declare because if time stands still, all fall back into and to singularity. Singularity is one being $k^0 = a^3 / T^2k$ and if T^0 then all the other factors would follow the same path.

The top has space but there is a space in which the top spins that covers the time part of space. That is the time part Einstein identified (1) as coming from (2) being at and (3) going to and the position holding singularity is represented by the entire body that holds all the space of the spinning top. As soon as the spinning of the top commences, the time aspect releases space in which the top spins from the space holding the time of the Earth and the rest of the Earth within that time. It is this space in time that becomes so hot when the aircraft is speeding because the motion takes the time back to what the time was when time was nearer to the Big Bang. By receiving space, singularity received a value from where it was in eternity Π^0 to just one point outside eternity as Π. But the motion of seven relating to ten brought about gravity as Π^2

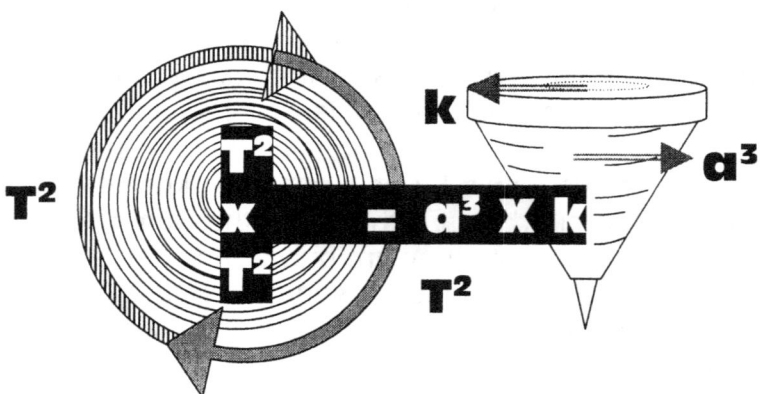

With everything in a cube or a circle or a potential of the two, brings about the implication of eternity in a form of singularity or the point of creation. Removing the radius of a circle does not remove the circle, because the circle is there, securing the ring. If the line (or imaginary line if you wish) holding the value of Π^0 = 1there has to be a point where the circle is no longer in infinity but claims existing outside the imaginary. At that point the radius may be lightly more than infinity, but to all calculating purposes it still remain as infinity. The spin was going on for eternity because the spin does not apply, it has a value of zero and zero is another expression for eternity. The full square of the motion $T^2 X\ T^2$ = is equal to the full space $a^3\ X\ k$ created by the motion and that relevance became the atom.

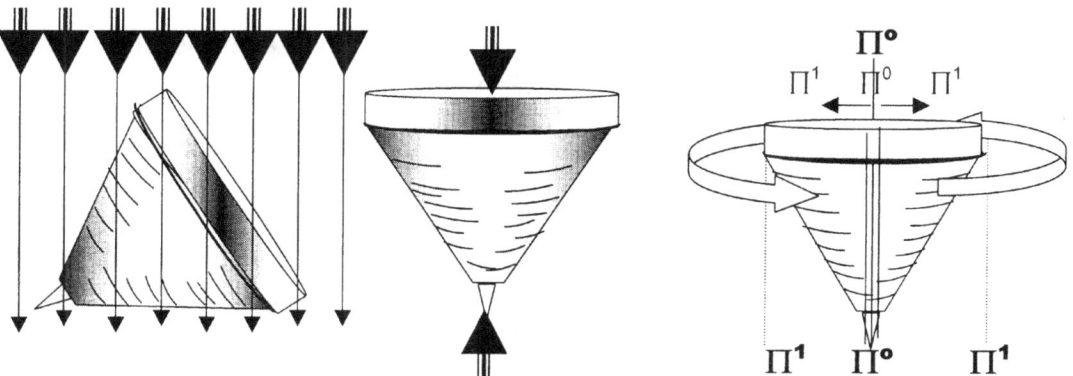

Singularity form lines running towards the centre of the Earth.

Singularity by Time

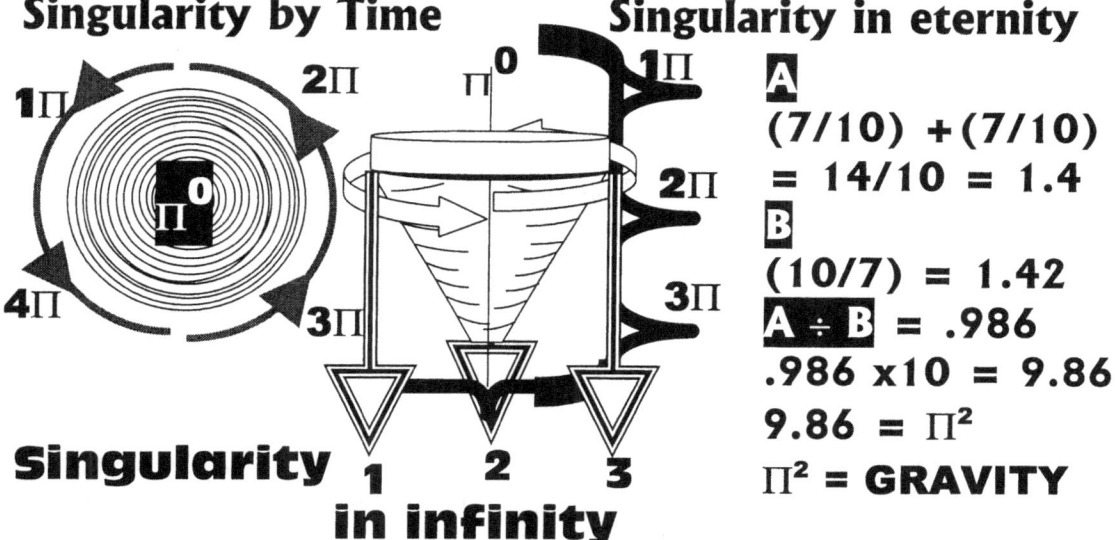

1Π 2Π Π^0 1Π

2Π

4Π 3Π 3Π

Singularity
1 2 3
in infinity

Singularity in eternity

A
$(7/10) + (7/10)$
$= 14/10 = 1.4$
B
$(10/7) = 1.42$
A ÷ B $= .986$
$.986 \times 10 = 9.86$
$9.86 = \Pi^2$
$\Pi^2 = $ **GRAVITY**

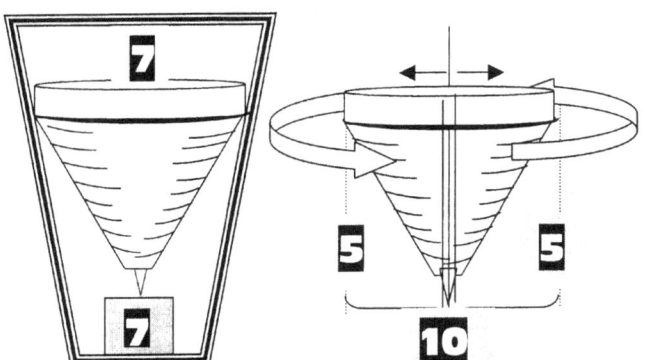

Having edges where Π^0 duplicate to present the edges singularity lost the value of Π^0 to the value of Π^1 with the same value singularity had being Π^1 to the one side and Π^1 to the other side, Π^0 must be the point splitting singularity into two parts of eternity, the eternal value of the first dimension outside eternity. It was the square of Π^1 being Π^{1+1}.

That was the first dimension outside singularity Π^0 where singularity has a value of Π^1 in the form of $\Pi^{1+1} = 2$. The first claim to space had a value of Π^2. This applied to both sides of the claim to space outside singularity, and the double proton became the dominant factor on matter.

The seven is part of infinity parting eternity. Infinity is a point that one find inside any and all solid spheres and the point is where all lines cross at a 180^0 as well as 90^0. The points form as rotating motion establish a charged line that take control of space –time. Then where singularity at seven points ends and the eighth point holding singularity begin another sector comes into action. This holds points in time eternity at eight nine and ten. The points are physical but needs to be generated while it never actually is there.

In they're using of such logic makes science appear foolish. Since the time of Newton, the arguments made by those in the time of Newton tarnished from being brilliant to clever to fair too poor and a hundred years ago it reached the point of being stupid.   That is what Kepler's formula is all about?   That is what Kepler indicated with his formula $a^3 = T^2 k$. The space of an object (a^3) is equal to the time (T^2), which it is in, in every given instant (**k**). If the space becomes smaller, the time duration becomes longer every instant of time's progress. The motion follows the graph in relation to motion and time.

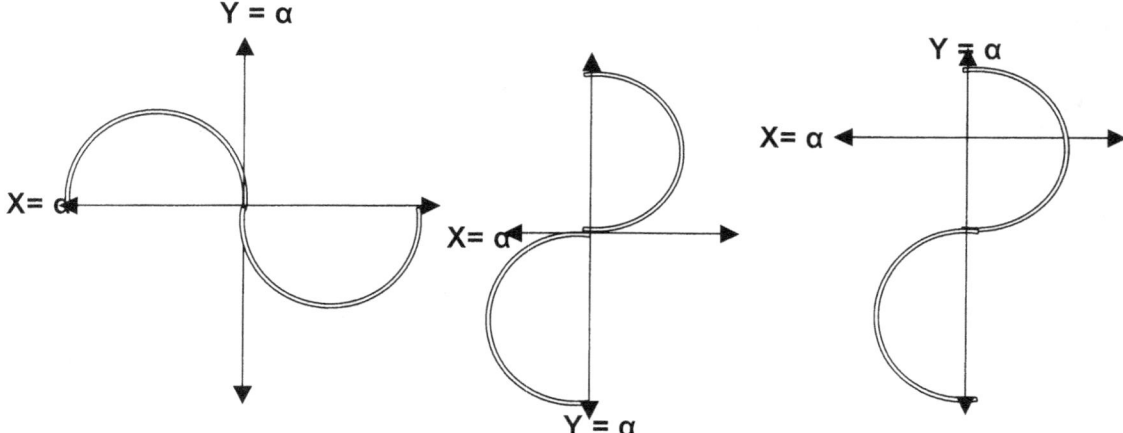

From the graph one can establish the link in the circle's rotation around a conforming unit being singularity. Saying that one therefore has to admit that the smallest spot has to hold space because the most insignificant dot can transmit light and being able to accomplish that, one must accept it to carry a value of something. If that spot had the value of nothing, it means that spot was not there to begin with. If the graph connected by zero that would mean there is no connection at all. With no connection the graph would be a mathematical tool with no value or use in any way.

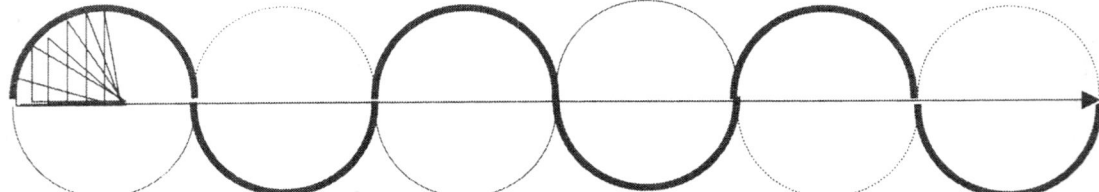

Holding space-time one should return to the original formula indicating space-time in as much as $a^3 = T^2 k$ where a = R and T = T. Being time it has to alternate positions and that can therefore only apply to **k** where **k** will indicate a relation to the space-time in question or the relevancy to singularity being $k^0 = 1$. By receiving k on top of the already $k^0 = 1$ that is in place the top becomes an atom by erecting the line of singularity from $k^0 = 1$ to $k^0 = a^3 / T^2 k$

It started with a dot, because that is the only form, size and dimension mathematical logic will allow our brain to accept. From the one dot had to come a second dot and a third dot. The dynamics of such a dot is smaller than we can understand because such a dot is in negative relation to what we see Π to be, and the deeper we delve in finding the smallest fragment where space started, in the spot where time is still eternal as much as we can accept eternity to be.

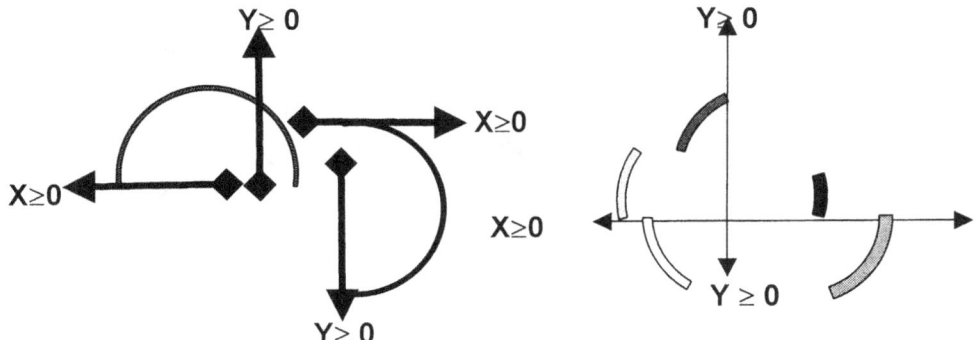

The graph with no connation between points because zero or nothing connects the points will render the use there of in mathematical terms quite obsolete.

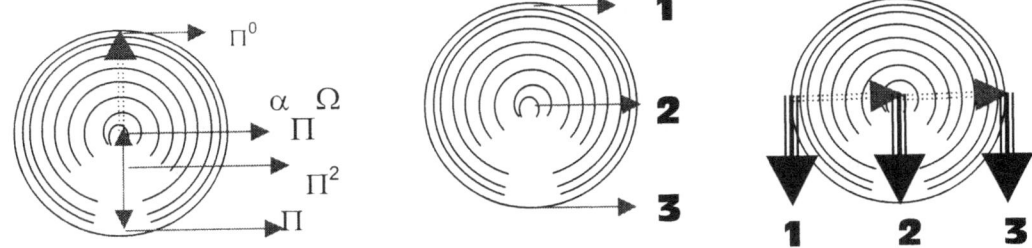

The reason why we should first locate the spot is because we can only work from that point forward. By working forward we have to work backwards to locate where we are heading. The cosmos started at a point and where such a point is, we will find the Universe. Every one knows where the Universe is, because we can see where the Universe is, but if we can see where the Universe is, then we should find the centre of the Universe in that spot. Einstein theoretically positioned the point of beginning at a place he indicated where singularity should be.

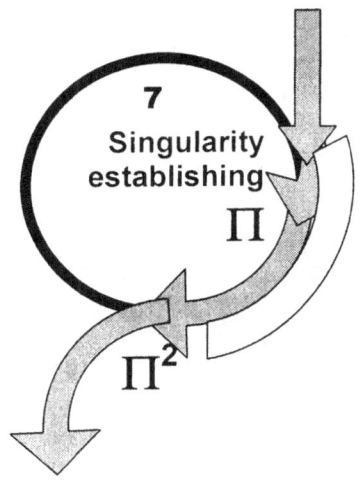

7

Singularity establishing

Π

Π^2

We not only have both as we now wish to see the Universe. By the duplication it therefore insists on relevancy because without relevancy there can be no motion and no motion means no space. The strongest proof there is about this is the manner in which the Coanda principle applies the reproducing of space taking shape from a round object and involving motion to produce such duplication. The relation forming the duplication of singularity is a duplication but applies as a dimensional forming of Π and placing 7 in relation to 10 forming Π^2 The liquid applying motion forms the 10 disciplines. No motion leaves no Coanda as well as no gravity because gravity is motion that duplicate singularity

The Coanda principle which in fact should be seen as a law because it is that strong is the principle of gravity duplicating with the motion that provides such duplication a relation of the particles having the seven factor and such a factor of seven produces through motion another three dimension. This total that material fill while in motion is ten and when ten crosses the line of singularity too duplicate the seven in the other side of the Universe the crossing cuts singularity in two as much as it puts singularity in the square. But it involves the motion of concentrating space to be or hold fluids around solids that may or may not move. In this must be a solid, a round basis Π, fluids concentrated in space and motion applying to one or all of the factors.

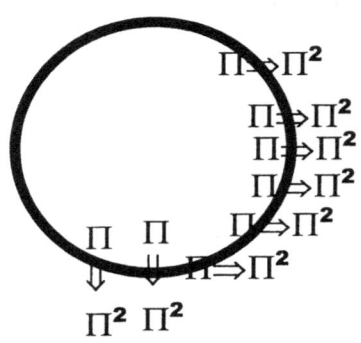

Conditions that prescribes the enactment of the Coanda effect is that the one surface has to duplicate singularity by establishing Π as a form. The round surface Π will bring about the shape of singularity Π that becomes enticed by the action of the motion of the liquid or of the solid or the motion of both around Π, which then establish and confirm singularity by form. The next factor is the presence of liquid. Air or atmosphere is liquid and water is liquid. Heat is liquid. The third factor being just as important is the motion establishing Π^2 by duplicating singularity as singularity becomes relevant through the applied motion that produces gravity from the singularity spot that provides the form.

That too forms the answer about the question concerning the Titius Bode gravity implicating of cosmology. The seven sides are linked by rotation nothing changes because there is a steady linking to the inside centre of the sphere. But it is to the outside that this rotation brings about dimensional complications. There are five T_1 points moving to five T_2 making contact with five moving points. The moving non fixed points is the point before reducing by five to the point after reducing by five that bring along the ten points in stead of the five to one point as it is the case with the Lagrangian system.

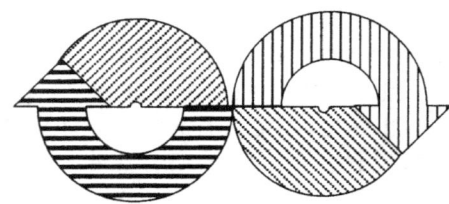

The heat brings about expanding singularity from a one sided affair to filling a volumetric Universe. But all of it is a relevancy where ten positions will sacrifice individuality and compromise singularity in order to secure two positions in singularity. The spin that comes about from such expanding and the duplication has the end as the Coanda principle where in the same motion of the same unit opposition spin forms both factors in gravity. It is this quality that forms the Coanda effect where two forms (solid and liquid) bong as a unit. The solid in relation to the liquid substantiate the contradicting nature there is in circular motion. The liquid will ale\ways substitute the solid because the liquid always contradicts the solid.

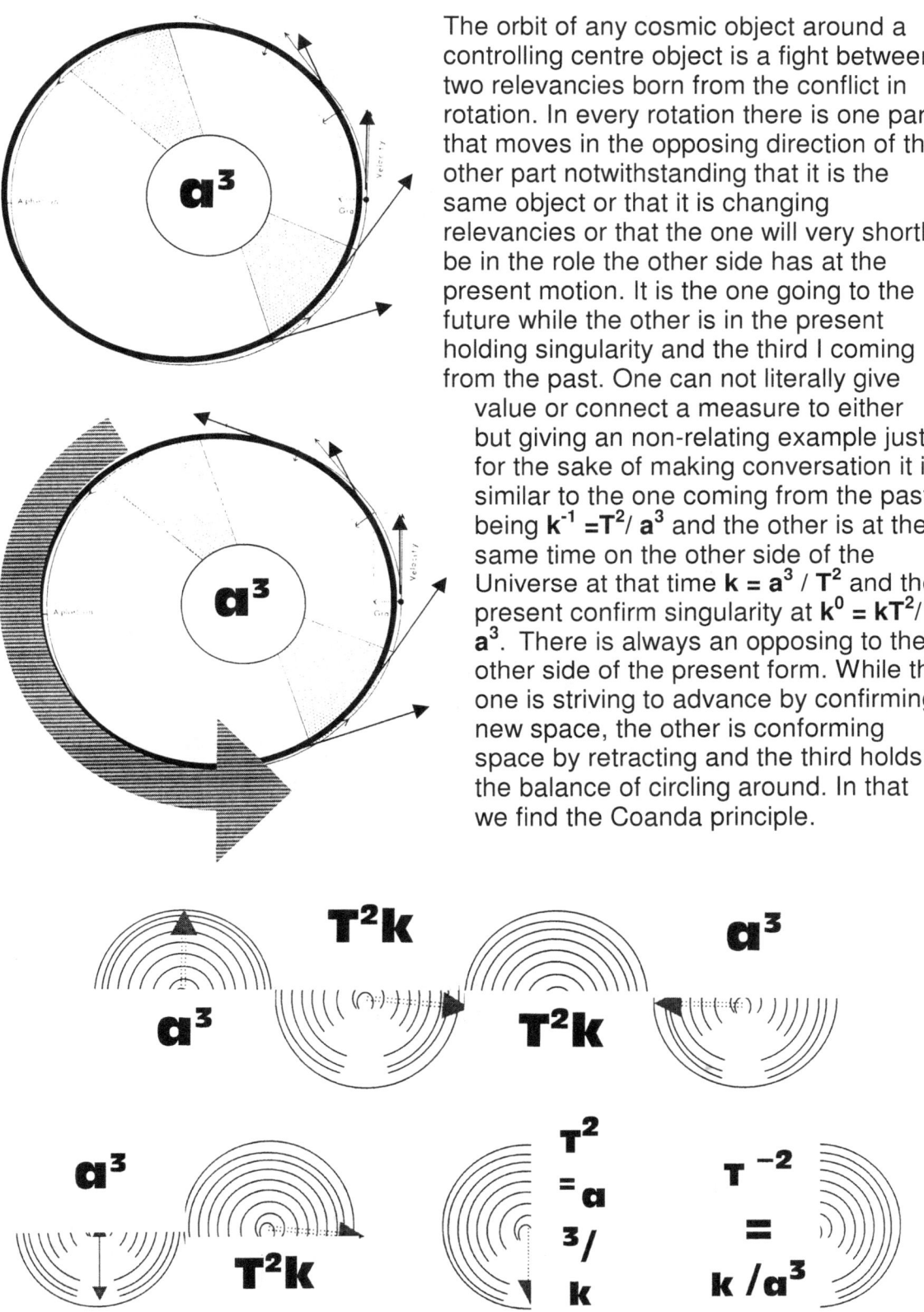

The orbit of any cosmic object around a controlling centre object is a fight between two relevancies born from the conflict in rotation. In every rotation there is one part that moves in the opposing direction of the other part notwithstanding that it is the same object or that it is changing relevancies or that the one will very shortly be in the role the other side has at the present motion. It is the one going to the future while the other is in the present holding singularity and the third I coming from the past. One can not literally give value or connect a measure to either but giving an non-relating example just for the sake of making conversation it is similar to the one coming from the past being $k^{-1} = T^2 / a^3$ and the other is at the same time on the other side of the Universe at that time $k = a^3 / T^2$ and the present confirm singularity at $k^0 = kT^2 / a^3$. There is always an opposing to the other side of the present form. While the one is striving to advance by confirming new space, the other is conforming space by retracting and the third holds the balance of circling around. In that we find the Coanda principle.

There are always the conflicting sides, which is a built in characteristic of rotation. The one side forms the space while the next is in motion of going away or coming towards at the same time. In that we find the Coanda effect producing gravity by applying opposing directions in the same unit as a result of the spin contracting as well as expanding simultaneously but on different sides of the divide.

The lot is more evidently moving further apart

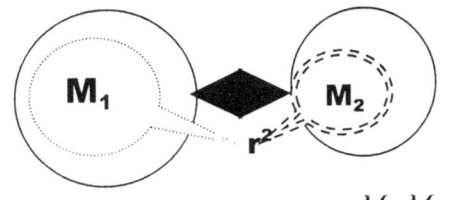

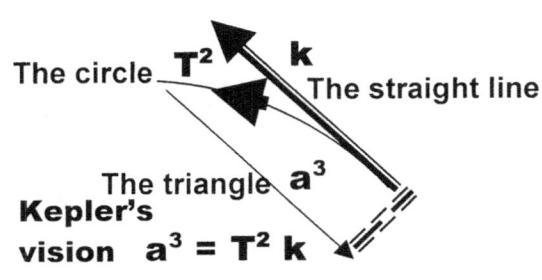

Newton's vision $F = \dfrac{M_1 M}{r^2} G$

Kepler's vision $a^3 = T^2 k$

$$\dfrac{M_1 M}{r^2} G$$

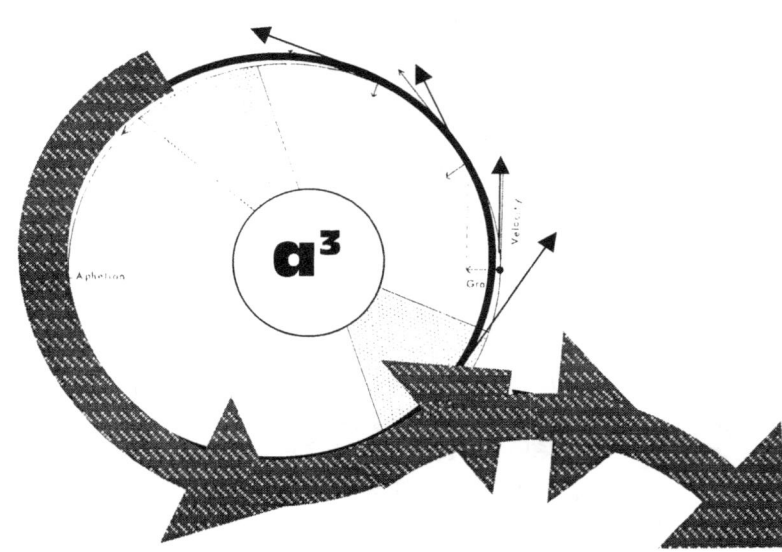

We have this tendency of a dual in rotating action that is the principle that is bringing about parts of the same unit orbiting other part of the same unit and still being in conflict with the other part of the same purpose and that is to rotate a centre.

If we look at the rotation we find from a human aspect that we put much claim to the circle. The circle holds importance but however important the circle may be, the circle forms part of eternity, which is the perfect part of creation.

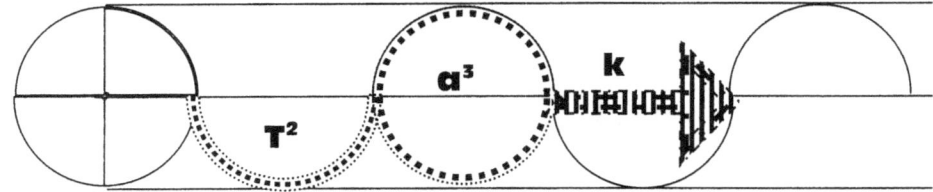

From the graph one can establish the link in the circle's rotation around a conforming unit being singularity.

Newton said that the rotation brings no influence and that he accomplished by allowing time to stand still. What he said was quite true because if time stood still all the Universe will tumble down into singularity or into the centre where from our perspective we find that **dJ / dt = 0.** This however will be a totally destructed cosmos

When $k = k^0 = 1$ then at the time also $a^3 = T^2$ and $a^3 = 1$ leaving $T^2 = 1$

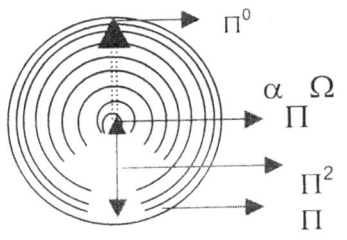

When Π^0 expands to Π we think of the rotation coming from the expanding. That is very much true but there is another aspect every one is missing. While the circle comes about time also shifts the centre to another location. In the duplication time brings the motion forward and although "forward" would not be a direction that is possible to point at, so is singularity not a point visible. Time in infinity does bring about the expanding of the sphere in the infinitive number of circles but the motion in time in eternity repositions such location to a new relevancy where the entirety of all the Universe rematch to find altogether new relevancy all over.

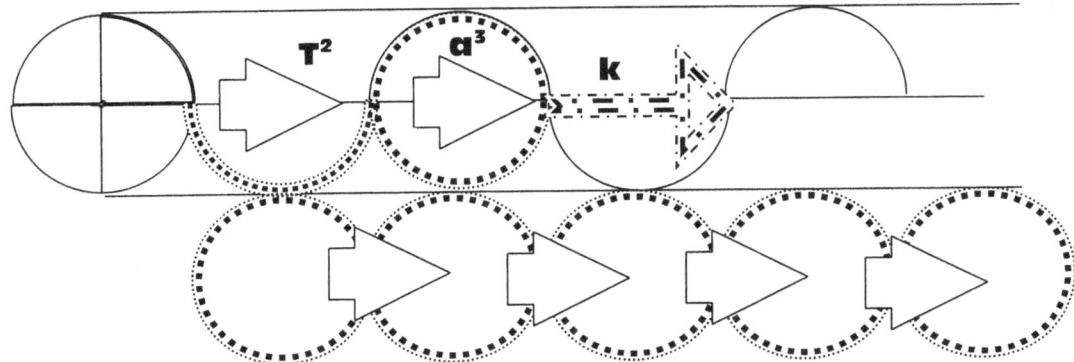

Matter or material is the concentration of heat that has gone dense because of time delay. I do not wish to elaborate that at this moment but there is a book out in which I explain this statement in much detail. It is the time delay of the shift in time from the past running through the present into the future that produces the material density we call the Universe in matter space and time. The duplication of the material is so extensive that the entirety of the Universe has to be demolished and again re-established to allow motion to take place. The one atom pulls the other atom while the one atom pushes the other atom to take its place the very next instant. Singularity cannot move but is totally rigid. Look at the centre of the top activated and one can see how immovable singularity truly is.

The entirety of time that we think of as the Universe is singularity holding every possibility that it can become when generated by heat. The circle remains the circle but the circle as a unit relocates while the circle complete the circle motion. By the way, that black stuff we see at night is not black stuff but it is light so bright our eyes kill of the brightness to allow us to see by daylight. That black stuff we see at night is heat and heat is time in motion.

From the past   In the present   Onto the future

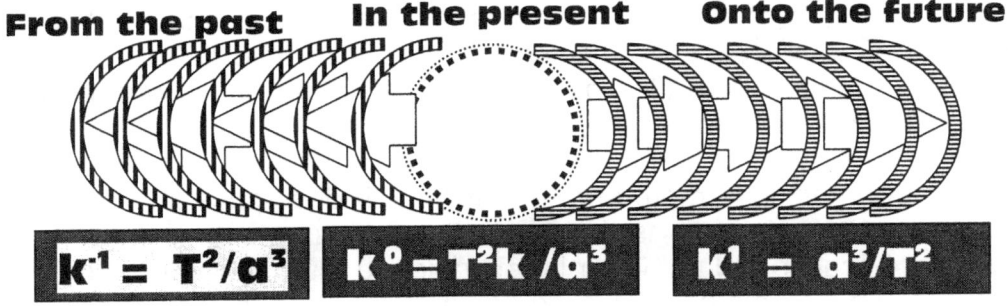

$$k^{-1} = T^2/a^3 \qquad k^0 = T^2k/a^3 \qquad k^1 = a^3/T^2$$

The material complete a circle by rotation and thereby confirm the heat that is sealed in the unit as the unit and that forms the unit we call material. That however is half the story where the other half is the directional redirecting of all that has spun into new allocated locations where each fin a relevancy that only apply at that specific moment. The motion redirecting movement plays as much a critical role as does the circle confirming the rotation and sealing the heat into the confined space that confirms material.

As the material duplicate and the duplication forms motion the material drags what it left behind as much as it pushes what it is catching in the future. The atom is following the other atom but to fill the time slot in the future the following atom has to relinquish the position it holds in the next one's future. In order to move into the next position in the future pretender the atom has to relinquish the claim it has

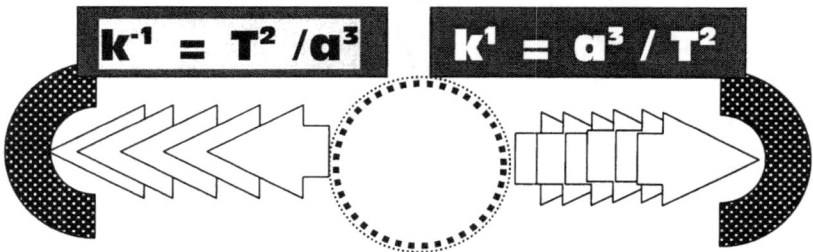

on the present location. The only easy way to do that is to drag the next tenant into the vacant to be slot the atom at that instant fills. To move it has to push the atom in front in to a new location while dragging the one behind to fill what it wishes to vacate. With that action we find t hat the relevancy k goes minus and goes plus which allow space to grow as time decline or to allow space to shrink as time becomes more prevalent.

Apparently according to informed sources we will find the Sun viewed from each planet indicated by name, as the photo would suggest. From Mercury the Sun seems large therefore the Sun is close and from Pluto the Sun seems small which we think of as further.

That way of thinking is very indicative of Newtonian thinking with a very explicit Earthly connotation. It is far from cosmology.

When an object such as one of the photos suggest shift further away by getting smaller, it is not the distance we should consider because the distance has no meaning in cosmology. It is the time it would take to reach the object that has to be considered.

The "further away" the object seems the longer it would take to reach the object and the "closer" the object seems the less time it would take to reach the object. There is more time between the Sun and Pluto than the time being between the Sun and Mercury.

In the case of Mercury time is more and therefore it divides space into a smaller factor $100k = a^3 / 100 T^2$ where as the space the Sun has seems to be more space at Mercury because $k = a^3 / T^2$ or in relevance to Pluto the Sun must be a **100** times bigger with the time being a hundred times less of a factor

$k = 100a^3 / T^2$. That mathematically with the aid of Kepler proves that the Black stuff being the Biblical light that was mentioned at the explaining of creation is time and not space. Material fills space but the filled space of material move through time we incorrectly think of as space.

When the one is pulling the other is in opposing mode not only by spin but also by having a singularity to protect.

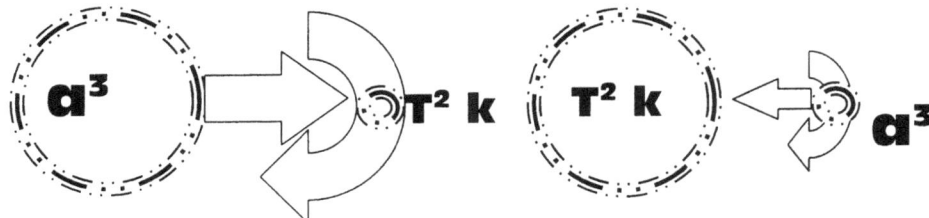

In the cosmos there is no big or small but only singularity generating space-time, Looking at the Sun and any planet we see two points holding singularity where in both cases singularity is equal at $k^0 = a^3 / T^2$ k =1. On the one side of the divide the one factor holding singularity take prominence and then crossing the divide the other point holding singularity holds dominance. Crossing the divide of singularity puts one of the two in dominance as far as controlling time and controlling space. The one holding lesser space would affectively control lesser time and the one controlling more space holds more time in control.

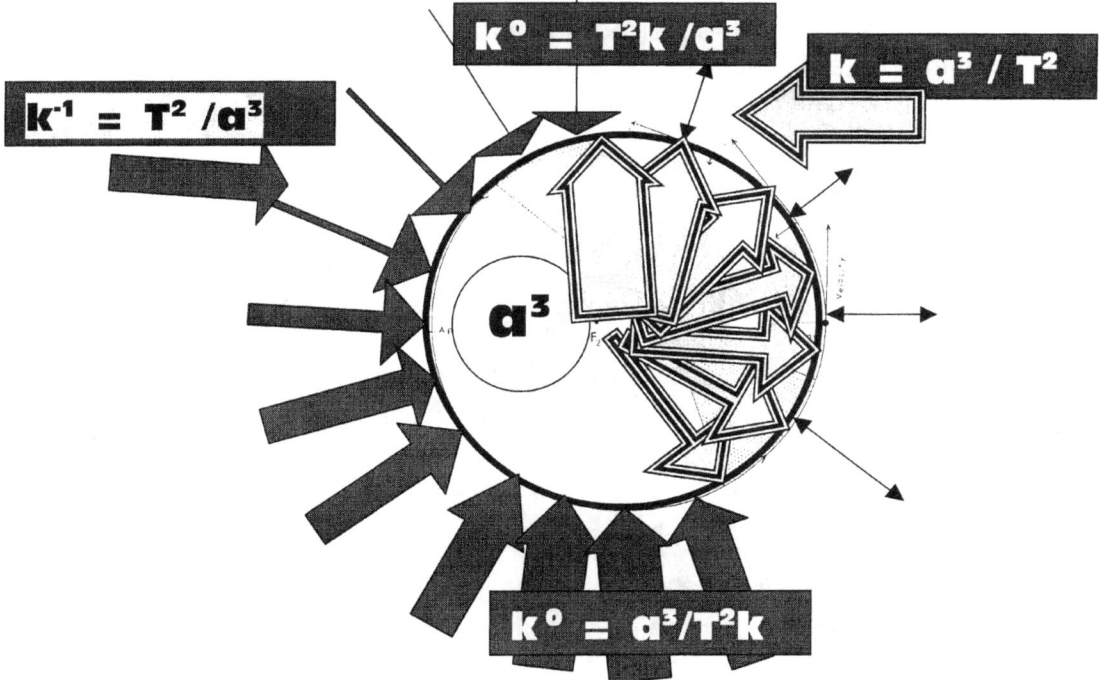

As I mentioned a while ago, material is the delay of time bringing about heat where the motion contain the heat in confirming the location of the preserved heat. The material fill time with space by relocating singularity while singularity generate material filling time in a specific allocated location. Through the motion there will be some area where the expanding $k = a^3 / T^2$ is more frivolous and where the material is pushing the next out of position to fill the next position. The there is the other where the delay becomes more because the delay is caused by matter not generating space filling fast enough so that the follow on filling can come about becoming motion or heat relocation of time.

In this changing of the relevancies we find that at one point the Sun gives the planet a nudge onwards and the next part where the planet go past the center singularity the Sun drags the planet onwards. The one part of the circle that the planet form in orbit is smaller than the other part that is bigger but small and big is no issue. It has more time concentrated by motion effort or less time concentrated by motion effort. In that way the Coanda effect forms gravity and locates the time intervals as structures orbit a center of concentration.

Before time moved there was a spot that repeated by being perfect.
Then the imperfect entered creation and the spot moved to a dot.
►By returning to the spot the dot advanced to form a new dot.

The motion of parting from the perfect to move into a position held by being imperfect gave time opportunity to part from time in eternity.

in infinity the

The returning to the previous position was always there and is part of the eternal perfect. Moving in one direction brought about a future that institutionalised time by implementing heat. It was the departing from the position the perfect held that is time but that time is a delay caused by time going imperfect. As time becomes more imperfect so would the relevancy of repeating the perfect in ratio of the imperfect grow and space filled with material will compact from that to become more, denser and dominating the entirety of the Universe.

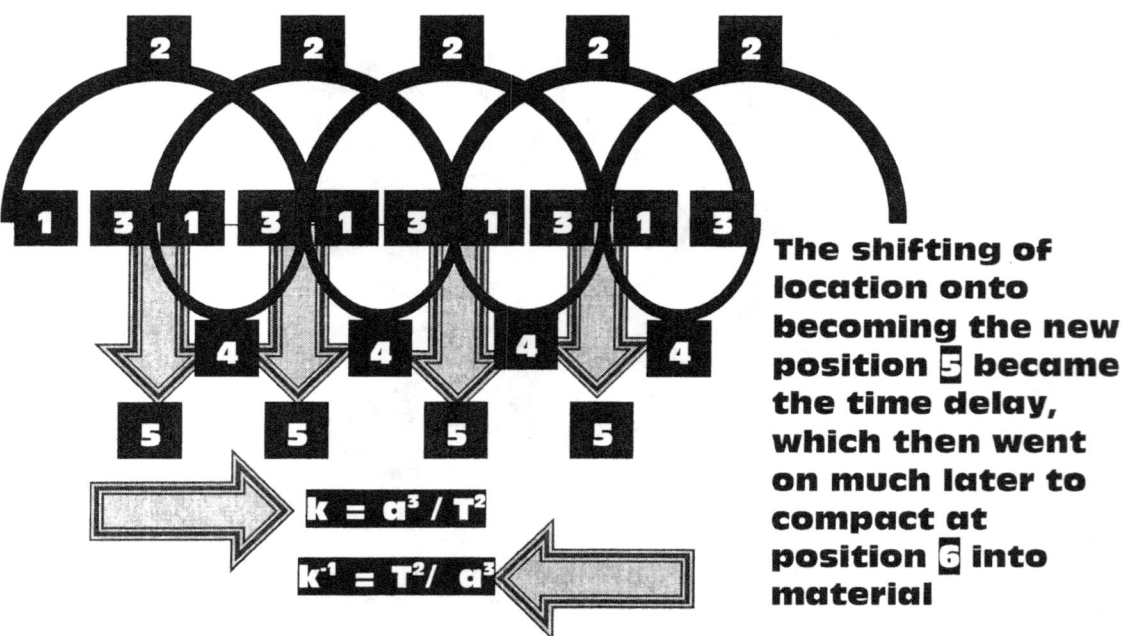

$$k = a^3 / T^2$$

$$k^{-1} = T^2 / a^3$$

The shifting of location onto becoming the new position 5 became the time delay, which then went on much later to compact at position 6 into material

Even high and low tides has nothing to do with "the pull of gravity". If it did have anything to do with the pull of gravity then there was no reason for cyclic change since the mass of both the Earth and the moon remains at a constant. The changing of tides is the rotational cyclic contradicting nature that motion has and the Moon crosses the divide twice daily where the cycle then begins to oppose what it was before. It remains just the crossing of singularity where singularity indicates the divide the motion establishes.

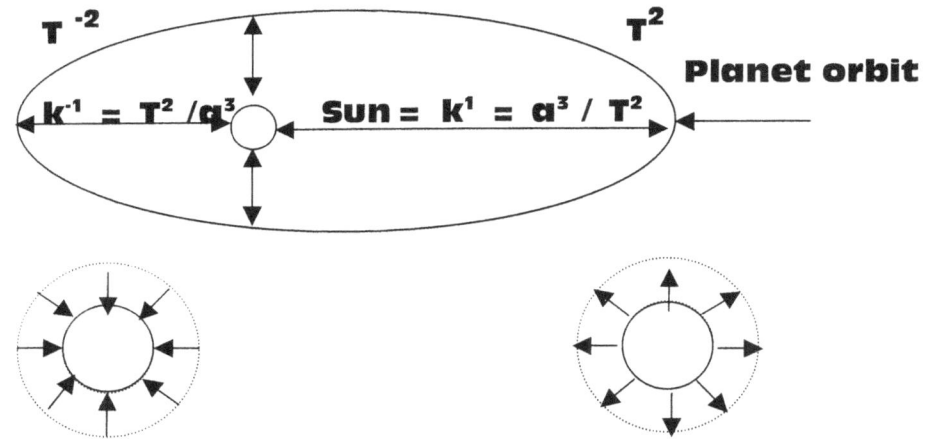

Gravity is about reducing space

Expanding is all about heating. Heating takes up more space and gravity reduces space.

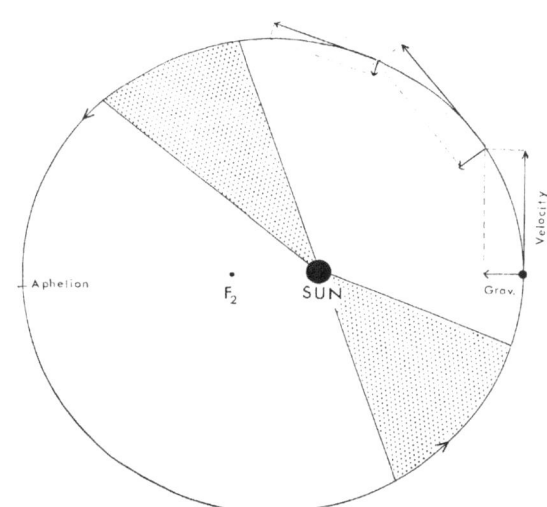

The same pattern is still very much visible in the way structures follow the centre of contraction. We still have the relevancy shifting from one to the other with not one point holding singularity absolutely domineering, We still have the one trying to expand while the other is trying to preserve and the relevancies does not go all out the way of the controlling structure. We find in this manner that the straight line goes bended by 7^0, and the curve follow the guidelines of singularity by measure of Π.

The second part of this book follows immediately from page 332 going as:

The second part of this book follows immediately from page 332 going as:

ISBN-13: 978-1533473844

ISBN-10: 1533473846